MATRIX ALGEBRA FOR
LINEAR MODELS

MATRIX ALGEBRA FOR LINEAR MODELS

MARVIN H. J. GRUBER
School of Mathematical Sciences
Rochester Institute of Technology
Rochester, NY

Published by John Wiley & Sons, Inc., Hoboken, New Jersey
Published simultaneously in Canada

For general information on our other products and services or for technical support, please contact
our Customer Care Department within the United States at (800) 762-2974, outside the United States
at (317) 572-3993 or fax (317) 572-4002.

Wiley also publishes its books in a variety of electronic formats. Some content that appears in print may
not be available in electronic formats. For more information about Wiley products, visit our web site at
www.wiley.com.

Library of Congress Cataloging-in-Publication Data:

Gruber, Marvin H. J., 1941–
    Matrix algebra for linear models / Marvin H. J. Gruber, Department of Mathematical Sciences,
Rochester Institute of Technology, Rochester, NY.
        pages   cm
    Includes bibliographical references and index.
    ISBN 978-1-118-59255-7 (cloth)
1. Linear models (Statistics)   2. Matrices.   I. Title.
    QA279.G78 2013
    519.5′36–dc23

                                                                        2013026537

ISBN: 9781118592557

10 9 8 7 6 5 4 3 2 1

To the memory of my parents, Adelaide Lee Gruber and Joseph George Gruber, who were always there for me while I was growing up and as a young adult.

CONTENTS

PREFACE

This is a book about matrix algebra with examples of its application to statistics, mostly the linear statistical model. There are 5 parts and 24 sections.

Part I (Sections 1–6) reviews topics in undergraduate linear algebra such as matrix operations, determinants, vector spaces, and solutions to systems of linear equations. In addition, it includes some topics frequently not covered in a first course that are of interest to statisticians. These include the Kronecker product of two matrices and inverses of partitioned matrices.

Part II (Sections 7–11) tells how to find the eigenvalues of a matrix and takes up the singular value decomposition and its generalizations. The applications studied include principal components and the multicollinearity problem.

Part III (Sections 12–16) deals with generalized inverses. This includes what they are and examples of how they are useful. It also considers different kinds of generalized inverses such as the Moore–Penrose inverse, minimum norm generalized inverses, and least square generalized inverses. There are a number of results about how to represent generalized inverses using nonsingular matrices and using the singular value decomposition. Results about least square estimators for the less than full rank case are given, which employ the properties of generalized inverses. Some of the results are applied in Parts IV and V.

The use of quadratic forms in the analysis of variance is the subject of Part IV (Sections 17–20). The distributional properties of quadratic forms of normal random variables are studied. The results are applied to the analysis of variance for a full rank regression model, the one- and two-way classification, the two-way classification with interaction, and a nested model. Testing the general linear hypothesis is also taken up.

Part V (Sections 21–24) is about the minimization of a second-degree form. Cases taken up are unconstrained minimization and minimization with respect to linear and quadratic constraints. The applications taken up include the least square estimator, canonical correlation, and ridge-type estimators.

Each part has an introduction that provides a more detailed overview of its contents, and each section begins with a brief overview and ends with a summary.

The book has numerous worked examples and most illustrate the important results with numerical computations. The examples are titled to inform the reader what they are about.

At the end of each of the 24 sections, there are exercises. Some of these are proof type; many of them are numerical. Answers are given at the end for almost all of the numerical examples and solutions, or partial solutions, are given for about half of the proof-type problems. Some of the numerical exercises are a bit cumbersome, and readers are invited to use a computer algebra system, such as Mathematica, Maple, and Matlab, to help with the computations. Many of the exercises have more than one right answer, so readers may, in some instances, solve a problem correctly and get an answer different from that in the back of the book.

The author has prepared a solutions manual with solutions to all of the exercises, which is available from Wiley to instructors who adopt this book as a textbook for a course.

The end of an example is denoted by the symbol □, the end of a proof by ■, and the end of a formal definition by ●.

The book is, for the most part, self-contained. However, it would be helpful if readers had a first course in matrix or linear algebra and some background in statistics.

There are a number of other excellent books on this subject that are given in the references. This book takes a slightly different approach to the subject by making extensive use of the singular value decomposition. Also, this book actually shows some of the statistical applications of the matrix theory; for the most part, the other books do not do this. Also, this book has more numerical examples than the others. Hopefully, it will add to what is out there on the subject and not necessarily compete with the other books.

MARVIN H. J. GRUBER

ACKNOWLEDGMENTS

There are a number of people who should be thanked for their help and support. I would like to thank three of my teachers at the University of Rochester, my thesis advisor, Poduri S.R.S. Rao, Govind Mudholkar, and Reuben Gabriel (may he rest in peace) for introducing me to many of the topics taken up in this book. I am very grateful to Steve Quigley for his guidance in how the book should be organized, his constructive criticism, and other kinds of help and support. I am also grateful to the other staff of John Wiley & Sons, which include the editorial assistant, Sari Friedman, the copy editor, Yassar Arafat, and the production editor, Stephanie Loh.

On a personal note, I am grateful for the friendship of Frances Johnson and her help and support.

PART I

BASIC IDEAS ABOUT MATRICES AND SYSTEMS OF LINEAR EQUATIONS

This part of the book reviews the topics ordinarily covered in a first course in linear algebra. It also introduces some other topics usually not covered in the first course that are important to statistics, in particular to the linear statistical model.

The first of the six sections in this part gives illustrations of how matrices are useful to the statistician for summarizing data. The basic operations of matrix addition, multiplication of a matrix by a scalar, and matrix multiplication are taken up. Matrices have some properties that are similar to real numbers and some properties that they do not share with real numbers. These are pointed out.

Section 2 is an informal review of the evaluation of determinants. It shows how determinants can be used to solve systems of equations. Cramer's rule and Gauss elimination are presented.

Section 3 is about finding the inverse of a matrix. The adjoint method and the use of elementary row and column operations are considered. In addition, the inverse of a partitioned matrix is discussed.

Special matrices important to statistical applications are the subject of Section 4. These include combinations of the identity matrix and matrices consisting of ones, orthogonal matrices in general, and some orthogonal matrices useful to the analysis of variance, for example, the Helmert matrix. The Kronecker product, also called the direct product of matrices, is presented. It is useful in the representation sums of squares in the analysis of variance. This section also includes a discussion of differentiation of matrices which proves useful in solving constrained optimization problems in Part V.

Matrix Algebra for Linear Models, First Edition. Marvin H. J. Gruber.
© 2014 John Wiley & Sons, Inc. Published 2014 by John Wiley & Sons, Inc.

Vector spaces are taken up in Section 5 because they are important to understanding eigenvalues, eigenvectors, and the singular value decomposition that are studied in Part II. They are also important for understanding what the rank of a matrix is and the concept of degrees of freedom of sums of squares in the analysis of variance. Inner product spaces are also taken up and the Cauchy–Schwarz inequality is established.

The Cauchy–Schwarz inequality is important for the comparison of the efficiency of estimators.

The material on vector spaces in Section 5 is used in Section 6 to explain what is meant by the rank of a matrix and to show when a system of linear equations has one unique solution, infinitely many solutions, and no solution.

SECTION 1

WHAT MATRICES ARE AND SOME BASIC OPERATIONS WITH THEM

1.1 INTRODUCTION

This section will introduce matrices and show how they are useful to represent data. It will review some basic matrix operations including matrix addition and multiplication. Some examples to illustrate why they are interesting and important for statistical applications will be given. The representation of a linear model using matrices will be shown.

1.2 WHAT ARE MATRICES AND WHY ARE THEY INTERESTING TO A STATISTICIAN?

Matrices are rectangular arrays of numbers. Some examples of such arrays are

$$A = \begin{bmatrix} 4 & -2 & 1 & 0 \\ 0 & 5 & 3 & -7 \end{bmatrix}, B = \begin{bmatrix} 1 \\ -2 \\ 6 \end{bmatrix}, \quad \text{and} \quad C = \begin{bmatrix} 0.2 & 0.5 & 0.6 \\ 0.7 & 0.1 & 0.8 \\ 0.9 & 0.4 & 0.3 \end{bmatrix}.$$

Often data may be represented conveniently by a matrix. We give an example to illustrate how.

Matrix Algebra for Linear Models, First Edition. Marvin H. J. Gruber.
© 2014 John Wiley & Sons, Inc. Published 2014 by John Wiley & Sons, Inc.

Example 1.1 Representing Data by Matrices

An example that lends itself to statistical analysis is taken from the Economic Report of the President of the United States in 1988. The data represent the relationship between a dependent variable Y (personal consumption expenditures) and three other independent variables X_1, X_2, and X_3. The variable X_1 represents the gross national product, X_2 represents personal income (in billions of dollars), and X_3 represents the total number of employed people in the civilian labor force (in thousands). Consider this data for the years 1970–1974 in Table 1.1.

TABLE 1.1 **Consumption expenditures in terms of gross national product, personal income, and total number of employed people**

Obs	Year	Y	X_1	X_2	X_3
1	1970	640.0	1015.5	831.8	78,678
2	1971	691.6	1102.7	894.0	79,367
3	1972	757.6	1212.8	981.6	82,153
4	1973	837.2	1359.3	1101.7	85,064
5	1974	916.5	1472.4	1210.1	86,794

The dependent variable may be represented by a matrix with five rows and one column. The independent variables could be represented by a matrix with five rows and three columns. Thus,

$$Y = \begin{bmatrix} 640.0 \\ 691.6 \\ 757.6 \\ 837.2 \\ 916.5 \end{bmatrix} \quad \text{and} \quad X = \begin{bmatrix} 1015.5 & 831.8 & 78,678 \\ 1102.7 & 894.0 & 79,367 \\ 1212.8 & 981.6 & 82,153 \\ 1359.3 & 1101.7 & 85,064 \\ 1472.8 & 1210.1 & 86,794 \end{bmatrix}.$$

A matrix with m rows and n columns is an $m \times n$ matrix. Thus, the matrix Y in Example 1.1 is 5×1 and the matrix X is 5×3. A square matrix is one that has the same number of rows and columns. The individual numbers in a matrix are called the elements of the matrix. $\qquad\Box$

We now give an example of an application from probability theory that uses matrices.

Example 1.2 A "Musical Room" Problem

Another somewhat different example is the following. Consider a triangular-shaped building with four rooms one at the center, room 0, and three rooms around it numbered 1, 2, and 3 clockwise (Fig. 1.1).

There is a door from room 0 to rooms 1, 2, and 3 and doors connecting rooms 1 and 2, 2 and 3, and 3 and 1. There is a person in the building. The room that he/she is

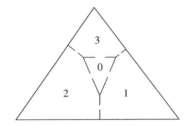

FIGURE 1.1   Building with four rooms.

in is the state of the system. At fixed intervals of time, he/she rolls a die. If he/she is in room 0 and the outcome is 1 or 2, he/she goes to room 1. If the outcome is 3 or 4, he/she goes to room 2. If the outcome is 5 or 6, he/she goes to room 3. If the person is in room 1, 2, or 3 and the outcome is 1 or 2, he/she advances one room in the clockwise direction. If the outcome is 3 or 4, he/she advances one room in the counterclockwise direction. An outcome of 5 or 6 will cause the person to return to room 0. Assume the die is fair.

Let p_{ij} be the probability that the person goes from room i to room j. Then we have the table of transitions

room	0	1	2	3
0	0	$\frac{1}{3}$	$\frac{1}{3}$	$\frac{1}{3}$
1	$\frac{1}{3}$	0	$\frac{1}{3}$	$\frac{1}{3}$
2	$\frac{1}{3}$	$\frac{1}{3}$	0	$\frac{1}{3}$
3	$\frac{1}{3}$	$\frac{1}{3}$	$\frac{1}{3}$	0

that indicates

$$p_{00} = p_{11} = p_{22} = p_{33} = 0$$

$$p_{01} = p_{02} = p_{03} = \frac{1}{3}$$

$$p_{12} = p_{13} = p_{10} = \frac{1}{3}$$

$$p_{21} = p_{23} = p_{20} = \frac{1}{3}$$

$$p_{31} = p_{32} = p_{30} = \frac{1}{3}.$$

Then the transition matrix would be

$$P = \begin{bmatrix} 0 & \frac{1}{3} & \frac{1}{3} & \frac{1}{3} \\ \frac{1}{3} & 0 & \frac{1}{3} & \frac{1}{3} \\ \frac{1}{3} & \frac{1}{3} & 0 & \frac{1}{3} \\ \frac{1}{3} & \frac{1}{3} & \frac{1}{3} & 0 \end{bmatrix}.$$

□

Matrices turn out to be handy for representing data. Equations involving matrices are often used to study the relationship between variables.

More explanation of how this is done will be offered in the sections of the book that follow.

The matrices to be studied in this book will have elements that are real numbers. This will suffice for the study of linear models and many other topics in statistics. We will not consider matrices whose elements are complex numbers or elements of an arbitrary ring or field.

We now consider some basic operations using matrices.

1.3 MATRIX NOTATION, ADDITION, AND MULTIPLICATION

We will show how to represent a matrix and how to add and multiply two matrices.

The elements of a matrix A are denoted by a_{ij} meaning the element in the ith row and the jth column. For example, for the matrix

$$C = \begin{bmatrix} 0.2 & 0.5 & 0.6 \\ 0.7 & 0.1 & 0.8 \\ 0.9 & 0.4 & 1.3 \end{bmatrix}$$

$c_{11} = 0.2$, $c_{12} = 0.5$, and so on. Three important operations include matrix addition, multiplication of a matrix by a scalar, and matrix multiplication. Two matrices A and B can be added only when they have the same number of rows and columns. For the matrix $C = A + B$, $c_{ij} = a_{ij} + b_{ij}$; in other words, just add the elements algebraically in the same row and column. The matrix $D = \alpha A$ where α is a real number has elements $d_{ij} = \alpha a_{ij}$; just multiply each element by the scalar. Two matrices can be multiplied only when the number of columns of the first matrix is the same as the number of rows of the second one in the product. The elements of the $n \times p$ matrix $E = AB$, assuming that A is $n \times m$ and B is $m \times p$, are

$$e_{ij} = \sum_{k=1}^{m} a_{ik} b_{kj}, \quad 1 \leq i \leq m, \ 1 \leq j \leq p.$$

Example 1.3 Illustration of Matrix Operations

Let $A = \begin{bmatrix} 1 & -1 \\ 2 & 3 \end{bmatrix}, B = \begin{bmatrix} -1 & 2 \\ 3 & 4 \end{bmatrix}.$

Then

$$C = A + B = \begin{bmatrix} 1+(-1) & -1+2 \\ 2+3 & 3+4 \end{bmatrix} = \begin{bmatrix} 0 & 1 \\ 5 & 7 \end{bmatrix},$$

$$D = 3A = \begin{bmatrix} 3 & -3 \\ 6 & 9 \end{bmatrix},$$

and

$$E = AB = \begin{bmatrix} 1(-1)+(-1)(3) & 1(2)+(-1)(4) \\ 2(-1)+3(3) & 2(2)+3(4) \end{bmatrix} = \begin{bmatrix} -4 & -2 \\ 7 & 16 \end{bmatrix}. \qquad \square$$

Example 1.4 Continuation of Example 1.2

Suppose that elements of the row vector $\pi^{(0)} = \begin{bmatrix} \pi_0^{(0)} & \pi_1^{(0)} & \pi_2^{(0)} & \pi_3^{(0)} \end{bmatrix}$ where $\sum_{i=0}^{3} \pi_i^{(0)} = 1$ represent the probability that the person starts in room i. Then $\pi^{(1)} = \pi^{(0)}P$. For example, if

$$\pi^{(0)} = \begin{bmatrix} \frac{1}{2} & \frac{1}{6} & \frac{1}{12} & \frac{1}{4} \end{bmatrix}$$

the probabilities the person is in room 0 initially are 1/2, room 1 1/6, room 2 1/12, and room 3 1/4, then

$$\pi^{(1)} = \begin{bmatrix} \frac{1}{2} & \frac{1}{6} & \frac{1}{12} & \frac{1}{4} \end{bmatrix} \begin{bmatrix} 0 & \frac{1}{3} & \frac{1}{3} & \frac{1}{3} \\ \frac{1}{3} & 0 & \frac{1}{3} & \frac{1}{3} \\ \frac{1}{3} & \frac{1}{3} & 0 & \frac{1}{3} \\ \frac{1}{3} & \frac{1}{3} & \frac{1}{3} & 0 \end{bmatrix} = \begin{bmatrix} \frac{1}{6} & \frac{5}{18} & \frac{11}{36} & \frac{1}{4} \end{bmatrix}.$$

Thus, after one transition given the initial probability vector above the probabilities that the person ends up in room 0, room 1, room 2, or room 3 after one transition are 1/6, 5/18, 11/36, and 1/4, respectively. This example illustrates a discrete Markov chain. The possible transitions are represented as elements of a matrix.

Suppose we want to know the probabilities that a person goes from room i to room j after two transitions. Assuming that what happens at each transition is independent, we could multiply the two matrices. Then

$$P^2 = P \cdot P = \begin{bmatrix} 0 & \frac{1}{3} & \frac{1}{3} & \frac{1}{3} \\ \frac{1}{3} & 0 & \frac{1}{3} & \frac{1}{3} \\ \frac{1}{3} & \frac{1}{3} & 0 & \frac{1}{3} \\ \frac{1}{3} & \frac{1}{3} & \frac{1}{3} & 0 \end{bmatrix} \begin{bmatrix} 0 & \frac{1}{3} & \frac{1}{3} & \frac{1}{3} \\ \frac{1}{3} & 0 & \frac{1}{3} & \frac{1}{3} \\ \frac{1}{3} & \frac{1}{3} & 0 & \frac{1}{3} \\ \frac{1}{3} & \frac{1}{3} & \frac{1}{3} & 0 \end{bmatrix} = \begin{bmatrix} \frac{1}{3} & \frac{2}{9} & \frac{2}{9} & \frac{2}{9} \\ \frac{2}{9} & \frac{1}{3} & \frac{2}{9} & \frac{2}{9} \\ \frac{2}{9} & \frac{2}{9} & \frac{1}{3} & \frac{2}{9} \\ \frac{2}{9} & \frac{2}{9} & \frac{2}{9} & \frac{1}{3} \end{bmatrix}.$$

Thus, for example, if the person is in room 1, the probability that he/she returns there after two transitions is 1/3. The probability that he/she winds up in room 3 is 2/9. Also when $\pi^{(0)}$ is the initial probability vector, we have that $\pi^{(2)}=\pi^{(1)}P=\pi^{(0)}P^2$. The reader is asked to find $\pi^{(2)}$ in Exercise 1.17.   □

   Two matrices are equal if and only if their corresponding elements are equal. More formally, $A=B$ if and only if $a_{ij}=b_{ij}$ for all $1\le i\le m$ and $1\le j\le n$.

   Most, but not all, of the rules for addition and multiplication of real numbers hold true for matrices. The associative and commutative laws hold true for addition. The zero matrix is the matrix with all of the elements zero. An additive inverse of a matrix A would be $-A$, the matrix whose elements are $(-1)a_{ij}$. The distributive laws hold true.

   However, there are several properties of real numbers that do not hold true for matrices. First, it is possible to have divisors of zero. It is not hard to find matrices A and B where $AB=0$ and neither A or B is the zero matrix (see Example 1.4).

   In addition the cancellation rule does not hold true. For real nonzero numbers a, b, c, $ba=ca$ would imply that $b=c$. However (see Example 1.5) for matrices, $BA=CA$ may not imply that $B=C$.

   Not every matrix has a multiplicative inverse. The identity matrix denoted by I has all ones on the longest (main) diagonal $(a_{ij}=1)$ and zeros elsewhere $(a_{ij}=0,\ i\ne j)$. For a matrix A, a multiplicative inverse would be a matrix such that $AB=I$ and $BA=I$. Furthermore, for matrices A and B, it is not often true that $AB=BA$. In other words, matrices do not satisfy the commutative law of multiplication in general.

   The transpose of a matrix A is the matrix A' where the rows and the columns of A are exchanged. For example, for the matrix A in Example 1.3,

$$A' = \begin{bmatrix} 1 & 2 \\ -1 & 3 \end{bmatrix}.$$

A matrix A is symmetric when $A=A'$. If $A=-A'$, the matrix is said to be skew symmetric. Symmetric matrices come up often in statistics.

Example 1.5 Two Nonzero Matrices Whose Product Is Zero

Consider the matrix

$$A = \begin{bmatrix} 1 & 2 \\ 1 & 2 \end{bmatrix}, B = \begin{bmatrix} 2 & -2 \\ -1 & 1 \end{bmatrix}.$$

Notice that

$$AB = \begin{bmatrix} 1(2)+2(-1) & 1(-2)+2(1) \\ 1(2)+2(-1) & 1(-2)+2(1) \end{bmatrix} = \begin{bmatrix} 0 & 0 \\ 0 & 0 \end{bmatrix}.$$

□

Example 1.6 The Cancellation Law for Real Numbers Does Not Hold for Matrices

Consider matrices A, B, C where

$$A = \begin{bmatrix} 1 & 2 \\ 1 & 2 \end{bmatrix}, B = \begin{bmatrix} 5 & 4 \\ 7 & 3 \end{bmatrix}, \text{ and } C = \begin{bmatrix} 3 & 6 \\ 8 & 2 \end{bmatrix}.$$

Now

$$BA = CA = \begin{bmatrix} 9 & 18 \\ 10 & 20 \end{bmatrix}$$

but $B \neq C$.   □

Matrix theory is basic to the study of linear models. Example 1.7 indicates how the basic matrix operations studied so far are used in this context.

Example 1.7 The Linear Model

Let Y be an n-dimensional vector of observations, an $n \times 1$ matrix. Let X be an $n \times m$ matrix where each column has the values of a prediction variable. It is assumed here that there are m predictors. Let β be an $m \times 1$ matrix of parameters to be estimated. The prediction of the observations will not be exact. Thus, we also need an n-dimensional column vector of errors ε. The general linear model will take the form

$$Y = X\beta + \varepsilon. \tag{1.1}$$

Suppose that there are five observations and three prediction variables. Then $n = 5$ and $m = 3$. As a result, we would have the multiple regression equation

$$Y_i = \beta_0 + \beta_1 X_{1i} + \beta_2 X_{2i} + \beta_3 X_{3i} + \varepsilon_i, \quad 1 \leq i \leq 5. \tag{1.2}$$

Equation (1.2) may be represented by the matrix equation

$$\begin{bmatrix} y_1 \\ y_2 \\ y_3 \\ y_4 \\ y_5 \end{bmatrix} = \begin{bmatrix} 1 & X_{11} & X_{21} & X_{31} \\ 1 & X_{12} & X_{22} & X_{32} \\ 1 & X_{13} & X_{23} & X_{33} \\ 1 & X_{14} & X_{24} & X_{34} \\ 1 & X_{15} & X_{25} & X_{35} \end{bmatrix} \begin{bmatrix} \beta_0 \\ \beta_1 \\ \beta_2 \\ \beta_3 \end{bmatrix} + \begin{bmatrix} \varepsilon_1 \\ \varepsilon_2 \\ \varepsilon_3 \\ \varepsilon_4 \\ \varepsilon_5 \end{bmatrix}. \tag{1.3}$$

In experimental design models, the matrix is frequently zeros and ones indicating the level of a factor. An example of such a model would be

$$\begin{bmatrix} Y_{11} \\ Y_{12} \\ Y_{13} \\ Y_{14} \\ Y_{21} \\ Y_{22} \\ Y_{23} \\ Y_{31} \\ Y_{32} \end{bmatrix} = \begin{bmatrix} 1 & 1 & 0 & 0 \\ 1 & 1 & 0 & 0 \\ 1 & 1 & 0 & 0 \\ 1 & 1 & 0 & 0 \\ 1 & 0 & 1 & 0 \\ 1 & 0 & 1 & 0 \\ 1 & 0 & 1 & 0 \\ 1 & 0 & 0 & 1 \\ 1 & 0 & 0 & 1 \end{bmatrix} \begin{bmatrix} \mu \\ \alpha_1 \\ \alpha_2 \\ \alpha_3 \end{bmatrix} + \begin{bmatrix} \varepsilon_{11} \\ \varepsilon_{12} \\ \varepsilon_{13} \\ \varepsilon_{14} \\ \varepsilon_{21} \\ \varepsilon_{22} \\ \varepsilon_{23} \\ \varepsilon_{31} \\ \varepsilon_{32} \end{bmatrix}. \tag{1.4}$$

This is an unbalanced one-way analysis of variance (ANOVA) model where there are three treatments with four observations of treatment 1, three observations of treatment 2, and two observations of treatment 3. Different kinds of ANOVA models will be studied in Part IV. □

1.4 SUMMARY

We have accomplished the following. First, we have explained what matrices are and illustrated how they can be used to summarize data. Second, we defined three basic matrix operations: addition, scalar multiplication, and matrix multiplication. Third, we have shown how matrices have some properties similar to numbers and do not share some properties that numbers have. Fourth, we have given some applications to probability and to linear models.

EXERCISES

1.1 Let

$$A = \begin{bmatrix} 3 & -1 & 2 \\ 4 & 6 & 0 \end{bmatrix} \quad \text{and} \quad B = \begin{bmatrix} 1 & 3 \\ -3 & -1 \\ 2 & 4 \end{bmatrix}.$$

Find AB and BA.

1.2 Let

$$C = \begin{bmatrix} 1 & 1 \\ 1 & 1 \end{bmatrix} \quad \text{and} \quad D = \begin{bmatrix} 1 & -1 \\ -1 & 1 \end{bmatrix}.$$

A. Show that $CD = DC = 0$.
B. Does C or D have a multiplicative inverse? If yes, find it. If not, why not?

1.3   Let

$$E = \begin{bmatrix} 1 & 2 \\ 3 & 4 \end{bmatrix} \quad \text{and} \quad F = \begin{bmatrix} 4 & 3 \\ 2 & 1 \end{bmatrix}.$$

Show that $EF \neq FE$.

1.4   Let

$$G = \begin{bmatrix} 5 & 1 \\ 1 & 5 \end{bmatrix} \quad \text{and} \quad H = \begin{bmatrix} 7 & -1 \\ -1 & 7 \end{bmatrix}.$$

A. Show that $GH = HG$.
B. Does this commutativity hold when

$$G = \begin{bmatrix} a & b \\ b & a \end{bmatrix}, H = \begin{bmatrix} c & -b \\ -b & c \end{bmatrix}$$

1.5   A diagonal matrix is a square matrix for which all of the elements that are not in the main diagonal are zero. Show that diagonal matrices commute.

1.6   Let P be a matrix with the property that $PP' = I$ and $P'P = I$. Let D_1 and D_2 be diagonal matrices. Show that the matrices $P'D_1P$ and $P'D_2P$ commute.

1.7   Show that any matrix is the sum of a symmetric matrix and a skew-symmetric matrix.

1.8   Show that in general for any matrices A and B that
   A. $A'' = A$.
   B. $(A + B)' = A' + B'$.
   C. $(AB)' = B'A'$.

1.9   Show that if A and B commute, then
$$A'B' = B'A'.$$

1.10   Determine whether the matrices

$$A = \begin{bmatrix} \dfrac{5}{2} & -\dfrac{3}{2} \\ -\dfrac{3}{2} & \dfrac{5}{2} \end{bmatrix} \quad \text{and} \quad B = \begin{bmatrix} \dfrac{13}{2} & -\dfrac{5}{2} \\ -\dfrac{5}{2} & \dfrac{13}{2} \end{bmatrix}$$

commute.

1.11   For the model (1.3), write the entries of $X'X$ using the appropriate sum notation.

1.12   For the data on gross national product in Example 1.1
   A. What is the X matrix? The Y matrix?
   B. Write the system of equations $X'X\beta = X'Y$ with numbers.
   C. Find the values of the β parameters that satisfy the system.

1.13 For the model in (1.4)

    A. Write out the nine equations represented by the matrix equation.

    B. Find X'X.

    C. What are the entries of X'Y? Use the appropriate sum notation.

    D. Write out the system of four equations X'Xα=X'Y.

    E. Let

$$G = \begin{bmatrix} 0 & 0 & 0 & 0 \\ 0 & \dfrac{1}{4} & 0 & 0 \\ 0 & 0 & \dfrac{1}{3} & 0 \\ 0 & 0 & 0 & \dfrac{1}{2} \end{bmatrix}.$$

Show that α=GX'Y satisfies the system of equations in D. The matrix G is an example of a generalized inverse. Generalized inverses will be studied in Part III.

1.14 Show that for any matrix X, X'X, and XX' are symmetric matrices.

1.15 Let A and B be two 2×2 matrices where the rows and the columns add up to 1. Show that AB has this property.

1.16 Consider the linear model

$$\begin{aligned} y_{11} &= \mu + \alpha_1 + \beta_1 + \varepsilon_{11} \\ y_{12} &= \mu + \alpha_1 + \beta_2 + \varepsilon_{12} \\ y_{21} &= \mu + \alpha_2 + \beta_1 + \varepsilon_{21} \\ y_{22} &= \mu + \alpha_2 + \beta_2 + \varepsilon_{22}. \end{aligned}$$

    A. Tell what the matrices should be for a model in the form

$$Y = X\gamma + \varepsilon$$

where Y is 4×1, X is 4×5, γ is 5×1, and ε is 4×1.

    B. Find X'X.

1.17   **A.** Find P^3 for the transition matrix in Example 1.2.

    B. Given the initial probability vector in Example 1.4, find

$$\pi^{(2)}, \pi^{(3)} = \pi^{(2)}P.$$

1.18 Suppose in Example 1.2 two coins are flipped instead of a die. A person in room 0 goes to room 1 if no heads are obtained, room 2 if one head is obtained, and room 3 if two heads are obtained. A person in rooms 1, 2, or 3 advances one room in the clockwise direction if no heads are obtained, goes to room 0 if one head is obtained, and advances one room in the counterclockwise direction if two heads are obtained. Assume that the coins are fair.

A. Find the transition matrix P.

B. Find P, P^2, and P^3.

C. Given $\pi^{(0)} = [1/2, \ 1/4, \ 1/16, \ 3/16]$, find $\pi^{(1)}, \pi^{(2)}, \pi^{(3)}$.

1.19 Give examples of nonzero 3×3 matrices whose product is zero and for which the cancellation rule fails.

1.20 A matrix A is nilpotent if there is an n such that $A^n = 0$. Show that

$$A = \begin{bmatrix} 0 & 0 & 0 & 0 \\ 1 & 0 & 0 & 0 \\ 0 & 2 & 0 & 0 \\ 0 & 0 & 3 & 0 \end{bmatrix}$$

is a nilpotent matrix.

SECTION 2

DETERMINANTS AND SOLVING A SYSTEM OF EQUATIONS

2.1 INTRODUCTION

This section will review informally how to find determinants of matrices and their use in solving systems of equations. We give the definition of determinants and show how to evaluate them by expanding along rows and columns. Some tricks for evaluating determinants are given that are based on elementary row and column operations on a matrix. We show how to solve systems of linear equations by Cramer's rule and Gauss elimination.

2.2 DEFINITION OF AND FORMULAE FOR EXPANDING DETERMINANTS

Let A be an $n \times n$ matrix. Let A_{ij} be the $(n-1) \times (n-1)$ submatrix formed by deleting the ith row and the jth column. Then formulae for expanding determinants are

$$\det(A) = \sum_{j=1}^{n} (-1)^{i+j} a_{ij} A_{ij}, \quad 1 \le i \le n \tag{2.1a}$$

and

$$\det(A) = \sum_{i=1}^{n} (-1)^{i+j} a_{ij} A_{ij}, \quad 1 \le j \le n. \tag{2.1b}$$

Matrix Algebra for Linear Models, First Edition. Marvin H. J. Gruber.
© 2014 John Wiley & Sons, Inc. Published 2014 by John Wiley & Sons, Inc.

These are the formulae used to compute determinants. The actual definition of a determinant is

$$\det(A) = \sum_{\text{overall permutations } \sigma} \text{sign}(\sigma) a_{1\sigma(1)} a_{2\sigma(2)} \cdots a_{n\sigma(n)}. \tag{2.2}$$

If $D = \prod_{i<j}(x_i - x_j)$, then depending on σ, $D' = \prod_{i<j}(x_{\sigma(i)} - x_{\sigma(j)}) = \pm D$, and sign$(\sigma)$ is the sign in front of D. The proof of the equality of all expressions in (2.1) and (2.2) is available in other books (see, e.g., Harville (2008)).

For example, suppose that $n = 3$. Then

$$\det(A) = \begin{vmatrix} a_{11} & a_{12} & a_{13} \\ a_{21} & a_{22} & a_{23} \\ a_{31} & a_{32} & a_{33} \end{vmatrix}$$

$$= a_{11}A_{11} - a_{12}A_{12} + a_{13}A_{13}$$

$$= a_{11}(a_{22}a_{33} - a_{23}a_{32}) - a_{12}(a_{21}a_{33} - a_{23}a_{31}) + a_{13}(a_{21}a_{32} - a_{22}a_{31})$$

$$= a_{11}a_{22}a_{33} + a_{12}a_{23}a_{31} + a_{13}a_{21}a_{32}$$

$$\quad - a_{11}a_{23}a_{32} - a_{12}a_{21}a_{33} - a_{13}a_{22}a_{31}$$

$$= -a_{12}(a_{21}a_{33} - a_{23}a_{31}) + a_{22}(a_{11}a_{33} - a_{13}a_{31}) - a_{32}(a_{11}a_{23} - a_{13}a_{21})$$

$$= -a_{12}A_{12} + a_{22}A_{22} - a_{32}A_{32}.$$

There are four other equalities that may be established. This is left to the reader. The above formulae indicate that a determinant may be expanded along any of the rows or columns.

Example 2.1 Calculation of a Determinant

Expanding along the first row, we have

$$\begin{vmatrix} 1 & 1 & 1 \\ 1 & 2 & 4 \\ 1 & 3 & 9 \end{vmatrix} = 1 \begin{vmatrix} 2 & 4 \\ 3 & 9 \end{vmatrix} - 1 \begin{vmatrix} 1 & 4 \\ 1 & 9 \end{vmatrix} + 1 \begin{vmatrix} 1 & 2 \\ 1 & 3 \end{vmatrix}$$

$$= ((2)(9) - (4)(3)) - ((1)(9) - (4)(1)) + ((1)(3) - (2)(1))$$

$$= 6 - 5 + 1 = 2.$$

Also, for example, expanding along the second column,

$$\begin{vmatrix} 1 & 1 & 1 \\ 1 & 2 & 4 \\ 1 & 3 & 9 \end{vmatrix} = (-1) \begin{vmatrix} 1 & 4 \\ 1 & 9 \end{vmatrix} + 2 \begin{vmatrix} 1 & 1 \\ 1 & 9 \end{vmatrix} - 3 \begin{vmatrix} 1 & 1 \\ 1 & 4 \end{vmatrix}$$

$$= (-1)((1)(9) - (4)(1)) + 2((1)(9) - (1)(1)) - 3((1)(4) - (1)(1))$$

$$= -5 + 16 - 9 = 2. \qquad \square$$

The reader may, if he or she wishes, calculate the determinant expanding along the remaining rows and columns and verify that the answer is indeed the same.

2.3 SOME COMPUTATIONAL TRICKS FOR THE EVALUATION OF DETERMINANTS

There are a few properties of determinants that are easily verified and applied that make their expansion easier. Instead of formal proofs, we give some illustrations to illustrate how these rules are true:

1. The determinant of a square matrix with two or more identical rows or columns is zero. Notice that, for example,

$$
\begin{vmatrix} a_{11} & a_{12} & a_{13} \\ c & c & c \\ c & c & c \end{vmatrix} = a_{11}\begin{vmatrix} c & c \\ c & c \end{vmatrix} - a_{12}\begin{vmatrix} c & c \\ c & c \end{vmatrix} + a_{13}\begin{vmatrix} c & c \\ c & c \end{vmatrix} = 0.
$$

2. If a row or column is multiplied by a number, the value of the determinant is multiplied by that number. For example, expanding along the first column,

$$
\begin{vmatrix} ka_{11} & a_{12} & a_{13} \\ ka_{21} & a_{22} & a_{23} \\ ka_{31} & a_{32} & a_{33} \end{vmatrix} = ka_{11}\begin{vmatrix} a_{22} & a_{23} \\ a_{32} & a_{33} \end{vmatrix} - ka_{21}\begin{vmatrix} a_{12} & a_{13} \\ a_{32} & a_{33} \end{vmatrix} + ka_{31}\begin{vmatrix} a_{12} & a_{13} \\ a_{22} & a_{23} \end{vmatrix}
$$

$$
= k\left(a_{11}\begin{vmatrix} a_{22} & a_{23} \\ a_{32} & a_{33} \end{vmatrix} - a_{21}\begin{vmatrix} a_{12} & a_{13} \\ a_{32} & a_{33} \end{vmatrix} + a_{31}\begin{vmatrix} a_{12} & a_{13} \\ a_{22} & a_{23} \end{vmatrix}\right) = k\begin{vmatrix} a_{11} & a_{12} & a_{13} \\ a_{21} & a_{22} & a_{23} \\ a_{31} & a_{32} & a_{33} \end{vmatrix}.
$$

3. If a row or column is exchanged with another row or column, the absolute value of the determinant is the same, but the sign changes. For example, in a three-by-three determinant, exchange rows one and three. Now expand along row two and obtain

$$
D = \begin{vmatrix} a_{31} & a_{32} & a_{33} \\ a_{21} & a_{22} & a_{23} \\ a_{11} & a_{12} & a_{13} \end{vmatrix} = -a_{21}\begin{vmatrix} a_{32} & a_{33} \\ a_{12} & a_{13} \end{vmatrix} + a_{22}\begin{vmatrix} a_{31} & a_{33} \\ a_{11} & a_{13} \end{vmatrix} - a_{23}\begin{vmatrix} a_{31} & a_{33} \\ a_{11} & a_{13} \end{vmatrix}.
$$

Observe that, for example,

$$
\begin{vmatrix} a_{32} & a_{33} \\ a_{12} & a_{13} \end{vmatrix} = a_{32}a_{13} - a_{33}a_{12} = -(a_{12}a_{33} - a_{13}a_{32}) = -\begin{vmatrix} a_{12} & a_{13} \\ a_{32} & a_{33} \end{vmatrix}.
$$

Similarly, each of the other two-by-two determinants is the negative of the determinant of the matrix with the two rows exchanged so that

$$D = - \begin{vmatrix} a_{11} & a_{12} & a_{13} \\ a_{21} & a_{22} & a_{23} \\ a_{31} & a_{32} & a_{33} \end{vmatrix}.$$

4. If a constant multiplied by a row or column of a determinant is added to another row or column, the value of the determinant is unchanged. For example,

$$\begin{vmatrix} a_{11}+ka_{21} & a_{12}+ka_{22} & a_{13}+ka_{23} \\ a_{21} & a_{22} & a_{23} \\ a_{32} & a_{32} & a_{33} \end{vmatrix} = \begin{vmatrix} a_{11} & a_{12} & a_{13} \\ a_{21} & a_{22} & a_{23} \\ a_{32} & a_{32} & a_{33} \end{vmatrix} + k \begin{vmatrix} a_{21} & a_{22} & a_{23} \\ a_{21} & a_{22} & a_{23} \\ a_{32} & a_{32} & a_{33} \end{vmatrix}$$

after expanding the determinant along the first row, applying the distributive law, and rewriting the three-by-three determinants and observing that the second determinant of the right-hand side is zero by Rule 1.

The fourth property stated above is particularly useful for expanding a determinant. The objective is to add multiples of rows or columns in such a way to get as many zeros as possible in a particular row and column. The determinant is then easily expanded along that row or column.

Example 2.2 Continuation of Example 2.1

Consider the determinant D in Example 2.1. When applying the rules, above a goal is to get some zeros in a row or column and then expand the determinant along that row or column. For the determinant D below, we subtract the first row from the second and third row obtaining two zeros in the first column. We then expand along that column to obtain

$$D = \begin{vmatrix} 1 & 1 & 1 \\ 1 & 2 & 4 \\ 1 & 3 & 9 \end{vmatrix} = \begin{vmatrix} 1 & 1 & 1 \\ 0 & 1 & 3 \\ 0 & 1 & 5 \end{vmatrix} = \begin{vmatrix} 1 & 3 \\ 1 & 5 \end{vmatrix} = \begin{vmatrix} 1 & 3 \\ 0 & 2 \end{vmatrix} = (1)(2) - (3)(0) = 2.$$

□

Example 2.3 Determinant of a Triangular Matrix

One possibility for the expansion of a determinant is to use the rules to put it in upper or lower triangular form. A matrix is in upper triangular form if all of the elements below the main diagonal are zero. Likewise a matrix is in lower triangular form if all of the elements above the main diagonal are zero. The resulting determinant is then the product of the elements in the main diagonal.

Consider the determinant

$$M = \begin{vmatrix} 3 & 5 & 9 & 11 \\ 2 & 3 & 8 & 2 \\ 3 & 6 & 9 & 12 \\ 6 & 2 & 8 & 4 \end{vmatrix}.$$

The following steps are one way to reduce the matrix to upper triangular form:

1. Factor 3 from the third row and 2 from the fourth row.
2. Subtract 2/3 times the first row from the second row and 1/3 times the first row from the third row and the first row from the fourth row obtaining three zeros in the first column.
3. Add the second row to the third row and add 12 times the third row to the fourth row.
4. Multiply the second row by three and the fourth row by 2 and expand the upper triangular matrix.

Thus,

$$M = \begin{vmatrix} 3 & 5 & 9 & 11 \\ 2 & 3 & 8 & 2 \\ 3 & 6 & 9 & 12 \\ 6 & 2 & 8 & 4 \end{vmatrix} = (3)(2)\begin{vmatrix} 3 & 5 & 9 & 11 \\ 2 & 3 & 8 & 2 \\ 1 & 2 & 3 & 4 \\ 3 & 1 & 4 & 2 \end{vmatrix} = 6\begin{vmatrix} 3 & 5 & 9 & 11 \\ 0 & -\frac{1}{3} & 2 & -\frac{16}{3} \\ 0 & \frac{1}{3} & 0 & \frac{1}{3} \\ 0 & -4 & -5 & -9 \end{vmatrix}$$

$$= 6\begin{vmatrix} 3 & 5 & 9 & 11 \\ 0 & -\frac{1}{3} & 2 & -\frac{16}{3} \\ 0 & 0 & 2 & -5 \\ 0 & 0 & -5 & -5 \end{vmatrix} = \begin{vmatrix} 3 & 5 & 9 & 11 \\ 0 & -1 & 6 & -16 \\ 0 & 0 & 2 & -5 \\ 0 & 0 & 0 & -35 \end{vmatrix} = 210.$$

□

If A is a square matrix $\det(A') = \det A$, expanding A using, say, the ith row will give the same result as expanding A' using the ith column.

An important property of determinants is that

$$\det(AB) = \det(A)\det(B).$$

A sketch of a proof of this important fact will be given in Subsection 4.5.

2.4   SOLUTION TO LINEAR EQUATIONS USING DETERMINANTS

Let A be a square $n \times n$ matrix with a nonzero determinant. A system of equations may be written

$$Ax = b.$$

The matrix A is called the coefficient matrix. The matrix [A, b] is called the augmented matrix.

For example, the system

$$2x + 3y = 12$$
$$3x - 4y = 1$$

can be written

$$\begin{bmatrix} 2 & 3 \\ 3 & -4 \end{bmatrix} \begin{bmatrix} x \\ y \end{bmatrix} = \begin{bmatrix} 12 \\ 1 \end{bmatrix}.$$

The system

$$x - y = 2$$
$$y - z = 2$$
$$x + y + z = 9$$

may be written

$$\begin{bmatrix} 1 & -1 & 0 \\ 0 & 1 & -1 \\ 1 & 1 & 1 \end{bmatrix} \begin{bmatrix} x \\ y \\ z \end{bmatrix} = \begin{bmatrix} 2 \\ 2 \\ 9 \end{bmatrix}.$$

There are some different methods of solving these equations. These include Cramer's method, Gauss elimination, and the use of the inverse of the matrix A. We will take up Cramer's rule first in this subsection. The next subsection will take up Gauss elimination, and the use of the inverse of a matrix will be discussed in Section 3.

Consider two equations in two unknowns, say

$$a_{11}x_1 + a_{12}x_2 = b_1$$
$$a_{21}x_1 + a_{22}x_2 = b_2.$$

Following a procedure that is generally taught in high school, we eliminate the x_2 variable by multiplying the first equation by a_{22} and the second equation by $-a_{12}$.

We then add the results to obtain

$$(a_{11}a_{22} - a_{12}a_{21})x_1 = b_1a_{22} - b_2a_{12}$$

so that

$$x_1 = \frac{\begin{vmatrix} b_1 & a_{12} \\ b_2 & a_{22} \end{vmatrix}}{\begin{vmatrix} a_{11} & a_{12} \\ a_{21} & a_{22} \end{vmatrix}}.$$

Similarly, the variable x_1 could be eliminated, and we would obtain

$$x_2 = \frac{\begin{vmatrix} a_{11} & b_1 \\ a_{21} & b_2 \end{vmatrix}}{\begin{vmatrix} a_{11} & a_{12} \\ a_{21} & a_{22} \end{vmatrix}}.$$

This is Cramer's rule for two equations in two unknowns. It is an easy formula to remember because the matrix in the denominator is the coefficients of x_1 and x_2 in the two equations and the matrix in the numerator is the same with the numbers on the right-hand side of the equation replacing the first column for x_1 and the second column for x_2. This idea generalizes to n equations in n unknowns. We now give a more formal derivation of Cramer's rule.

Let $C_{ij} = (-1)^{i+j} A_{ij}$ where A_{ij} is the $(n-1) \times (n-1)$ submatrix formed by deleting the ith row and the jth column. The C_{ij} are called cofactors. Let adj(A) (the adjoint matrix of A) be the matrix with elements C_{ji}, the transpose of the matrix of cofactors. The elements of the matrix

$$\text{adj}(A)A \text{ are} \sum_{k=1}^{n} a_{ik} C_{kj} = \begin{cases} \det(A) & i = j \\ 0 & i \neq j \end{cases}. \tag{2.3}$$

Thus,

$$\text{adj}(A)A = \det(A)I. \tag{2.4}$$

Let A_j represent the $n \times n$ matrix with the elements of A everywhere but the jth column. The jth column consists of the n-dimensional column vector b. The determinant of A_j is the scalar product of the jth row of adj(A) and b. Then since $Ax = b$,

$$\text{adj}(A)b = \text{adj}(A)Ax = \det(A)Ix, \tag{2.5}$$

and

$$x = \frac{1}{\det(A)} \text{adj}(A)b = \frac{1}{\det(A)} \begin{bmatrix} \sum_{j=1}^{n} b_j C_{j1} \\ \vdots \\ \sum_{j=1}^{n} b_j C_{jn} \end{bmatrix} = \frac{1}{\det(A)} \begin{bmatrix} \det(A_1) \\ \vdots \\ \det(A_n) \end{bmatrix} \tag{2.6}$$

Example 2.4 Solution of a System of Two Equations in Two Unknowns by Cramer's Rule

Consider the system of equations

$$2x + 3y = 12$$
$$3x - 4y = 1.$$

Observe that for x the vector $\begin{bmatrix} 12 \\ 1 \end{bmatrix}$ is in the first column of the determinant in the numerator and for y the same vector is in the second column of the numerator. Thus,

$$x = \frac{\begin{vmatrix} 12 & 3 \\ 1 & -4 \end{vmatrix}}{\begin{vmatrix} 2 & 3 \\ 3 & -4 \end{vmatrix}} = \frac{12(-4) - (3)(1)}{2(-4) - (3)(3)} = \frac{-51}{-17} = 3$$

$$y = \frac{\begin{vmatrix} 2 & 12 \\ 3 & 1 \end{vmatrix}}{\begin{vmatrix} 2 & 3 \\ 3 & -4 \end{vmatrix}} = \frac{(2)(1) - (12)(3)}{-17} = \frac{-34}{-17} = 2.$$

□

Example 2.5 Solution of a System of Equations in Matrix Form

Consider

$$\begin{bmatrix} 1 & -1 & 0 \\ 0 & 1 & -1 \\ 1 & 1 & 1 \end{bmatrix} \begin{bmatrix} x \\ y \\ z \end{bmatrix} = \begin{bmatrix} 2 \\ 2 \\ 9 \end{bmatrix}.$$

Observe that

$$\begin{vmatrix} 1 & -1 & 0 \\ 0 & 1 & -1 \\ 1 & 1 & 1 \end{vmatrix} = \begin{vmatrix} 1 & -1 & 0 \\ 0 & 1 & -1 \\ 1 & 2 & 0 \end{vmatrix} = \begin{vmatrix} 1 & -1 \\ 1 & 2 \end{vmatrix} = 3$$

and that

$$x = \frac{\begin{vmatrix} 2 & -1 & 0 \\ 2 & 1 & -1 \\ 9 & 1 & 1 \end{vmatrix}}{3} = 5,$$

$$y = \frac{\begin{vmatrix} 1 & 2 & 0 \\ 0 & 2 & -1 \\ 1 & 9 & 1 \end{vmatrix}}{3} = 3,$$

$$z = \frac{\begin{vmatrix} 1 & -1 & 2 \\ 0 & 1 & 2 \\ 1 & 1 & 9 \end{vmatrix}}{3} = 1.$$

□

2.5   GAUSS ELIMINATION

Gauss elimination reduces the coefficient matrix of a system of linear equations to an upper triangular form using elementary row operations giving the answer for the last of the variables. The other variables may be found progressively as linear combinations. First, write down the matrix consisting of the coefficients of the variables and the numbers on the right-hand side. The matrices will yield equivalent equations with the same solution if:

1. Two rows are interchanged.
2. A row is multiplied by a number.
3. A row is multiplied by a number and added to another row. Here the objective is to make the leading coefficient zero.

The same operations may be performed on the columns. These operations are called elementary row and column operations.

The goal is to get an equivalent system of equations where the bottom equation has one variable (for purposes of this section), the next equation up has at most two variables, the equation second from the bottom has at most three variables, and so on. The examples below provide illustrations.

Example 2.6 Two Equations in Two Unknowns

$$3x - 2y = 5$$
$$4x + y = 3$$

The coefficient matrix is the first matrix below. To get the second matrix, subtract 4/3 times the first row from the second row. Now multiply the second row by 3/11:

$$\begin{bmatrix} 3 & -2 & 5 \\ 4 & 1 & 3 \end{bmatrix} \rightarrow \begin{bmatrix} 3 & -2 & 5 \\ 0 & \frac{11}{3} & -\frac{11}{3} \end{bmatrix} \rightarrow \begin{bmatrix} 3 & -2 & 5 \\ 0 & 1 & -1 \end{bmatrix}.$$

The equivalent system of equations is

$$3x - 2y = 5$$
$$y = -1.$$

Then $3x - 2(-1) = 5$ and $3x = 3$ so $x = 1$.   □

Example 2.7 Three Equations in Three Unknowns

Consider the system in Example 2.5

$$x - y = 2$$
$$y - z = 2$$
$$x + y + z = 9.$$

Now we have

$$\begin{bmatrix} 1 & -1 & 0 & 2 \\ 0 & 1 & -1 & 2 \\ 1 & 1 & 1 & 9 \end{bmatrix} \rightarrow \begin{bmatrix} 1 & 1 & 1 & 9 \\ 0 & 1 & -1 & 2 \\ 1 & -1 & 0 & 2 \end{bmatrix} \rightarrow \begin{bmatrix} 1 & 1 & 1 & 9 \\ 0 & 1 & -1 & 2 \\ 0 & -2 & -1 & -7 \end{bmatrix}$$

$$\rightarrow \begin{bmatrix} 1 & 1 & 1 & 9 \\ 0 & 1 & -1 & 1 \\ 0 & 0 & -3 & -3 \end{bmatrix} \rightarrow \begin{bmatrix} 1 & 1 & 1 & 9 \\ 0 & 1 & -1 & 1 \\ 0 & 0 & 1 & 1 \end{bmatrix}.$$

The steps here are:

1. Exchange row 1 and row 3.
2. Subtract row 1 from row 3.
3. Multiply row 2 by 2 and add it to row 3.
4. Multiply row 3 by $-1/3$.

The resulting equivalent system of equations is

$$x + y + z = 9$$
$$y - z = 2$$
$$z = 1.$$

Then $y - 1 = 2$ so $y = 3$ and $x + 3 + 1 = 9$ so $x = 5$. $\qquad\square$

Not all systems of linear equations have a unique solution. They may not have a solution or they may have infinitely many solutions. We illustrate this in Examples 2.8 and 2.9.

Example 2.8 A System with Infinitely Many Solutions

Consider the system of equations

$$x + y = 4$$
$$x + z = 3$$
$$x + w = 2$$
$$3x + y + z + w = 9.$$

The augmented matrix of coefficients in this system is

$$\begin{bmatrix} 1 & 1 & 0 & 0 & 4 \\ 1 & 0 & 1 & 0 & 3 \\ 1 & 0 & 0 & 1 & 2 \\ 3 & 1 & 1 & 1 & 9 \end{bmatrix} \rightarrow \begin{bmatrix} 1 & 1 & 0 & 0 & 4 \\ 0 & -1 & 1 & 0 & -1 \\ 0 & 0 & -1 & 1 & -1 \\ 0 & 0 & 0 & 0 & 0 \end{bmatrix}.$$

Here the sum of rows one, two, and three was subtracted from row 4, row 2 was subtracted from row three, and row one was subtracted from row 2. The equivalent system of equations is

$$x + y = 4$$
$$-y + z = -1$$
$$--z + w = -1$$
$$0x + 0y + 0z + 0w = 0.$$

We wind up with only three equations in four unknowns with nonzero coefficients. Such a system can be solved but only in terms of one of the variables. Let us solve the system in terms of w. Then

$$z = w + 1$$
$$y = w + 2$$
$$x = 2 - w$$

Thus, there is a different solution for each w. For $w = 0$ $x = 2$, $y = 2$, $z = 1$. For $w = 1$ $x = 1$, $y = 3$ and $z = 2$.   □

Example 2.9 A System of Equations with No Solutions

Suppose we have the same system of equations as in Example 2.8 except that the last one is slightly different. We have

$$x + y = 4$$
$$-y + z = -1$$
$$-z + w = -1$$
$$3x + y + z + w = 10$$

Following the same elimination process as before, we get

$$\begin{bmatrix} 1 & 1 & 0 & 0 & 4 \\ 1 & 0 & 1 & 0 & 3 \\ 1 & 0 & 0 & 1 & 2 \\ 3 & 1 & 1 & 1 & 10 \end{bmatrix} \rightarrow \begin{bmatrix} 1 & 1 & 0 & 0 & 4 \\ 0 & -1 & 1 & 0 & -1 \\ 0 & 0 & -1 & 1 & -1 \\ 0 & 0 & 0 & 0 & 1 \end{bmatrix}.$$

This time the equivalent system of equations is

$$x + y = 4$$
$$-y + z = -1$$
$$-z + w = -1$$
$$0x + 0y + 0z + 0w = 1,$$

a system with no solution. Such a system is inconsistent. More will be said about this later when conditions are obtained for systems equations to have one solution, no solution, or infinitely many solutions.   □

Example 2.10 Two Equations in Two Unknowns with Literal Coefficients

$$a_{11}x_1 + a_{12}x_2 = b_1$$
$$a_{21}x_1 + a_{22}x_2 = b_2$$

Using Gauss elimination,

$$\begin{bmatrix} a_{11} & a_{12} & b_1 \\ a_{21} & a_{22} & b_2 \end{bmatrix} \rightarrow \begin{bmatrix} a_{11} & a_{12} & b_1 \\ 0 & a_{22} - a_{11}^{-1}a_{21}a_{12} & b_2 - a_{11}^{-1}a_{21}b_1 \end{bmatrix} \rightarrow \begin{bmatrix} a_{11} & a_{12} & b_1 \\ 0 & 1 & \dfrac{a_{22}a_{11} - a_{21}a_{12}}{b_2 a_{11} - a_{21}b_1} \end{bmatrix}$$

so that we have

$$a_{11}x_1 + a_{12}x_2 = b_1$$

$$x_2 = \frac{b_2 a_{11} - a_{21}b_1}{a_{22}a_{11} - a_{21}a_{12}}$$

$$= \frac{\begin{vmatrix} a_{11} & b_1 \\ a_{21} & b_2 \end{vmatrix}}{\begin{vmatrix} a_{11} & a_{12} \\ a_{21} & a_{22} \end{vmatrix}}$$

and

$$x_1 = a_{11}^{-1}b_1 - \frac{a_{11}^{-1}a_{12}(b_2 a_{11} - a_{21}b_1)}{a_{22}a_{11} - a_{21}a_{12}}$$

$$= \frac{b_1(a_{22}a_{11} - a_{21}a_{12}) - a_{12}(b_2 a_{11} - a_{21}b_1)}{a_{11}(a_{22}a_{11} - a_{21}a_{12})}$$

$$= \frac{a_{11}(b_1 a_{22} - b_2 a_{12})}{a_{11}(a_{22}a_{11} - a_{21}a_{12})}$$

$$= \frac{(b_1 a_{22} - b_2 a_{12})}{(a_{22}a_{11} - a_{21}a_{12})} = \frac{\begin{vmatrix} b_1 & a_{12} \\ b_2 & a_{22} \end{vmatrix}}{\begin{vmatrix} a_{11} & a_{12} \\ a_{21} & a_{22} \end{vmatrix}}.$$

For any system of n equations in n unknowns, a general algorithm would have the following steps:

1. Divide each row by the element in its first column.
2. Subtract the first row from each of rows 2 to n. You now have zeros in the first column below the first row.
3. Divide each of rows 2 to n by the element in the second column.
4. Subtract the second row from each of rows 3 to n. You now have zeros from the third row down.
5. Repeat the process until you have an upper triangular matrix. Then divide the nth row by the element in the nth column.

If the nth row has a nonzero element in the nth column, the system has a unique solution. If the nth row has all zeros, the system has infinitely many solutions. If for the nth row there are zeros in the first n columns and a non-zero element in the $n+1$th column, the system is inconsistent and has no solutions. □

Example 2.11 An Illustration of the Suggested Algorithm Above

Consider the system of equations

$$2x + y + 2z = 10$$
$$3x - 2y + z = 2$$
$$6x + 4y + 5z = 29.$$

Following the algorithm above, we have

$$
\begin{bmatrix} 2 & 1 & 2 & 10 \\ 3 & -2 & 1 & 2 \\ 6 & 4 & 5 & 29 \end{bmatrix}
\rightarrow
\begin{bmatrix} 1 & \frac{1}{2} & 1 & 5 \\ 1 & -\frac{2}{3} & \frac{1}{3} & \frac{2}{3} \\ 1 & \frac{2}{3} & \frac{5}{6} & \frac{29}{6} \end{bmatrix}
\rightarrow
\begin{bmatrix} 1 & \frac{1}{2} & 1 & 5 \\ 0 & -\frac{7}{6} & -\frac{2}{3} & -\frac{13}{3} \\ 0 & \frac{1}{6} & -\frac{1}{6} & -\frac{1}{6} \end{bmatrix}
\rightarrow
\begin{bmatrix} 1 & \frac{1}{2} & 1 & 5 \\ 0 & 1 & \frac{4}{7} & \frac{26}{7} \\ 0 & 1 & -1 & -1 \end{bmatrix}
$$

$$
\rightarrow
\begin{bmatrix} 1 & \frac{1}{2} & 1 & 5 \\ 0 & 1 & \frac{4}{7} & \frac{26}{7} \\ 0 & 0 & -\frac{11}{7} & -\frac{33}{7} \end{bmatrix}
\rightarrow
\begin{bmatrix} 1 & \frac{1}{2} & 1 & 5 \\ 0 & 1 & \frac{4}{7} & \frac{26}{7} \\ 0 & 0 & 1 & 3 \end{bmatrix}.
$$

The equivalent system of equations is

$$x + \tfrac{1}{2}y + z = 5$$
$$y + \tfrac{4}{7}z = \tfrac{26}{7}$$
$$z = 3.$$

Then by substitution into the equations above it,

$$y + \tfrac{4}{7}(3) = \tfrac{26}{7}$$
$$y = 2$$
$$x + \tfrac{1}{2}(2) + 3 = 5$$
$$x = 1. \qquad \square$$

2.6 SUMMARY

We have shown how to evaluate determinants and explained how to use them for the solution of systems of linear equations. We took up solution by Cramer's rule and Gauss elimination. We also gave illustrations of inconsistent systems of equations with no solution and systems of equations with infinitely many solutions.

EXERCISES

2.1 Find

A. $\begin{vmatrix} 6 & 3 & 3 \\ 3 & 3 & 0 \\ 3 & 0 & 3 \end{vmatrix}$

B. $\begin{vmatrix} 1 & 1 & 1 & 1 \\ 1 & 3 & 9 & 27 \\ 1 & 5 & 25 & 125 \\ 1 & 9 & 81 & 729 \end{vmatrix}$

C. $\begin{vmatrix} 1 & 2 & 0 & 0 \\ 3 & 4 & 0 & 0 \\ 0 & 0 & 1 & 2 \\ 0 & 0 & 3 & 4 \end{vmatrix}$

D. $\begin{vmatrix} 1 & 2 & 1 & 1 \\ 3 & 4 & 1 & 1 \\ 1 & 1 & 1 & 2 \\ 1 & 1 & 3 & 4 \end{vmatrix}$

E. $\begin{vmatrix} 0 & \frac{1}{3} & \frac{1}{3} & \frac{1}{3} \\ \frac{1}{3} & 0 & \frac{1}{3} & \frac{1}{3} \\ \frac{1}{3} & \frac{1}{3} & 0 & \frac{1}{3} \\ \frac{1}{3} & \frac{1}{3} & \frac{1}{3} & 0 \end{vmatrix}$

2.2   **A.** Show that

$$\begin{vmatrix} 1 & x_1 & x_1^2 & x_1^3 \\ 1 & x_2 & x_2^2 & x_2^3 \\ 1 & x_3 & x_3^2 & x_3^3 \\ 1 & x_4 & x_4^2 & x_4^2 \end{vmatrix} = \prod_{1 \le i < j \le 4} (x_i - x_j).$$

   B. Recalculate the determinant in 2.1B using the formula in Part A.

2.3   Verify the answers in Example 2.4 by evaluating the determinants.

2.4   Give an example to show that $\det(A+B) \ne \det A + \det B$.

2.5   Later we will show that in general for square matrices $\det(AB) = \det(A)\det(B)$. Show by direct calculation that this holds true when

$$A = \begin{bmatrix} 1 & 2 & 3 & 4 \\ 0 & 1 & 2 & 3 \\ 0 & 0 & 1 & 2 \\ 0 & 0 & 0 & 1 \end{bmatrix}, B = \begin{bmatrix} 0 & 0 & 0 & 1 \\ 0 & 0 & 1 & 2 \\ 0 & 1 & 2 & 3 \\ 1 & 2 & 3 & 4 \end{bmatrix}.$$

2.6   Solve the systems of equations by Cramer's rule and by Gauss elimination.

   A. $\begin{aligned} 3x - 4y &= 5 \\ x + 2y &= 9 \end{aligned}$

   B. $\begin{aligned} 3x + 4y + 5z &= 6 \\ 2x - 6y + 4z &= -14 \\ x + 3y - 5z &= 1 \end{aligned}$

   C. $\begin{aligned} x - y + z &= 3 \\ x + z + w &= 7 \\ x + 2y + 3w &= 13 \\ y + z + w &= 6 \end{aligned}$

2.7   **A.** Find all values of λ so that $\det(A-\lambda I)=0$ where I is the identity matrix containing 1 on the main diagonal and zero elsewhere and $A = \begin{bmatrix} 3 & 2 \\ 2 & 4 \end{bmatrix}$.

   B. Show that

$$A^2 - 7A + 8I = 0.$$

2.8   Let

$$A = \begin{bmatrix} a_{11} & a_{12} \\ a_{21} & a_{22} \end{bmatrix}, I = \begin{bmatrix} 1 & 0 \\ 0 & 1 \end{bmatrix}, trA = a_{11} + a_{12}.$$

Show that

$$A^2 - (trA)A + (\det A)I = 0.$$

2.9   Let A be as in 2.8 with all real entries. Show that the solutions to $\det(A - \lambda I) = 0$ are real

A. If and only if

$$\frac{(trA)^2}{\det A} \geq 4$$

B. If A is symmetric

C. When a_{12} and a_{21} have the same sign

2.10 For the system

$$a_{11}x_1 + a_{12}x_2 = b_1$$
$$a_{21}x_1 + a_{22}x_2 = b_2$$

studied at the beginning of Subsection 2.3, show how to obtain x_2 in the formula for Cramer's rule by the elimination process mentioned there.

SECTION 3

THE INVERSE OF A MATRIX

3.1 INTRODUCTION

Two methods of finding the inverse of a matrix will be taken up in this section. The first uses the adjoint of a matrix and gives a formula in terms of it. The second is based on elementary row operations. We will then present a method of solving a system of equations that uses a matrix inverse. We will show how to find the inverse of a partitioned matrix and in so doing derive an important matrix identity. Finally, we show how to find the least square estimator by using the matrix operations discussed here and in Sections 1 and 2.

3.2 THE ADJOINT METHOD OF FINDING THE INVERSE OF A MATRIX

Let A be a square matrix (one with the same number of rows and columns) where $\det(A) \neq 0$. Matrix B is the inverse of A if $AB = I$ and $BA = I$. The inverse matrix is denoted by A^{-1}. The method of finding the inverse of a matrix to be studied in this sub-section uses the adjoint and the determinant of a matrix.

From (2.4) recall that

$$\text{adj}(A)A = \det(A)I. \tag{3.1}$$

Then

$$A^{-1} = \frac{1}{\det(A)} \text{adj}(A). \tag{3.2}$$

Matrix Algebra for Linear Models, First Edition. Marvin H. J. Gruber.
© 2014 John Wiley & Sons, Inc. Published 2014 by John Wiley & Sons, Inc.

Thus, the solution for B in the equation $AB=I$ is A^{-1} as given in (3.2). Post-multiplication of (3.2) by A and multiplication by $det(A)$ give

$$A\,adj(A) = I\det(A) = \det(A)I. \tag{3.3}$$

Thus, for a square nonsingular matrix, only one of the conditions $AB=I$ and $BA=I$ must be shown to hold true to say that B is an inverse of A.

Example 3.1 The Inverse of a Two-by-Two Matrix by the Adjoint Method

Let

$$A = \begin{bmatrix} 1 & 3 \\ 2 & 4 \end{bmatrix}.$$

Then $det(A)=(1)(4)-(3)(2)=-2$. Also

$$adj(A) = \begin{bmatrix} 4 & -3 \\ -2 & 1 \end{bmatrix}.$$

Then

$$A^{-1} = \begin{bmatrix} -2 & \frac{3}{2} \\ 1 & -\frac{1}{2} \end{bmatrix}. \qquad \square$$

Example 3.2 The Inverse of a Three-by-Three Matrix

Let

$$G = \begin{bmatrix} 1 & 2 & 2 \\ 2 & 1 & 2 \\ 2 & 2 & 1 \end{bmatrix}.$$

The reader may verify that $det(G)=5$. Then

$$G^{-1} = \frac{1}{5}\begin{bmatrix} -3 & 2 & 2 \\ 2 & -3 & 2 \\ 2 & 2 & -3 \end{bmatrix}$$

where the matrix of cofactors is multiplied by the scalar 1/5. $\qquad \square$

3.3 USING ELEMENTARY ROW OPERATIONS

Another method is to write the matrix to be inverted down with the identity matrix to the right of it. Do elementary row operations to reduce the matrix on the left to the identity matrix. The matrix on the right will be the inverse.

Example 3.3 Inverse of a Two-by-Two Matrix Using Row Operations

Using the matrix of Example 3.1, we have

$$\begin{bmatrix} 1 & 3 & | & 1 & 0 \\ 2 & 4 & | & 0 & 1 \end{bmatrix} \rightarrow \begin{bmatrix} 1 & 2 & | & 1 & 0 \\ 0 & -2 & | & -2 & 1 \end{bmatrix} \rightarrow \begin{bmatrix} 1 & 0 & | & -2 & \frac{3}{2} \\ 0 & -2 & | & -2 & 1 \end{bmatrix} \rightarrow \begin{bmatrix} 1 & 0 & | & -2 & \frac{3}{2} \\ 0 & 1 & | & 1 & -\frac{1}{2} \end{bmatrix}.$$

The inverse is that obtained in Example 3.1. The steps were:

1. Multiply the first row by −2 and add it to the second row. The matrix being inverted is now upper triangular.
2. Multiply the second row by 3/2 and add it to the first row.
3. Divide the second row by −2.

The inverse now appears on the right, namely,

$$A^{-1} = \begin{bmatrix} -2 & \frac{3}{2} \\ 1 & -\frac{1}{2} \end{bmatrix}.$$

□

Example 3.4 Inverting a Three-by-Three Matrix by Row Operations

Consider the matrix from Example 3.2. The steps required to obtain its inverse follow:

$$\begin{bmatrix} 1 & 2 & 2 & | & 1 & 0 & 0 \\ 2 & 1 & 2 & | & 0 & 1 & 0 \\ 2 & 2 & 1 & | & 0 & 0 & 1 \end{bmatrix} \rightarrow \begin{bmatrix} 1 & 2 & 2 & | & 1 & 0 & 0 \\ 0 & -3 & -2 & | & -2 & 1 & 0 \\ 0 & -2 & -3 & | & -2 & 0 & 1 \end{bmatrix} \rightarrow \begin{bmatrix} 1 & 2 & 2 & | & 1 & 0 & 0 \\ 0 & -3 & -2 & | & -2 & 1 & 0 \\ 0 & 0 & -\frac{5}{3} & | & -\frac{2}{3} & -\frac{2}{3} & 1 \end{bmatrix}$$

$$\rightarrow \begin{bmatrix} 1 & 2 & 2 & | & 1 & 0 & 0 \\ 0 & -3 & -2 & | & -2 & 1 & 0 \\ 0 & 0 & 1 & | & \frac{2}{5} & \frac{2}{5} & -\frac{3}{5} \end{bmatrix} \rightarrow \begin{bmatrix} 1 & 2 & 2 & | & 1 & 0 & 0 \\ 0 & -3 & 0 & | & -\frac{6}{5} & \frac{9}{5} & -\frac{6}{5} \\ 0 & 0 & 1 & | & \frac{2}{5} & \frac{2}{5} & -\frac{3}{5} \end{bmatrix}$$

$$\rightarrow \begin{bmatrix} 1 & 2 & 2 & | & 1 & 0 & 0 \\ 0 & 1 & 0 & | & \frac{2}{5} & -\frac{3}{5} & \frac{2}{5} \\ 0 & 0 & 1 & | & \frac{2}{5} & \frac{2}{5} & -\frac{3}{5} \end{bmatrix} \rightarrow \begin{bmatrix} 1 & 0 & 2 & | & \frac{1}{5} & \frac{6}{5} & -\frac{4}{5} \\ 0 & 1 & 0 & | & \frac{2}{5} & -\frac{3}{5} & \frac{2}{5} \\ 0 & 0 & 1 & | & \frac{2}{5} & \frac{2}{5} & -\frac{3}{5} \end{bmatrix}$$

$$\rightarrow \begin{bmatrix} 1 & 0 & 0 & | & -\frac{3}{5} & \frac{2}{5} & \frac{2}{5} \\ 0 & 1 & 0 & | & \frac{2}{5} & -\frac{3}{5} & \frac{2}{5} \\ 0 & 0 & 1 & | & \frac{2}{5} & \frac{2}{5} & -\frac{3}{5} \end{bmatrix}.$$

The steps were:

1. Simultaneously subtract two times the first row from the second row and from the third row.
2. Multiply the second row by −2/3 and add it to the third row. The 3×3 matrix on the left is now in upper triangular form.

3. Multiply the third row by $-3/5$.
4. Multiply the third row by 2 and add it to the second row.
5. Multiply the second row by $-1/3$.
6. Multiply the third row by -2 and add it to the first row.

The matrix on the right in columns 3 to 6 is the inverse. ☐

3.4 USING THE MATRIX INVERSE TO SOLVE A SYSTEM OF EQUATIONS

The solution to a matrix equation $Ax = b$ where $\det(A) \neq 0$ is

$$x = A^{-1}y. \tag{3.4}$$

Example 3.5 Solution to Two Equations in Two Unknowns Using an Inverse

The system of equations

$$3x - 2y = 5$$
$$4x + y = 3$$

may be written in the form

$$\begin{bmatrix} 3 & -2 \\ 4 & 1 \end{bmatrix} \begin{bmatrix} x \\ y \end{bmatrix} = \begin{bmatrix} 5 \\ 3 \end{bmatrix}.$$

Pre-multiply both sides of equation by the inverse of the 2×2 matrix to obtain

$$\begin{bmatrix} x \\ y \end{bmatrix} = \frac{1}{11} \begin{bmatrix} 1 & 2 \\ -4 & 3 \end{bmatrix} \begin{bmatrix} 5 \\ 3 \end{bmatrix} = \begin{bmatrix} 1 \\ -1 \end{bmatrix}.$$

Thus, $x = 1$, $y = -1$. ☐

Example 3.6 Solution to Three Equations in Three Unknowns Using Inverses

Consider the system of equations in matrix form

$$\begin{bmatrix} 1 & 2 & 2 \\ 2 & 1 & 2 \\ 2 & 2 & 1 \end{bmatrix} \begin{bmatrix} x \\ y \\ z \end{bmatrix} = \begin{bmatrix} 4 \\ 9 \\ 2 \end{bmatrix}.$$

The solution is after inverting and premultiplying by the 3×3 matrix

$$\begin{bmatrix} x \\ y \\ z \end{bmatrix} = \frac{1}{5} \begin{bmatrix} -3 & 2 & 2 \\ 2 & -3 & 2 \\ 2 & 2 & -3 \end{bmatrix} \begin{bmatrix} 4 \\ 9 \\ 2 \end{bmatrix} = \begin{bmatrix} 2 \\ -3 \\ 4 \end{bmatrix}.$$

☐

3.5   PARTITIONED MATRICES AND THEIR INVERSES

Matrices may be partitioned into blocks of submatrices. For example, if M is an $m \times n$ matrix, it could be partitioned into four submatrices as shown:

$$M = \begin{bmatrix} M_{11} & M_{12} \\ M_{21} & M_{22} \end{bmatrix}$$

where M_{11} has dimensions $m_1 \times n_1$, M_{12} has dimensions $m_1 \times (n - n_1)$, M_{21} has dimensions $(m - m_1) \times n_1$, and M_{22} has dimensions $(m - m_1) \times (n - n_1)$. The partitioned matrices may be added or subtracted if both matrices have partitions of the same size. They may be multiplied considering the blocks as individual elements provided that the sizes of the matrices are such that matrix multiplication is possible.

Example 3.7 An Example of a Partitioned Matrix with Operations on It

Let

$$M = \left[\begin{array}{ccc|cc} 1 & 2 & 1 & 5 & 3 \\ 3 & 4 & 1 & 2 & 1 \\ \hline 1 & 6 & 7 & 1 & 2 \\ 2 & 4 & 8 & 3 & 4 \\ 5 & 3 & 2 & 5 & 6 \end{array}\right].$$

Then

$$M' = \left[\begin{array}{cc|ccc} 1 & 3 & 1 & 2 & 5 \\ 2 & 4 & 6 & 4 & 3 \\ 1 & 1 & 7 & 8 & 2 \\ \hline 5 & 2 & 1 & 3 & 5 \\ 3 & 1 & 2 & 4 & 6 \end{array}\right].$$

As a result,

$$M'M = \begin{bmatrix} \begin{bmatrix} 1 & 3 \\ 2 & 4 \\ 1 & 1 \end{bmatrix}\begin{bmatrix} 1 & 2 & 1 \\ 3 & 4 & 1 \end{bmatrix} + \begin{bmatrix} 1 & 2 & 5 \\ 6 & 4 & 3 \\ 7 & 8 & 2 \end{bmatrix}\begin{bmatrix} 1 & 6 & 7 \\ 2 & 4 & 8 \\ 5 & 3 & 2 \end{bmatrix} & \begin{bmatrix} 1 & 3 \\ 2 & 4 \\ 1 & 1 \end{bmatrix}\begin{bmatrix} 5 & 3 \\ 2 & 1 \end{bmatrix} + \begin{bmatrix} 1 & 2 & 5 \\ 6 & 4 & 3 \\ 7 & 8 & 2 \end{bmatrix}\begin{bmatrix} 1 & 2 \\ 3 & 4 \\ 5 & 6 \end{bmatrix} \\ \begin{bmatrix} 5 & 2 \\ 3 & 1 \end{bmatrix}\begin{bmatrix} 1 & 2 & 1 \\ 3 & 4 & 1 \end{bmatrix} + \begin{bmatrix} 1 & 3 & 5 \\ 2 & 4 & 6 \end{bmatrix}\begin{bmatrix} 1 & 6 & 7 \\ 2 & 4 & 8 \\ 5 & 3 & 2 \end{bmatrix} & \begin{bmatrix} 5 & 2 \\ 3 & 1 \end{bmatrix}\begin{bmatrix} 5 & 3 \\ 2 & 1 \end{bmatrix} + \begin{bmatrix} 1 & 3 & 5 \\ 2 & 4 & 6 \end{bmatrix}\begin{bmatrix} 1 & 2 \\ 3 & 4 \\ 5 & 6 \end{bmatrix} \end{bmatrix}$$

$$= \left[\begin{array}{ccc|cc} 40 & 43 & 37 & 43 & 46 \\ 43 & 81 & 86 & 51 & 56 \\ 37 & 86 & 119 & 48 & 62 \\ \hline 43 & 51 & 48 & 64 & 61 \\ 46 & 56 & 62 & 61 & 66 \end{array}\right].$$

□

The formulae for inverses of a partitioned matrix are useful for some of the ideas associated with least square estimators. We will derive a formula for the inverse of a symmetric matrix. Consider a symmetric matrix

$$M = \begin{bmatrix} A & B \\ B' & D \end{bmatrix}.$$

To find M^{-1}, solve the matrix equation

$$\begin{bmatrix} E & F \\ F' & G \end{bmatrix} \begin{bmatrix} A & B \\ B' & D \end{bmatrix} = \begin{bmatrix} I & 0 \\ 0 & I \end{bmatrix} \tag{3.5}$$

by solving the corresponding systems of equations

$$EA + FB' = I$$
$$EB + FD = 0 \tag{3.6a}$$

and

$$F'A + GB' = 0$$
$$F'B + GD = I. \tag{3.6b}$$

To do this, observe that from the first equation of (3.4),

$$E = A^{-1} - FB'A^{-1}. \tag{3.7}$$

Substitution of (3.7) into the second equation of (3.4)

$$(A^{-1} - FB'A^{-1}) B + FD = 0 \tag{3.8}$$

so that

$$-A^{-1}B = F(D - B'A^{-1}B). \tag{3.9}$$

Let

$$W = (D - B'A^{-1}B)^{-1}. \tag{3.10}$$

Then

$$F = -A^{-1}BW \tag{3.11}$$

and

$$E = A^{-1} + A^{-1}BWB'A^{-1}. \tag{3.12}$$

To obtain G, observe that from the first equation in (3.6b),

$$F' = -GB'A^{-1}. \tag{3.13}$$

Substitute (3.13) into the second equation of (3.6b) to obtain

$$G(D - B'A^{-1}B) = I. \tag{3.14}$$

Thus, G=W. Thus,

$$
\begin{aligned}
M^{-1} &= \begin{bmatrix} A^{-1} + A^{-1}BWB'A^{-1} & -A^{-1}BW \\ -WB'A^{-1} & W \end{bmatrix} \\
&= \begin{bmatrix} A^{-1} & 0 \\ 0 & 0 \end{bmatrix} + \begin{bmatrix} -A^{-1}B \\ I \end{bmatrix} W \begin{bmatrix} -B'A^{-1} & I \end{bmatrix}.
\end{aligned} \tag{3.15}
$$

Alternatively, we could have solved the system of equations

$$\begin{bmatrix} A & B \\ B' & D \end{bmatrix} \begin{bmatrix} E & F \\ F' & G \end{bmatrix} = \begin{bmatrix} I & 0 \\ 0 & I \end{bmatrix}. \tag{3.16}$$

Then we have the systems

$$
\begin{aligned}
AE + BF' &= I \\
B'E + DF' &= 0
\end{aligned} \tag{3.17}
$$

and

$$
\begin{aligned}
AF + BG &= 0 \\
B'F + DG &= I.
\end{aligned} \tag{3.18}
$$

From the system of equations in (3.17), we have

$$E = A^{-1}(I - BF') \tag{3.19}$$

and

$$F' = -D^{-1}B'E. \tag{3.20}$$

Then substituting for F in the first of equations (3.17), we get

$$AE - BD^{-1}B'E = I$$

so

$$E = (A - BD^{-1}B')^{-1}.$$

Then let

$$Z = (A - BD^{-1}B')^{-1}$$

so that $E = Z$ and

$$F = -DB'Z.$$

From (3.18),

$$G = D^{-1}(I - B'F)$$

and

$$F = -A^{-1}BG.$$

Then

$$G = D^{-1}(I + B'A^{-1}BG)$$

yielding

$$G = (D - B'A^{-1}B)^{-1} = W \tag{3.21}$$

and

$$F = -A^{-1}BW. \tag{3.22}$$

Equating the two different expressions for E, we get

$$(A - BD^{-1}B')^{-1} = A^{-1} + A^{-1}B(D - B'A^{-1}B)^{-1}B'A^{-1}. \tag{3.23}$$

Let $C = -D^{-1}$. Then (3.23) becomes

$$(A + BCB')^{-1} = A^{-1} - A^{-1}B(C^{-1} + B'A^{-1}B)^{-1}B'A^{-1}. \tag{3.24}$$

The identity in (3.24) will be important in Part V.

Example 3.8 Inverting a Partitioned Matrix

Consider the matrix

$$M = \begin{bmatrix} 2 & 1 & 1 & 1 \\ 1 & 2 & 1 & 1 \\ 1 & 1 & 3 & 1 \\ 1 & 1 & 1 & 3 \end{bmatrix}.$$

Then

$$A = \begin{bmatrix} 2 & 1 \\ 1 & 2 \end{bmatrix}, B = \begin{bmatrix} 1 & 1 \\ 1 & 1 \end{bmatrix}, D = \begin{bmatrix} 3 & 1 \\ 1 & 3 \end{bmatrix}$$

and

$$
W = \left[\begin{bmatrix} 3 & 1 \\ 1 & 3 \end{bmatrix} - \begin{bmatrix} 1 & 1 \\ 1 & 1 \end{bmatrix} \begin{bmatrix} \frac{2}{3} & -\frac{1}{3} \\ -\frac{1}{3} & \frac{2}{3} \end{bmatrix} \begin{bmatrix} 1 & 1 \\ 1 & 1 \end{bmatrix} \right]^{-1} = \begin{bmatrix} \frac{7}{16} & -\frac{1}{16} \\ -\frac{1}{16} & \frac{7}{16} \end{bmatrix}.
$$

Then

$$
M^{-1} = \begin{bmatrix} \frac{2}{3} & -\frac{1}{3} & 0 & 0 \\ -\frac{1}{3} & \frac{2}{3} & 0 & 0 \\ 0 & 0 & 0 & 0 \\ 0 & 0 & 0 & 10 \end{bmatrix} + \begin{bmatrix} -\frac{1}{3} & -\frac{1}{3} \\ -\frac{1}{3} & -\frac{1}{3} \\ 1 & 0 \\ 0 & 1 \end{bmatrix} \begin{bmatrix} \frac{7}{16} & -\frac{1}{16} \\ -\frac{1}{16} & \frac{7}{16} \end{bmatrix} \begin{bmatrix} -\frac{1}{3} & -\frac{1}{3} & 1 & 0 \\ -\frac{1}{3} & -\frac{1}{3} & 0 & 1 \end{bmatrix}
$$

$$
= \begin{bmatrix} \frac{3}{4} & -\frac{1}{4} & -\frac{1}{8} & -\frac{1}{8} \\ -\frac{1}{4} & \frac{3}{4} & -\frac{1}{8} & -\frac{1}{8} \\ -\frac{1}{8} & -\frac{1}{8} & \frac{7}{16} & -\frac{1}{16} \\ -\frac{1}{8} & \frac{1}{8} & -\frac{1}{16} & \frac{7}{16} \end{bmatrix}.
$$

$\square$

3.6 FINDING THE LEAST SQUARE ESTIMATOR

Consider a linear model of the form

$$
Y = X\beta + \varepsilon. \tag{3.25}
$$

The objective is to minimize the sum of the squares of the differences between the observed values Y and the predicted values $X\beta$. This will be accomplished by finding the vector β that minimizes

$$
F(\beta) = (Y - X\beta)'(Y - X\beta). \tag{3.26}
$$

The usual way of obtaining the least square estimator is to differentiate $F(\beta)$ with respect to the components of the vector β, set the result equal to zero, and solve the matrix equation. We will do this in Section 4 when we study the differentiation of matrices and vectors. Right now we give an algebraic derivation by completing the square. Notice that

$$
\begin{aligned}
&(Y - X\beta)'(Y - X\beta) \\
&= Y'Y - \beta'X'Y - Y'X\beta + \beta'X'X\beta \\
&= Y'Y - \beta'(X'X)(X'X)^{-1}X'Y - Y'X(X'X)^{-1}(X'X)\beta \\
&\quad + Y'X(X'X)^{-1}(X'X)(X'X)^{-1}X'Y - Y'X(X'X)^{-1}(X'X)(X'X)^{-1}X'Y + \beta'X'X\beta \\
&= Y'(I - X(X'X)^{-1}X')Y + (\beta - (X'X)^{-1}X'Y)'(X'X)(\beta - (X'X)^{-1}X'Y).
\end{aligned} \tag{3.27}
$$

The expression in (3.27) is minimized when its last term is zero or when

$$\beta = (X'X)^{-1}X'Y. \qquad (3.28)$$

Example 3.9 Numerical Example of Finding Least Square Estimates

Consider the example of economic data that was given in Subsection 1.1. In order to include the intercept term in the model

$$y = \beta_0 + \beta_1 x_1 + \beta_2 x_2 + \beta_3 x_3 + \varepsilon,$$

we add a column of ones at the left end of the X matrix. Thus,

$$X = \begin{bmatrix} 1 & 1015.5 & 831.8 & 78678 \\ 1 & 1102.7 & 894.0 & 79367 \\ 1 & 1212.8 & 981.6 & 82153 \\ 1 & 1359.3 & 1101.7 & 85064 \\ 1 & 1472.8 & 1210.1 & 86794 \end{bmatrix}.$$

Then

$$X'X = \begin{bmatrix} 5 & 6163.1 & 5019.2 & 412056 \\ 6163.1 & 7.73491 \times 10^6 & 6.30077 \times 10^6 & 5.10508 \times 10^8 \\ 5019.2 & 6.30077 \times 10^6 & 5.13275 \times 10^6 & 4.15784 \times 10^8 \\ 412056 & 5.10508 \times 10^8 & 4.15784 \times 10^8 & 3.40075 \times 10^{10} \end{bmatrix},$$

$$(X'X)^{-1} = \begin{bmatrix} 5997.5 & 1.62547 & 0.323355 & -0.101024 \\ 1.62547 & 0.00385599 & -0.00401386 & -0.0000285055 \\ 0.323355 & -0.00401386 & 0.00496308 & -4.34325 \times 10^6 \\ -0.101024 & -0.0000285055 & -4.34325 \times 10^6 & 1.70511 \times 10^{-6} \end{bmatrix},$$

$$X'Y = \begin{bmatrix} 3842.9 \\ 4.81919 \times 10^6 \\ 3.9257 \times 10^6 \\ 3.18246 \times 10^8 \end{bmatrix},$$

and

$$b = \begin{bmatrix} 302.356 \\ 0.326265 \\ 0.428298 \\ -0.00443968 \end{bmatrix}.$$

If this calculation had been done by hand, it would have proved to be very tedious. In this day and age, most regression calculations are done with a computer. The software package Mathematica was used.   □

Example 3.10 One Variable Regression

Suppose that there is only one prediction variable. Then the linear model would look like

$$
\begin{bmatrix} y_1 \\ y_2 \\ \vdots \\ y_n \end{bmatrix} = \begin{bmatrix} 1 & x_1 \\ 1 & x_2 \\ \vdots & \vdots \\ 1 & x_n \end{bmatrix} \begin{bmatrix} b_0 \\ b_1 \end{bmatrix} + \varepsilon.
$$

Then, since

$$
X = \begin{bmatrix} 1 & x_1 \\ 1 & x_2 \\ \vdots & \vdots \\ 1 & x_n \end{bmatrix},
$$

multiplication of the transpose of X, X′ by X gives

$$
X'X = \begin{bmatrix} n & \sum_{i=1}^{n} x_i \\ \sum_{i=1}^{n} x_i & \sum_{i=1}^{n} x_i^2 \end{bmatrix}.
$$

The inverse of X′X is

$$
(X'X)^{-1} = \frac{1}{n\sum_{i=1}^{n} x_i^2 - \left(\sum_{i=1}^{n} x_i\right)^2} \begin{bmatrix} \sum_{i=1}^{n} x_i^2 & -\sum_{i=1}^{n} x_i \\ -\sum_{i=1}^{n} x_i & n \end{bmatrix}.
$$

Now

$$
X'Y = \begin{bmatrix} \sum_{i=1}^{n} y_i \\ \sum_{i=1}^{n} x_i y_i \end{bmatrix}
$$

and thus,

$$
\begin{bmatrix} b_0 \\ b_1 \end{bmatrix} = \frac{1}{n\sum_{i=1}^{n} x_i^2 - \left(\sum_{i=1}^{n} x_i\right)^2} \begin{bmatrix} \sum_{i=1}^{n} x_i^2 \sum_{i=1}^{n} y_i - \sum_{i=1}^{n} x_i \sum_{i=1}^{n} x_i y_i \\ -\sum_{i=1}^{n} x_i \sum_{i=1}^{n} y_i + n\sum_{i=1}^{n} x_i y_i \end{bmatrix}. \tag{3.29}
$$

$\square$

Example 3.11 A Simple Linear Regression Numerical Example

Year	y	x
1975	1012.2	1598.4
1976	1129.2	1782.8
1977	1257.2	1990.5
1978	1403.5	2249.7
1979	1566.8	2508.2
1980	1732.6	2732.0
1981	1915.1	3052.6
1982	2050.7	3166.0
1983	2234.5	3405.7
1984	2430.5	3772.2

We consider consumption expenditures as a linear function of gross national product. The scatter plot in Figure 3.1 indicates that fitting a linear equation appears to be reasonable.

Now

$$
X = \begin{bmatrix} 1 & 1598.4 \\ 1 & 1782.8 \\ 1 & 1990.5 \\ 1 & 2249.7 \\ 1 & 2508.2 \\ 1 & 2732.0 \\ 1 & 3052.6 \\ 1 & 3166.0 \\ 1 & 3405.7 \\ 1 & 3772.2 \end{bmatrix}, Y = \begin{bmatrix} 1012.2 \\ 1129.2 \\ 1257.2 \\ 1403.5 \\ 1566.8 \\ 1732.6 \\ 1915.1 \\ 2050.7 \\ 2234.5 \\ 2430.5 \end{bmatrix},
$$

$$
X'X = \begin{bmatrix} 10 & 25898.1 \\ 25898.1 & 7.13591 \times 10^7 \end{bmatrix},
$$

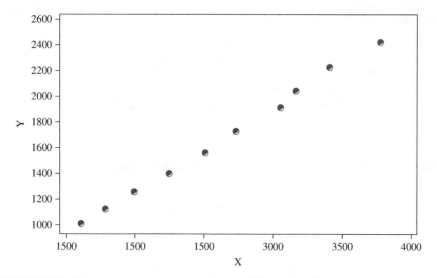

FIGURE 3.1 Personal consumption expenditures (billions of dollars y) versus gross national product.

$$(X'X)^{-1} = \begin{bmatrix} 1.66418 & -0.000603976 \\ -0.000603976 & 2.33212 \times 10^{-7} \end{bmatrix},$$

and

$$X'Y = \begin{bmatrix} 16732.3 \\ 4.70712 \times 10^{7} \end{bmatrix}.$$

Thus,

$$\begin{bmatrix} a \\ b \end{bmatrix} = \begin{bmatrix} 1.66418 & -0.000603976 \\ -0.000603976 & 2.33212 \times 10^{-7} \end{bmatrix} \begin{bmatrix} 16732.3 \\ 4.70712 \times 10^{7} \end{bmatrix}$$

$$= \begin{bmatrix} -66.2844 \\ 0.662468 \end{bmatrix}$$

and the regression equation is $y = -66.2844 + 0.662468x$.   □

Example 3.12 Regression Estimators Using the Corrected Sums of Squares

Consider a multiple regression model for an $n \times m$ matrix where the first column is 1_n. The linear model takes the form

$$Y = \begin{bmatrix} 1_n & Z \end{bmatrix} \begin{bmatrix} \alpha \\ \beta_1 \end{bmatrix} + \varepsilon. \tag{3.30}$$

Then the regression coefficients

$$b = \begin{bmatrix} a \\ \tilde{b} \end{bmatrix} = \begin{bmatrix} n & 1'_n Z \\ Z'1_n & Z'Z \end{bmatrix}^{-1} \begin{bmatrix} 1'_n Y \\ Z'Y \end{bmatrix}. \tag{3.31}$$

Using the formula for the inverse of a partitioned matrix,

$$\begin{bmatrix} n & 1'_n Z \\ Z'1_n & Z'Z \end{bmatrix}^{-1} = \begin{bmatrix} \frac{1}{n} + \frac{1}{n}1'_n ZS^{-1}Z'1_n \frac{1}{n} & -\frac{1}{n}1'_n ZS^{-1} \\ -S^{-1}Z'1_n \frac{1}{n} & S^{-1} \end{bmatrix} \tag{3.32}$$

where $S = Z'(I - (1/n)J)Z$. Let $\bar{z}' = (1/n)1'_n Z$ be the row vector consisting of the means of the columns of X_1. Let $\bar{y} = (1/n)1'_n Y$ denote the mean of the elements of Y. Then

$$\begin{aligned} a &= \bar{y} - \bar{z}'\tilde{b} \\ \tilde{b} &= S^{-1}(Z'Y - n\bar{y}\bar{z}') \end{aligned} \tag{3.33}$$

Let $\tilde{X} = Z - (1/n)JZ$. Then

$$\tilde{b} = (\tilde{X}\tilde{X})^{-1}\tilde{X}Y \tag{3.34}$$

$$a = \bar{y} - \tilde{b}\bar{z}. \tag{3.35}$$

$\square$

Example 3.13 For the data given in Subsection 1.1, we have

$$Y = \begin{bmatrix} 640.0 \\ 691.6 \\ 757.6 \\ 837.2 \\ 916.5 \end{bmatrix}, Z = \begin{bmatrix} 1015.5 & 831.8 & 78678 \\ 1102.7 & 894 & 79367 \\ 1212.8 & 981.6 & 82153 \\ 1356.3 & 1101.7 & 85064 \\ 1472.8 & 1210.1 & 86794 \end{bmatrix}.$$

Now

$$\tilde{X} = \begin{bmatrix} 1015.5 & 831.8 & 78678 \\ 1102.7 & 894 & 79367 \\ 1212.8 & 981.6 & 82153 \\ 1356.3 & 1101.7 & 85064 \\ 1472.8 & 1210.1 & 86794 \end{bmatrix} - \frac{1}{5}\begin{bmatrix} 1 & 1 & 1 & 1 & 1 \\ 1 & 1 & 1 & 1 & 1 \\ 1 & 1 & 1 & 1 & 1 \\ 1 & 1 & 1 & 1 & 1 \\ 1 & 1 & 1 & 1 & 1 \end{bmatrix}\begin{bmatrix} 1015.5 & 831.8 & 78678 \\ 1102.7 & 894 & 79367 \\ 1212.8 & 981.6 & 82153 \\ 1356.3 & 1101.7 & 85064 \\ 1472.8 & 1210.1 & 86794 \end{bmatrix}$$

$$= \begin{bmatrix} -216.52 & -172.04 & -3733.2 \\ -129.32 & -109.84 & -3044.2 \\ -19.22 & -22.24 & -258.2 \\ 124.28 & 97.86 & 2652.8 \\ 240.78 & 206.26 & 4382.8 \end{bmatrix}$$

Substitution in (3.33) yields

$$\tilde{b} = \begin{bmatrix} 0.379615 \\ 0.357472 \\ -0.00411033 \end{bmatrix}.$$

Now

$$\bar{y} = 768.58$$
$$\bar{z}' = \begin{bmatrix} 1232.02 & 1003.84 & 82411.20 \end{bmatrix}$$

and

$$a = 768.58 - \begin{bmatrix} 1232.02 & 1003.84 & 82411.20 \end{bmatrix} \begin{bmatrix} 0.379615 \\ 0.357472 \\ -0.00411033 \end{bmatrix}$$

$$= 280.779. \qquad \qquad \Box$$

3.7 SUMMARY

We have shown how to find the inverse of a matrix by two methods, using the adjoint of the matrix and by elementary row operations. We have also shown how to find the inverse of a partitioned matrix. In so doing, we derived an important matrix identity. We demonstrated how these methods are applied when finding the least square estimator.

EXERCISES

3.1 Find the inverse of each of the matrices in A through F both by the adjoint method and by using elementary row operations. In part G invert the matrix by using the formula for the inverse of a partitioned matrix after partitioning it into four two-by-two matrices.

A. $\begin{bmatrix} 2 & 1 \\ 1 & 2 \end{bmatrix}$

B. $\begin{bmatrix} \dfrac{5\sqrt{3}}{2} & \dfrac{5}{2} \\ -\dfrac{5}{2} & \dfrac{5\sqrt{3}}{2} \end{bmatrix}$

C. $\begin{bmatrix} \cos\theta & \sin\theta \\ -\sin\theta & \cos\theta \end{bmatrix}$

D. $\begin{bmatrix} 2 & 1 & 1 \\ 1 & 2 & 1 \\ 1 & 1 & 2 \end{bmatrix}$

E. $\begin{bmatrix} 1 & 0.9 & 0.9 \\ 0.9 & 1 & 0.9 \\ 0.9 & 0.9 & 1 \end{bmatrix}$

F. $\begin{bmatrix} 1 & 0.9 & 0.8 \\ 0.9 & 1 & 0.7 \\ 0.8 & 0.7 & 1 \end{bmatrix}$

G. $\begin{bmatrix} 1 & \rho & \rho & \rho \\ \rho & 1 & \rho & \rho \\ \rho & \rho & 1 & \rho \\ \rho & \rho & \rho & 1 \end{bmatrix}$

3.2 Solve the systems of equations in Exercise 2.6 by rewriting them in matrix form and inverting the coefficient matrix.

3.3 Show that the inverse of a symmetric matrix is also a symmetric matrix.

3.4 Let

$$R_\alpha = \begin{bmatrix} \cos\alpha & \sin\alpha \\ -\sin\alpha & \cos\alpha \end{bmatrix}, R_\beta = \begin{bmatrix} \cos\beta & \sin\beta \\ -\sin\beta & \cos\beta \end{bmatrix}.$$

Show that
A. $R_\alpha R_\beta = R_{\alpha+\beta}$.
B. $R_\alpha^{-1} = R_{-\alpha}$.
C. $R_\alpha R_\beta^{-1} = R_{\alpha-\beta}$.

3.5 Show that if A and B are nonsingular matrices, then

$$(A^{-1} + B^{-1})^{-1} = A(A+B)^{-1}B$$
$$= A - A(A+B)^{-1}A$$
$$= B - B(A+B)^{-1}B.$$

3.6   In standard textbooks, the formulae for the least square estimators in a one variable linear regression model are given by

$$b_0 = \bar{y} - b_1 \bar{x}$$

where $\bar{y} = (1/n) \sum_{i=1}^{n} y_i$ and $\bar{x} = (1/n) \sum_{i=1}^{n} x_i$ and

$$b_1 = \frac{n \sum_{i=1}^{n} x_i y_i - \left(\sum_{i=1}^{n} x_i\right)\left(\sum_{i=1}^{n} y_i\right)}{n \sum_{i=1}^{n} x_i^2 - \left(\sum_{i=1}^{n} x_i\right)^2}$$

Show how these formulae may be derived from the result in (3.29).

3.7   **A.** Consider the model

$$\begin{bmatrix} y_1 \\ y_2 \\ \vdots \\ y_n \end{bmatrix} = \begin{bmatrix} x_1 \\ x_2 \\ \vdots \\ x_n \end{bmatrix} b_1 + \varepsilon.$$

Derive a formula for the least square estimator of b_1.
B. Consider the model

$$\begin{bmatrix} y_1 \\ y_2 \\ \vdots \\ y_n \end{bmatrix} = \begin{bmatrix} 1 \\ 1 \\ \vdots \\ 1 \end{bmatrix} b_0 + \varepsilon.$$

Derive a formula for the least square estimator of b_0.

3.8   For the economic data in Section 1 consider the models
A. $y = b_0 + b_1 x_1 + \varepsilon.$
B. $y = b_0 + b_1 x_1 + b_2 x_2 + \varepsilon.$
Find the numerical estimates of the least square estimator. You might like to use a computer algebra system. What other models could you consider? (Hint: There are 13 more possibilities including that worked out in Example 3.8.)

3.9   Show that $A\,\text{adj}(A) = \det(A)I$.

SECTION 4

SPECIAL MATRICES AND FACTS ABOUT MATRICES THAT WILL BE USED IN THE SEQUEL

4.1 INTRODUCTION

We will consider some special kinds of matrices in this section. Of particular interest will be orthogonal matrices and matrices of the form $aI + bJ$ where I is the identity matrix and J is a matrix with all of its elements 1. We will also consider two important matrix operations, namely, finding the trace of a matrix and the direct product of two matrices. These will be especially important in Part IV where we consider analysis of variance models. We also study matrix differentiation and derive the least square estimator again using it. Matrix differentiation will be of particular importance in solving optimization problems in Part V.

4.2 MATRICES OF THE FORM $aI_n + bJ_n$

Matrices of the form $aI_n + bJ_n$ are very useful for quadratic forms that represent sums of squares. In what follows, the matrix I_n will denote the $n \times n$ identity matrix. It has ones on the main diagonal and zeros elsewhere. The matrix J_n will denote the $n \times n$ matrix of ones. Observe that $I_n \cdot I_n = I_n$ and that $J_n \cdot J_n = nJ_n$. The $n \times 1$ matrix 1_n will denote the one-dimensional column of ones. Notice that $1_n \cdot 1_n' = J_n$.

A matrix A is idempotent if $A \cdot A = A$. Consider the idempotent matrices of the form $A = aI_n + bJ_n$. Matrix A is idempotent if and only if

$$(aI_n + bJ_n)(aI_n + bJ_n) = a^2 I_n + 2abJ_n + nb^2 J_n$$
$$= aI_n + bJ_n.$$

(4.1)

Equation 4.1 implies that $a^2 = a$, so $a = 0$ or $a = 1$, and that $2ab + nb^2 = b$. Then for $a = 0$ $b = 1/n$ and for $a = 1$, $b = -1/n$. Thus, the only idempotent matrices are $(1/n)J_n$ and $I_n - \frac{1}{n}J_n$. Both matrices are important because, as we shall see below, they are matrices of quadratic forms that represent sums of squares commonly used in statistics.

Let x be an n-dimensional column vector (an $n \times 1$ matrix) with elements x_i, $1 \leq i \leq n$. A quadratic form is an expression of the form $x'Ax$ where A is an $n \times n$ matrix. Two important examples for expressions in statistics that involve sums of squares are $x'I_n x = \sum_{i=1}^{n} x_i^2$ and $x'J_n x = \left(\sum_{i=1}^{n} x_i\right)^2$. The quadratic form that uses the identity matrix yields the sum of squares, while the quadratic form that uses the matrix of ones yields the square of a sum.

The formula for the sample standard deviation may be expressed as a quadratic form containing I and J matrices. Indeed,

$$
\begin{aligned}
s^2 &= \frac{1}{n-1}\sum_{i=1}^{n}(x_i - \bar{x})^2 = \frac{1}{n-1}x'\left(I_n - \frac{1}{n}J_n\right)\left(I_n - \frac{1}{n}J_n\right)x \\
&= \frac{1}{n-1}x'\left(I_n - \frac{1}{n}J_n\right)x = \frac{1}{n-1}\left(x'I_n x - \frac{1}{n}x'J_n x\right) \\
&= \frac{1}{n-1}\left(\sum_{i=1}^{n}x_i^2 - \frac{1}{n}\left(\sum_{i=1}^{n}x_i\right)^2\right) = \frac{n\sum_{i=1}^{n}x_i^2 - \left(\sum_{i=1}^{n}x_i\right)^2}{n(n-1)}.
\end{aligned}
$$

(4.2)

Thus, we obtain an elegant expression for the sample standard deviation (the third expression in (4.2)) and an elegant proof of its algebraic equivalence to the computing formula.

The inverse of these matrices is also interesting. Observe that the product of two matrices $A = aI_n + bJ_n$ and $B = cI_n + dJ_n$ is

$$AB = (aI_n + bJ_n)(cI_n + dJ_n) = acI_n + (ad + bc + nbd)J_n.$$

(4.3)

The reader might like to verify that the matrices $A = aI_n + bJ_n$ and $B = cI_n + dJ_n$ commute. In order that $AB = I$, it follows that $ac = 1$ and $ad + bc + nbd = 0$. Then

$$ac = 1$$

(4.4)

and

$$ad + bc + nbd = 0.$$

(4.5)

Solving (4.4) and (4.5) for c and d in terms of a and b yields

$$c = \frac{1}{a} \qquad (4.6)$$

and

$$d = -\frac{b}{a(a + nb)}. \qquad (4.7)$$

Thus,

$$A^{-1} = \frac{1}{a}I_n - \frac{b}{a(a + nb)}J_n. \qquad (4.8)$$

Observe that the formula for the inverse in (4.8) is valid only when $a \ne 0$ and $a \ne -nb$. For example, the matrix $I - (1/n)J$ would not have an inverse.

Example 4.1 An Equicorrelation Matrix

Suppose you have n variables where any pair of two have the same correlation coefficient. Then the correlation matrix takes the form

$$\Sigma = (1 - \rho)I_n + \rho J_n$$

and substituting $a = 1 - \rho$ and $b = \rho$ into (4.8) yields the inverse

$$\Sigma^{-1} = \frac{1}{1 - \rho}\left[I_n - \frac{\rho}{1 + (n - 1)\rho}J_n\right]. \qquad \square$$

4.3 ORTHOGONAL MATRICES

A matrix A is said to be orthogonal if $A'A = I$ and $AA' = I$. For orthogonal matrices, both relationships hold true. If only $A'A = I$, then A is column orthogonal. If only $AA' = I$, then A is row orthogonal. For square matrices, the three conditions (i) $A'A = I$, (ii) $AA' = I$, (iii) $A' = A^{-1}$ are equivalent, so only one of the conditions needs to hold for a matrix to be orthogonal. A simple theorem about orthogonal matrices is stated below.

Theorem 4.1 If A and B are orthogonal matrices, then AB is orthogonal.

Proof. Observe that

$$(AB)(AB)' = (AB)(B'A') = A(BB')A' = AIA' = AA' = I$$

and

$$(AB)'(AB) = (B'A')(AB) = B'(A'A)B = B'IB = B'B = I. \qquad \blacksquare$$

Example 4.2 An Orthogonal Rotation Matrix

The matrix

$$\begin{bmatrix} \cos\theta & \sin\theta \\ -\sin\theta & \cos\theta \end{bmatrix}$$

is orthogonal. Indeed,

$$
\begin{bmatrix} \cos\theta & \sin\theta \\ -\sin\theta & \cos\theta \end{bmatrix}\begin{bmatrix} \cos\theta & -\sin\theta \\ \sin\theta & \cos\theta \end{bmatrix}
$$
$$
= \begin{bmatrix} \cos^2\theta + \sin^2\theta & -\cos\theta\sin\theta + \sin\theta\cos\theta \\ -\sin\theta\cos\theta + \cos\theta\sin\theta & (-\sin\theta)(-\sin\theta) + \cos^2\theta \end{bmatrix}
$$
$$
= \begin{bmatrix} 1 & 0 \\ 0 & 1 \end{bmatrix}
$$

and

$$
\begin{bmatrix} \cos\theta & -\sin\theta \\ \sin\theta & \cos\theta \end{bmatrix}\begin{bmatrix} \cos\theta & \sin\theta \\ -\sin\theta & \cos\theta \end{bmatrix}
$$
$$
= \begin{bmatrix} \cos^2\theta + (-\sin\theta)^2 & \cos\theta\sin\theta - \sin\theta\cos\theta \\ \sin\theta\cos\theta + (\cos\theta)(-\sin\theta) & (-\sin\theta)(-\sin\theta) + \cos^2\theta \end{bmatrix}
$$
$$
= \begin{bmatrix} 1 & 0 \\ 0 & 1 \end{bmatrix}.
$$

□

What are the orthogonal matrices of the form $aI_n + bJ_n$? These matrices are symmetric. Thus, they are orthogonal if

$$(aI_n + bJ_n)(aI_n + bJ_n) = I_n.$$

It follows that

$$a^2 I_n + 2abJ_n + nb^2 J_n = a^2 I_n + (2ab + nb^2)J_n = I_n.$$

This happens if and only if

$$a^2 = 1$$
$$2ab + nb^2 = 0.$$

Then either $a=1$ or $a=-1$ and $b=0$ so the matrix takes the form I_n or $-I_n$, or for $a=1$ $b=-2/n$ and for $a=-1$, $b=2/n$. Thus, the orthogonal matrices are $\pm I_n$ or

$$\pm\left(I_n - \left(\frac{2}{n}\right)J_n\right).$$

Example 4.3 An Orthogonal Matrix of the Form $aI + bJ$

If $n = 5$, an orthogonal matrix of the form $aI_5 + bJ_5$ would be

$$
A = \begin{bmatrix}
\dfrac{3}{5} & -\dfrac{2}{5} & -\dfrac{2}{5} & -\dfrac{2}{5} & -\dfrac{2}{5} \\
-\dfrac{2}{5} & \dfrac{3}{5} & -\dfrac{2}{5} & -\dfrac{2}{5} & -\dfrac{2}{5} \\
-\dfrac{2}{5} & -\dfrac{2}{5} & \dfrac{3}{5} & -\dfrac{2}{5} & -\dfrac{2}{5} \\
-\dfrac{2}{5} & -\dfrac{2}{5} & -\dfrac{2}{5} & \dfrac{3}{5} & -\dfrac{2}{5} \\
-\dfrac{2}{5} & -\dfrac{2}{5} & -\dfrac{2}{5} & -\dfrac{2}{5} & \dfrac{3}{5}
\end{bmatrix}.
$$

□

Another interesting orthogonal matrix is the Helmert matrix. The Helmert matrix H_n has rows

$$
\begin{aligned}
r_1 &= \frac{1}{\sqrt{n}}(1, 1, \ldots, 1) \\
r_k &= \frac{1}{\sqrt{k(k-1)}}(1, 1, \ldots, 1, 1-k, 0, \ldots, 0), \quad 2 \le k \le n.
\end{aligned}
\tag{4.9}
$$

Example 4.4 An Example of a Helmert Matrix

The 5×5 Helmert matrix would be

$$
H_5 = \begin{bmatrix}
\dfrac{1}{\sqrt{5}} & \dfrac{1}{\sqrt{5}} & \dfrac{1}{\sqrt{5}} & \dfrac{1}{\sqrt{5}} & \dfrac{1}{\sqrt{5}} \\
\dfrac{1}{\sqrt{2}} & -\dfrac{1}{\sqrt{2}} & 0 & 0 & 0 \\
\dfrac{1}{\sqrt{6}} & \dfrac{1}{\sqrt{6}} & -\dfrac{2}{\sqrt{6}} & 0 & 0 \\
\dfrac{1}{\sqrt{12}} & \dfrac{1}{\sqrt{12}} & \dfrac{1}{\sqrt{12}} & -\dfrac{3}{\sqrt{12}} & 0 \\
\dfrac{1}{\sqrt{20}} & \dfrac{1}{\sqrt{20}} & \dfrac{1}{\sqrt{20}} & \dfrac{1}{\sqrt{20}} & -\dfrac{4}{\sqrt{20}}
\end{bmatrix}.
$$

□

Consider the submatrix $\tilde{H}_n$ consisting of all but the first row of H_n. This submatrix is row orthogonal that is $\tilde{H}_n \tilde{H}_n = I_n$ but not column orthogonal. However, what is interesting is that

$$\tilde{H}_n\tilde{H}_n = I_n - \frac{1}{n}J_n. \tag{4.10}$$

Let $y = \tilde{H}_n x$. Then

$$\sum_{i=1}^{n-1} y_i^2 = y'y = x'\tilde{H}_n\tilde{H}_n x = x'\left(I_n - \frac{1}{n}J_n\right)x = \sum_{i=1}^{n}(x_i - \bar{x})^2 = (n-1)s^2.$$

Assuming that the x_i are independent observations from a normal population with mean μ and variance σ^2, it can be shown that the y_i/σ has a standard normal distribution and is independent so that $(n-1)s^2/\sigma^2$ has a chi-square distribution with $n-1$ degrees of freedom. If θ is an n-dimensional vector of parameters, then the $n-1$ equalities in $\tilde{H}_n\theta = 0$ are orthogonal contrasts frequently studied in Design of Experiments. More will be said about this in Part IV.

4.4   DIRECT PRODUCT OF MATRICES

The Kronecker or direct product of matrices will be very useful in the study of experimental design models in Part IV.

Let A be an $m \times n$ and B be a $p \times q$ matrix. The Kronecker product (also called the direct product) is given by

$$A \otimes B = a_{ij}B, \quad i = 1, 2, \ldots, m, \; j = 1, 2, \ldots, n.$$

Example 4.5 A Kronecker Product of Two Specific Matrices

If

$$A = \begin{bmatrix} 1 & 2 \\ 3 & 4 \end{bmatrix}, B = \begin{bmatrix} 2 & 1 \\ -1 & 0 \\ 3 & 2 \end{bmatrix},$$

then

$$A \otimes B = \begin{bmatrix} 2 & 1 & 4 & 2 \\ -1 & 0 & -2 & 0 \\ 3 & 2 & 6 & 4 \\ 6 & 3 & 8 & 4 \\ -3 & 0 & -4 & 0 \\ 9 & 6 & 12 & 8 \end{bmatrix}$$

and

$$B \otimes A = \begin{bmatrix} 2 & 4 & 1 & 2 \\ 6 & 8 & 3 & 4 \\ -1 & -2 & 0 & 0 \\ -3 & -4 & 0 & 0 \\ -3 & 6 & 2 & 4 \\ 9 & 12 & 6 & 8 \end{bmatrix}.$$

□

Some useful properties of the Kronecker product are given in Theorem 4.2.

Theorem 4.2 The following are properties of the Kronecker product:

1. Assume that A and B are the same size. Then
   a. $(A+B) \otimes C = A \otimes C + B \otimes C$.
   b. $C \otimes (A+B) = C \otimes A + C \otimes B$.
2. Assuming A, B, C, and D have appropriate dimensions so that AC and BD are defined, $(A \otimes B)(C \otimes D) = (AC \otimes BD)$.
3. For two nonsingular matrices A and B, $(A \otimes B)^{-1} = A^{-1} \otimes B^{-1}$.
4. The transpose $(A \otimes B)' = A' \otimes B'$.

The reader is asked to prove 1 in Exercise 4.7. To prove 2 observe that

$$AC \otimes BD = \sum_{k=1}^{n} a_{ik} c_{kj} BD = \sum_{k=1}^{n} a_{ik} Bc_{kj} D = (A \otimes B)(C \otimes D).$$

To prove 3 let A be an $m \times m$ square nonsingular matrix and let B be an $n \times n$ square nonsingular matrix. Notice that

$$(A \otimes B)(A^{-1} \otimes B^{-1}) = (AA^{-1} \otimes BB^{-1}) = I_m \otimes I_n = I_{mn}.$$

To complete the proof the reader may show by a similar argument that

$$(A^{-1} \otimes B^{-1})(A \otimes B) = I_{mn}.$$

∎

4.5 AN IMPORTANT PROPERTY OF DETERMINANTS

An important property of determinants is that

$$\det(AB) = \det(A)\det(B).$$

A sketch of a proof of this important fact will now be given. Corresponding to an elementary row operation is a matrix multiplication. For example, for a three-by-three matrix, swapping row 1 with row 3 would be the same as multiplying the matrix on the left by

$$P_{13} = \begin{bmatrix} 0 & 0 & 1 \\ 0 & 1 & 0 \\ 1 & 0 & 0 \end{bmatrix}.$$

Multiplying a row by a constant would correspond to a matrix where if, say, row k is being multiplied, then $a_{kk} = c$, the multiplier and the other elements of that row are zero. The ith row would have $a_{ii} = 1$ and zeros elsewhere. The matrix that adds, say, k times, row 2 to row 1 for a three-by-three matrix would be

$$M = \begin{bmatrix} 1 & 0 & 0 \\ 1 & k & 0 \\ 0 & 0 & 1 \end{bmatrix}.$$

Such matrices are called elementary matrices. The elementary matrices correspond to the elementary row operations used to reduce a determinant to a form where it is easier to evaluate. Any nonsingular matrix can be expressed as the product of elementary matrices, and for an elementary matrix E, it is not hard to show that
$\det(EA) = \det E \det A$. Since if, say,

$$A = E_1 E_2 \ldots E_m$$

and

$$B = F_1 F_2 \ldots F_n,$$

then

$$AB = E_1 E_2 \ldots E_m F_1 F_2 \ldots F_n.$$

Now $\det A = \prod_{i=1}^{m} \det E_i$, $\det B = \prod_{i=1}^{m} \det F_j$, and $\det(AB)$ are the product of these two quantities so the result follows.

Example 4.6 Inverse of a Matrix in Terms of Elementary Matrices

The inverse of a matrix may be expressed as the product of elementary matrices. For example, consider the application of elementary row operations to the matrix

$$M = \begin{bmatrix} 5 & 3 \\ 3 & 5 \end{bmatrix}$$

to obtain the identity matrix. Observe that

$$\begin{bmatrix} 5 & 3 \\ 3 & 5 \end{bmatrix} \rightarrow \begin{bmatrix} 1 & \frac{3}{5} \\ 3 & 5 \end{bmatrix} \rightarrow \begin{bmatrix} 1 & \frac{3}{5} \\ 0 & \frac{16}{5} \end{bmatrix} \rightarrow \begin{bmatrix} 1 & \frac{3}{5} \\ 0 & 1 \end{bmatrix} \rightarrow \begin{bmatrix} 1 & 0 \\ 0 & 1 \end{bmatrix}$$

corresponds to finding the product using pre-multiplication as shown below. We have

$$\begin{bmatrix} 1 & 0 \\ 0 & 1 \end{bmatrix} = \begin{bmatrix} 1 & -\frac{3}{5} \\ 0 & 1 \end{bmatrix}\begin{bmatrix} 1 & 0 \\ 0 & \frac{5}{16} \end{bmatrix}\begin{bmatrix} 1 & 0 \\ -3 & 1 \end{bmatrix}\begin{bmatrix} \frac{1}{5} & 0 \\ 0 & 1 \end{bmatrix}\begin{bmatrix} 5 & 3 \\ 3 & 5 \end{bmatrix}.$$

Now

$$M^{-1} = \begin{bmatrix} 1 & -\frac{3}{5} \\ 0 & 1 \end{bmatrix}\begin{bmatrix} 1 & 0 \\ 0 & \frac{5}{16} \end{bmatrix}\begin{bmatrix} 1 & 0 \\ -3 & 1 \end{bmatrix}\begin{bmatrix} \frac{1}{5} & 0 \\ 0 & 1 \end{bmatrix} = \begin{bmatrix} \frac{5}{16} & -\frac{3}{16} \\ -\frac{3}{16} & \frac{5}{16} \end{bmatrix}.$$

Also

$$M = \begin{bmatrix} 5 & 0 \\ 0 & 1 \end{bmatrix}\begin{bmatrix} 1 & 0 \\ 3 & 1 \end{bmatrix}\begin{bmatrix} 1 & 0 \\ 0 & \frac{16}{5} \end{bmatrix}\begin{bmatrix} 1 & 0 \\ \frac{3}{5} & 1 \end{bmatrix}.$$

$\square$

By the same token we can find the inverse of a matrix by doing analogous operations on columns. This would correspond to doing post-multiplication by elementary matrices. One alternative would be to reduce the matrix to upper triangular form by row operations and then use column operations to reduce it to diagonal form. Suppose that

$$P_1 \ldots P_r M Q_1 \ldots Q_s = PMQ = I,$$

where $P=P_1 \ldots P_r$ and $Q=Q_1 \ldots Q_s$. Since $PM=T$, an upper triangular matrix $PMQ = TQ=I$ so that $Q=T^{-1}=(PM)^{-1}=M^{-1}P^{-1}$ and $M^{-1}=QP$. Also $M=P^{-1}Q^{-1}$.

Example 4.7 Inverse of a Matrix Using Row and Column Operations

Consider

$$M = \begin{bmatrix} 1 & 1 & 2 \\ 1 & 4 & 5 \\ 2 & 5 & 6 \end{bmatrix}$$

now $PMQ=I$ where

$$P = \begin{bmatrix} 1 & 0 & 0 \\ 0 & 1 & 0 \\ 0 & -1 & 1 \end{bmatrix}\begin{bmatrix} 1 & 0 & 0 \\ 0 & 1 & 0 \\ -2 & 0 & 1 \end{bmatrix}\begin{bmatrix} 1 & 0 & 0 \\ -1 & 1 & 0 \\ 0 & 0 & 1 \end{bmatrix} = \begin{bmatrix} 1 & 0 & 0 \\ -1 & 1 & 0 \\ -1 & -1 & 1 \end{bmatrix}$$

and

$$Q = \begin{bmatrix} 1 & -1 & 0 \\ 0 & 1 & 0 \\ 0 & 0 & 1 \end{bmatrix} \begin{bmatrix} 1 & 0 & 0 \\ 0 & 1 & -1 \\ 0 & 0 & 1 \end{bmatrix} \begin{bmatrix} 1 & 0 & -2 \\ 0 & 1 & 0 \\ 0 & 0 & 1 \end{bmatrix} \begin{bmatrix} 1 & 0 & 0 \\ 0 & \frac{1}{3} & 0 \\ 0 & 0 & -1 \end{bmatrix} = \begin{bmatrix} 1 & -\frac{1}{3} & 1 \\ 0 & \frac{1}{3} & 1 \\ 0 & 0 & -1 \end{bmatrix}$$

so that

$$M^{-1} = QP = \begin{bmatrix} -\frac{1}{3} & -\frac{4}{3} & 1 \\ -\frac{4}{3} & -\frac{2}{3} & 1 \\ 1 & 1 & -1 \end{bmatrix}.$$

$\square$

4.6   THE TRACE OF A MATRIX

Let A be an $n \times n$ matrix. The trace of matrix A is the sum of the diagonal elements. Thus,

$$tr(A) = \sum_{i=1}^{n} a_{ii}.$$

A few important properties of the trace whose proofs will be left as exercises include:

1. If k is a scalar, then $tr(kA) = k\,trA$.
2. $tr(A+B) = tr(A) + tr(B)$.
3. $tr(A') = tr(A)$.

Another property of the trace is given in Theorem 4.3.

Theorem 4.3   For any $m \times n$ matrix A and $n \times m$ matrix B,

$$tr(AB) = tr(BA)$$

Proof. Observe that

$$trAB = \sum_{i=1}^{m}\sum_{j=1}^{n} a_{ij} b_{ji} = \sum_{j=1}^{n}\sum_{i=1}^{m} a_{ij} b_{ji} = \sum_{j=1}^{n}\sum_{i=1}^{m} b_{ji} a_{ij} = trBA.$$

■

Corollary 4.1 For matrices A, B, and C, assuming matrix multiplication is possible,

$$tr(ABC) = tr(CAB) = tr(BCA).$$

The next theorem gives the trace of a direct product.

■

Theorem 4.4 Let A be an $n \times n$ matrix and B be a $p \times p$ matrix. Then $\mathrm{tr}(A \otimes B) = \mathrm{tr} A \mathrm{tr} B$.

Proof. Observe that

$$\mathrm{tr}(A \otimes B) = \mathrm{tr} \begin{bmatrix} a_{11}B & a_{12}B & \cdots & a_{1n}B \\ a_{21}B & a_{22}B & \cdots & a_{2n}B \\ \vdots & \vdots & \vdots & \vdots \\ a_{n1}B & a_{n2}B & \cdots & a_{nn}B \end{bmatrix}$$

$$= \sum_{i=1}^{n} \left(a_{ii} \sum_{j=1}^{p} b_{jj} \right) = \sum_{i=1}^{n} a_{ii} \sum_{j=1}^{p} b_{jj} = \mathrm{tr} A \mathrm{tr} B. \qquad \blacksquare$$

4.7 MATRIX DIFFERENTIATION

The material in this section will be useful in Part V on optimization problems. Some more material on this topic will be presented at appropriate places in Parts III and IV. We shall differentiate matrix quantities with respect to elements of matrices. Suppose that y is an m-dimensional vector and x is an n-dimensional vector. Let

$$y = (y_1, y_2, \ldots, y_m) = (f_1(x_1, x_2, \ldots, x_n), f_2(x_1, x_2, \ldots, x_n), \ldots, f_m(x_1, x_2, \ldots, x_n)) = f(x)$$

be a "nice" vector function in the sense that all of the f_i are continuous and differentiable in each of the elements of the vector x. The symbol

$$\frac{\partial y}{\partial x} = \begin{bmatrix} \dfrac{\partial y_1}{\partial x_1} & \dfrac{\partial y_1}{\partial x_2} & \cdots & \dfrac{\partial y_1}{\partial x_n} \\ \dfrac{\partial y_2}{\partial x_1} & \dfrac{\partial y_2}{\partial x_2} & \cdots & \dfrac{\partial y_2}{\partial x_n} \\ \vdots & \vdots & \cdots & \vdots \\ \dfrac{\partial y_m}{\partial x_1} & \dfrac{\partial y_m}{\partial x_2} & \cdots & \dfrac{\partial y_m}{\partial x_n} \end{bmatrix}$$

denotes the $m \times n$ matrix of first-order partial derivatives. Such a matrix is called a Jacobian. The results that follow will help us in Part V in deriving optimal estimators.

Theorem 4.5 Let

$$y = Ax$$

where A is an $n \times m$ matrix of constants, y is an $m \times 1$ column vector, and x is an $n \times 1$ column vector. Then

$$\frac{\partial y}{\partial x} = A. \tag{4.11}$$

Proof. The ith element of y is given by

$$y_i = \sum_{i=1}^{n} a_{ik} x_k.$$

It follows that for all $i = 1, 2, \ldots, m$, $j = 1, 2, \ldots, n$ that

$$\frac{\partial y_i}{\partial x_j} = a_{ij}.$$

This establishes (4.11). ∎

We now consider an extension of Theorem 4.5 where x is a function of a vector z where A is independent of z. The formal statement is given in Theorem 4.5.

Theorem 4.6 Let y=Ax where again y is an $m \times 1$ column vector, x is an $n \times 1$ column vector, and A is an $n \times m$ matrix independent of z. Then

$$\frac{\partial y}{\partial z} = A \frac{\partial x}{\partial z}. \tag{4.12}$$

Proof. The ith element of y is given by

$$y_i = \sum_{i=1}^{n} a_{ik} x_k.$$

It follows that for all $i = 1, 2, \ldots, m$, $j = 1, 2, \ldots, n$,

$$\frac{\partial y_i}{\partial z_j} = \sum_{k=1}^{n} a_{ik} \frac{\partial x_k}{\partial z_j}.$$

However, this is the element of the $m \times n$ matrix $A(\partial x / \partial z)$ in the ith row and the jth column. Thus, we obtain 4.12. ∎

Some useful results about bilinear forms will now be obtained. When A is an $m \times n$ matrix and x is an m-dimensional column vector and y is an n-dimensional column vector, then $y'Ax$ is said to be a bilinear form. If A is a square matrix and $x = y$, then we have a quadratic form already mentioned.

Theorem 4.7 Consider the scalar

$$\alpha = y'Ax$$

where y is an m-dimensional column vector, x is an n-dimensional column vector, and A is an m×n matrix independent of x and y. Then

$$\frac{\partial \alpha}{\partial x} = y'A, \qquad (4.13a)$$

and

$$\frac{\partial \alpha}{\partial y} = x'A'. \qquad (4.13b)$$

Proof. Let $w' = y'A$. Then $\alpha = w'x$. From Theorem 4.5

$$\frac{\partial \alpha}{\partial x} = w' = y'A,$$

the result of 4.13. On the other hand

$$\alpha = \alpha' = x'A'y.$$

By application of Theorem 4.6, we obtain (4.13b).

We will now specialize the result of Theorem 4.7 to the special case of a quadratic form. This result will be most useful in Part V.   ∎

Theorem 4.8 Let $\alpha = x'Ax$ where x is n×1, A is n×n, and A does not depend on x; then

$$\frac{\partial \alpha}{\partial x} = x'(A + A'). \qquad (4.14a)$$

When A is a symmetric matrix,

$$\frac{\partial \alpha}{\partial x} = 2x'A. \qquad (4.14b)$$

Proof. Write the quadratic form out as a sum to obtain

$$\alpha = \sum_{i=1}^{n} a_{ii} x_i^2 + \sum_{i \neq j} a_{ij} x_i x_j.$$

Then

$$\frac{\partial \alpha}{\partial x_k} = 2a_{kk} x_k + \sum_{k \neq j} a_{kj} x_k + \sum_{k \neq j} a_{jk} x_k$$

$$= \sum_{k=1}^{n} (a_{kj} + a_{jk}) x_k.$$

These are the elements of (4.14a). For a symmetric matrix $a_{kj} = a_{jk}$, yielding the elements of (4.14b). ■

A chain rule for the differentiation of bilinear forms would be as follows.

Theorem 4.9 Let $\alpha = y'Ax$ where y, A, and x are as in Theorem 4.7. However, x and y are functions of another vector z. Then

$$\frac{\partial \alpha}{\partial z} = x'A'\frac{\partial y}{\partial z} + y'A\frac{\partial x}{\partial z}. \tag{4.15a}$$

For the quadratic form $\alpha = x'Ax$,

$$\frac{\partial \alpha}{\partial z} = x'(A + A')\frac{\partial x}{\partial z}. \tag{4.15b}$$

Proof. Let $v = Ax$. Then

$$\frac{\partial v}{\partial z} = \frac{\partial v}{\partial x} \cdot \frac{\partial x}{\partial z} = A\frac{\partial x}{\partial z}.$$

Now $\alpha = y'v$ so that

$$\frac{\partial \alpha}{\partial z} = y'\frac{\partial v}{\partial z} + \frac{\partial y'}{\partial z}v$$

$$= v'\frac{\partial y}{\partial z} + y'\frac{\partial v}{\partial z}$$

$$= x'A'\frac{\partial y}{\partial z} + y'A\frac{\partial x}{\partial z}. \qquad ■$$

Let A be an $m \times n$ matrix whose elements are functions of the scalar parameter α. Then the derivative of matrix A with respect to α is the $m \times n$ matrix of derivatives of the elements. Thus,

$$\frac{\partial A}{\partial \alpha} = \begin{bmatrix} \dfrac{\partial a_{11}}{\partial \alpha} & \dfrac{\partial a_{12}}{\partial \alpha} & \cdots & \dfrac{\partial a_{1n}}{\partial \alpha} \\ \dfrac{\partial a_{21}}{\partial \alpha} & \dfrac{\partial a_{22}}{\partial \alpha} & \cdots & \dfrac{\partial a_{2n}}{\partial \alpha} \\ \vdots & \vdots & & \vdots \\ \dfrac{\partial a_{m1}}{\partial \alpha} & \dfrac{\partial a_{m2}}{\partial \alpha} & \cdots & \dfrac{\partial a_{mn}}{\partial \alpha} \end{bmatrix}.$$

Theorem 4.10 Let γ and δ be scalars. Let A and B be two matrices of the same size. Then

$$\frac{\partial}{\partial \alpha}(\gamma A + \delta B) = \gamma \frac{\partial A}{\partial \alpha} + \delta \frac{\partial B}{\partial \alpha}. \tag{4.16}$$

Proof. See Exercise 4.19.   ■

A product rule for derivatives of matrices will be developed in Theorem 4.11.

Theorem 4.11 Let A be an $m \times k$ matrix. Let B be an $k \times n$. The elements of both matrices are functions of the scalar α. Let $C = AB$. Then

$$\frac{\partial C}{\partial \alpha} = A \frac{\partial B}{\partial \alpha} + \frac{\partial A}{\partial \alpha} B. \tag{4.17a}$$

Proof. The element

$$c_{ij} = \sum_{t=1}^{k} a_{it} b_{tj}.$$

Using the rules for derivatives of sums and products of scalar functions,

$$\frac{\partial c_{ij}}{\partial \alpha} = \sum_{t=1}^{k} \left(a_{it} \frac{\partial b_{tj}}{\partial \alpha} + \frac{\partial a_{it}}{\partial \alpha} b_{tj} \right), \quad i = 1, 2, \ldots, m, \ j = 1, 2, \ldots, n. \tag{4.17b}$$

The right-hand side of (4.17b) is the elements of (4.17a).   ■

Theorem 4.12 will be used to develop a formula for the derivative of the inverse of a nonsingular matrix.

Theorem 4.12 Let A be a nonsingular $m \times m$ matrix whose elements are functions of the scalar parameter α. Then

$$\frac{\partial A^{-1}}{\partial \alpha} = -A^{-1} \frac{\partial A}{\partial \alpha} A^{-1}. \tag{4.18a}$$

Proof. Using the definition of an inverse,

$$AA^{-1} = I.$$

Then by Theorem 4.11

$$A\frac{\partial A^{-1}}{\partial \alpha} + \frac{\partial A}{\partial \alpha}A^{-1} = 0. \tag{4.18b}$$

Multiplication of the left side of (4.18b) by A^{-1} and transposition of the second term yield (4.18a). ∎

4.8   THE LEAST SQUARE ESTIMATOR AGAIN

In Subsection 3.5 we gave a derivation of the least square estimator by completing the square. We now give the Calculus derivation using the matrix differentiation rules considered in Subsection 4.7. Again consider a linear model of the form

$$Y = X\beta + \varepsilon. \tag{4.19}$$

As before the objective is to minimize the sum of the squares of the differences between the observed values Y and the predicted values $X\beta$. This will be accomplished by finding the vector β that minimizes

$$F(\beta) = (Y - X\beta)'(Y - X\beta). \tag{4.20}$$

The usual way of obtaining the least square estimator is to differentiate $F(\beta)$ with respect to the components of the vector β, set the result equal to zero, and solve the matrix equation. We will do this now. First

$$\begin{aligned} F(\beta) &= (Y - X\beta)'(Y - X\beta) \\ &= Y'Y - \beta'X'Y - Y'X\beta + \beta'X'X\beta. \end{aligned} \tag{4.21}$$

From Theorems 4.4 and 4.7 and the fact that $\beta'X'Y = Y'X\beta$,

$$\frac{\partial F(\beta)}{\partial \beta} = -2X'Y + 2X'X\beta = 0. \tag{4.22}$$

Solving the matrix equation for $\hat{\beta}$ when X is of full rank, we again obtain

$$\hat{\beta} = (X'X)^{-1}X'Y. \tag{4.23}$$

4.9   SUMMARY

We considered two kinds of matrices used a lot in statistics, matrices of the form $aI_n + bJ_n$ and orthogonal matrices. We defined and gave properties of the Kronecker product of two matrices. We showed that the determinant of the product of two matrices is the product of their determinants and that the trace of the sum of two matrices is the sum of their traces. Then we defined and established properties of matrix derivatives and showed how they could be used to find the least square estimator.

EXERCISES

4.1   Let $A = aI_n + bJ_n$, $X = xI_n + yJ_n$ and $B = cI_n + dJ_n$. Give conditions on A so that the equation $AX = B$ has a solution and find the solution when these conditions hold true.

4.2   Show that $I_n - (2/n)J_n$ is its own inverse.

4.3   Find an inverse for $I_n - (g/n)J_n$ of the form $aI_n + bJ_n$.

4.4   Write out the matrix H_6 and show that it is an orthogonal matrix by direct computation.

4.5   Show using mathematical induction that the product of n orthogonal matrices is an orthogonal matrix.

4.6   Let A be a square matrix and D a diagonal matrix. Show that $\det(AD) = \det(DA) = \det A \det D$. (This can be done without the more general result that $\det(AB) = \det A \det B$).

4.7   **A.** Establish 1a and 1b in Theorem 4.2.
         B. Use 1a and 1b to show that

$$(A + B) \otimes (C + D) = (A \otimes C) + (A \otimes D) + (B \otimes C) + (B \otimes D).$$

         C. Complete the proof of 3 in Theorem 4.2. Also show that the Kronecker product of two symmetric matrices is a symmetric matrix.
         D. Prove 4 in Theorem 4.2.

4.8   Find $A \otimes B$ and $B \otimes A$ when

A. $A = \begin{bmatrix} 1 & 1 \\ 1 & 1 \end{bmatrix}$, $B = \begin{bmatrix} 1 & 0 \\ 0 & 1 \end{bmatrix}$.

B. $A = \begin{bmatrix} 1 & 6 \\ 2 & 4 \end{bmatrix}$, $B = \begin{bmatrix} -1 & 2 & 4 \\ 4 & -1 & 6 \end{bmatrix}$.

4.9   Given $A = \begin{bmatrix} 1 & 1 \\ 1 & 1 \end{bmatrix}$, $B = \begin{bmatrix} 1 & 0 \\ 0 & 1 \end{bmatrix}$, $C = \begin{bmatrix} 1 \\ 1 \end{bmatrix}$, find $A \otimes (B \otimes C)$ and $(A \otimes B) \otimes C$. Compare your answers. Another property of the direct product is the associative law

$$A \otimes (B \otimes C) = (A \otimes B) \otimes C.$$

A proof may be found in Harville (2008).

4.10 Let P and Q be orthogonal matrices. Show that $P \otimes Q$ is orthogonal.

4.11 Show that $(A \otimes B) = (A \otimes I)(I \otimes B)$.

4.12 **A.** Show that for nonsingular matrix A,

$$\det A^{-1} = \frac{1}{\det A}.$$

B. Show that a real orthogonal matrix has determinant 1 or −1.

4.13 Characterize all of the elementary 2×2 matrices.

4.14 Prove the following properties of the trace of a matrix:
A. If k is a scalar, then $\operatorname{tr}(kA) = k\operatorname{tr}A$.
B. $\operatorname{tr}(A+B) = \operatorname{tr}(A) + \operatorname{tr}(B)$.
C. $\operatorname{tr}(A') = \operatorname{tr}(A)$.

4.15 Prove the corollary to Theorem 4.3.

4.16 Let A be an $n \times n$ matrix. Let p be an n-dimensional column vector. Show that $\operatorname{tr}App' = p'Ap$.

4.17 Show that $\operatorname{tr}(A'A) \geq 0$.

4.18 Use elementary row operations and column operations to reduce each of the following matrices to the identity matrix. Find the corresponding product that gives the inverse of the matrices. Then express the original matrix as the product of elementary matrices:

A. $A = \begin{bmatrix} 4 & 2 \\ 3 & 1 \end{bmatrix}$.

B. $B = \begin{bmatrix} a & b \\ b & a \end{bmatrix}$, $a \neq b$.

C. $C = \begin{bmatrix} 0 & 1 & 1 \\ 1 & 0 & 1 \\ 1 & 1 & 0 \end{bmatrix}$.

D. $D = \begin{bmatrix} 1 & 2 & 3 \\ 3 & 1 & 2 \\ 2 & 3 & 1 \end{bmatrix}$.

4.19 Prove Theorem 4.10.

4.20 Show that

A. $\dfrac{\partial A'}{\partial \alpha} = \left(\dfrac{\partial A}{\partial \alpha} \right)'$.

B. $\dfrac{\partial (A'A)}{\partial \alpha} = A' \dfrac{\partial A}{\partial \alpha} + \left(\dfrac{\partial A}{\partial \alpha} \right)' A$.

4.21 Let

$$A = \begin{bmatrix} \cos\theta & \sin\theta \\ -\sin\theta & \cos\theta \end{bmatrix}.$$

A. Show that

$$A^2 = \begin{bmatrix} \cos 2\theta & \sin 2\theta \\ -\sin 2\theta & \cos 2\theta \end{bmatrix}.$$

B. Find $\partial A^2 / \partial \theta$ two ways.

C. Find $A'A$. Also find its derivative two ways.

D. Find the derivative of A^{-1} two ways.

4.22 Let A and B be nonsingular $m \times m$ matrices whose elements are functions of a scalar parameter. Show the following:

A. $\dfrac{\partial}{\partial \alpha} AB^{-1} = -AB^{-1} \dfrac{\partial B}{\partial \alpha} B^{-1} + \dfrac{\partial A}{\partial \alpha} B^{-1}.$

B. $\dfrac{\partial}{\partial \alpha} A^{-1}B = A^{-1} \dfrac{\partial B}{\partial \alpha} - A^{-1} \dfrac{\partial A}{\partial \alpha} A^{-1}B.$

C. If A and B are 1×1 matrices, then the elements of AB^{-1} and $A^{-1}B$ satisfy the quotient rule for derivatives from differential calculus.

4.23 Let A and B be matrices whose elements are functions of a scalar parameter. Show that

$$\frac{\partial}{\partial \alpha} (A \otimes B) = A \otimes \frac{\partial B}{\partial \alpha} + \frac{\partial A}{\partial \alpha} \otimes B.$$

4.24 Show that if A is a square nonsingular matrix, then the following three conditions are equivalent:

1. $A'A = I$
2. $AA' = I$
3. $A' = A^{-1}$

4.25 Using both row and column operations as was done in Example 4.7, find inverses of the matrices as a product of elementary matrices; that is, find matrices P and Q where $PMQ = I$ and $M^{-1} = QP$:

A. $\begin{bmatrix} 7 & 3 \\ 3 & 7 \end{bmatrix}$

B. $\begin{bmatrix} 1 & 2 & 1 \\ 2 & 1 & 1 \\ 1 & 1 & 2 \end{bmatrix}$

SECTION 5

VECTOR SPACES

5.1 INTRODUCTION

Vector spaces will now be taken up because the associated terminology will be important for understanding concepts like the rank of a matrix, eigenvalues and eigenvectors, and the singular value decomposition. Vector spaces generalize the notion of vectors to more abstract spaces than the plane or three-dimensional space.

We will explain what a vector space is and describe some of its properties. In order to talk about the dimension of a vector space, we will need the concepts of linear dependence and independence. We then define what is meant by an inner product space and prove two important inequalities, the triangle inequality and the Cauchy–Schwarz inequality. The Cauchy–Schwarz inequality will be important for results that compare the efficiency of estimators in Part V.

We conclude this section with a discussion of linear transformations. These are linear mappings between vector spaces. We will show how there is a matrix associated with each linear transformation.

5.2 WHAT IS A VECTOR SPACE?

A vector space consists of two kinds of objects, vectors and scalars. Vectors may be n tuples of numbers, functions, or polynomials depending on the nature of the problem for solution. For our purposes scalars will be real numbers. However, scalars can be complex numbers or elements of any field.

Matrix Algebra for Linear Models, First Edition. Marvin H. J. Gruber.
© 2014 John Wiley & Sons, Inc. Published 2014 by John Wiley & Sons, Inc.

There are two operations. They are the addition of vectors and the multiplication of a vector by a scalar. For example, for vectors in the Cartesian plane, two vectors might be denoted by $v = (v_1, v_2)$ and $w = (w_1, w_2)$.

The sum of the two vectors would be

$$v + w = (v_1 + w_1, v_2 + w_2).$$

The scalar multiple av would be (av_1, av_2). The sums and scalar multiples of vectors satisfy certain axioms. These include associate and commutative laws of addition; the existence of an identity element, for each of the vectors the existence of an additive inverse; the distributive law for scalar multiplication with respect to vector addition; and addition of real numbers and an identity element for scalar multiplication. Definition 5.1 is a more formal definition of a vector space.

Definition 5.1 A vector space V is a set of two kinds of objects vectors and scalars with binary operations of addition and multiplication of vectors by scalars. The sum of two vectors and the scalar multiples of a vector are contained in the vector space V. The following axioms must be satisfied:

1. For two vectors u and v,

$$u + v = v + u \, (\text{commutative law}).$$

2. Given three vectors u, v, and w,

$$u + (v + w) = (u + v) + w \, (\text{associative law}).$$

3. There is an additive identity, a zero vector 0.
4. Each vector has an additive inverse. For any element v, there exists an element denoted by $-v$ such that $v + (-v) = 0$.
5. a. $a(u + v) = au + av$ (distributive laws for scalar multiplication).
   b. $(a + b)u = au + bu$.
6. $(ab)u = a(bu)$ (compatibility of scalar multiplication with multiplication of real numbers).
7. There is a number denoted by 1 where $1v = v$. Thus, there is an identity element for scalar multiplication.

We define subtraction by saying that $v - w = v + (-w)$. ●

A subset of a vector space that is itself a vector space is called a subspace. If W is a subset of V such that for all vectors u and v that belong to W and all scalars, a $u - v$ belongs to W and au belongs to W, it is not hard to establish that W is a vector subspace.

5.3   THE DIMENSION OF A VECTOR SPACE

We now consider vectors that are linear combinations of other vectors. These vectors take the form

$$v = \sum_{i=1}^{n} a_i v_i \tag{5.1}$$

where the a_i are real numbers. If every vector in a vector space V is expressible in the form of (5.1), then the vectors v_i, $1 \le i \le n$ are said to span the space. Suppose that none of the vectors v_i, $1 \le i \le n$ is expressible as a linear combination of the other vectors. Then the vectors are said to be linearly independent. If a set of vectors spans a vector space and are linearly independent, they form a basis of the space. The number of elements in the basis is the dimension of the vector space. More formally we make the following definitions.

Definition 5.2 Let V be a vector space. If every element $v \in V$ may be expressed as a linear combination of vectors $v_1, v_2, \ldots, v_n$ using (5.1), then the vectors $v_1, v_2, \ldots, v_n$ are said to span V.   ●

Definition 5.3 Suppose n vectors $v_1, v_2, \ldots, v_n$ are such that there do not exist scalars $c_1, c_2, \ldots, c_n$. that are different from zero where

$$\sum_{i=1}^{n} c_i v_i = 0. \tag{5.2}$$

Then the vectors $v_1, v_2, \ldots, v_n$ are said to be linearly independent. On the other hand if there do exist nonzero scalars $c_1, c_2, \ldots, c_n$ where (5.2) does hold true, then the vectors are said to be linearly dependent.   ●

Definition 5.3 is a formal way of saying that if the n vectors are linearly independent, one of the v_i cannot be expressed as a linear combination of the others. On the other hand if the vectors are linearly dependent, one of the vectors can be expressed as a linear combination of the others.

Definition 5.4 If a set of n vectors $v_1, v_2, \ldots, v_n$

1. Span the vector space V
2. Are linearly independent

then the n vectors form a basis for the vector space V and the space V has dimension n.   ●

A vector space may have many bases. The basis is not unique. Only the number of elements in the basis is unique. An illustration of two different basis of a vector space will be considered below.

THE DIMENSION OF A VECTOR SPACE

Example 5.1 Two Different Basis of a Three-Dimensional Vector Space

Consider the vector space that consists of three tuples (a,b,c) where a, b, and c are any real numbers. Observe that

$$(a, b, c) = a(1, 0, 0) + b(0, 1, 0) + c(0, 0, 1)$$

and

$$(a, b, c) = \left(\frac{b+c-a}{2} \right)(0, 1, 1) + \left(\frac{a+c-b}{2} \right)(1, 0, 1) + \left(\frac{a+b-c}{2} \right)(1, 1, 0).$$

It is not difficult to show that $(1,0,0), (0,1,0), (0,0,1)$ and $(0,1,1), (1,0,1), (1,1,0)$ are each linearly independent vectors and, thus, basis of the vector space generated by the three tuples (a,b,c).

Subspaces of this vector space that are of dimension 2 would be the set of elements of the form (a, b, 0), the set of elements of the form (a, 0, c), and the set of elements of the form (0, b, c).   □

Example 5.2 A Vector Space Whose Elements Are Matrices

The set of $n \times m$ matrices where addition is matrix addition and scalar multiplication multiplies every element of a matrix by a scalar. A basis for this vector space would be the $n \times m$ matrices E_{ij}, $1 \le i \le n$, $1 \le j \le m$ with the entry 1 in the ith row and the jth column and zero elsewhere. The dimension of this vector space would be nm.   □

The sum of two vector spaces U and V, U+V, consists of the sum of elements $u+v$ where $u \in U$ and $v \in V$. Given bases of U and V, the vector space U+V may be generated by the linear combinations of the basis elements in U and V, in other words the span of the two sets of bases. Suppose $\dim(U) = m$ and $\dim(V) = n$. If $\dim(U+V) > n+m$, there would be at least $n+m+1$ linearly independent elements in the basis of U+V that would be linear combinations of the $n+m$ elements in the span of U+V. However, the vector space U+V would then require $m+n+1$ elements in a spanning set contradicting the definition. On the other hand as illustrated by Example 5.3, the $m+n$ elements may not be linearly independent so $\dim[U+V]$ can be less than $m+n$. Thus,

$$\dim[U + V] \le m + n.$$

In fact,

$$\dim[U + V] = \dim U + \dim V - \dim[U \cap V]. \tag{5.3}$$

A proof of (5.3) is available in Harville (2008).

When $U \cap V = \{0\}$, then the vector space U+V is a direct sum of two vector spaces denoted by $U \oplus V$. It is not hard to show that

$$\dim(U \oplus V) = \dim U + \dim V. \tag{5.4}$$

Example 5.3 Illustration of Result in (5.3)

Let U be the vector space with basis $\begin{bmatrix} 1 \\ 0 \\ 0 \\ 0 \end{bmatrix}$ and $\begin{bmatrix} 0 \\ 1 \\ 0 \\ 0 \end{bmatrix}$. Let V be the vector space with

basis $\begin{bmatrix} 0 \\ 1 \\ 0 \\ 0 \end{bmatrix}$ and $\begin{bmatrix} 0 \\ 0 \\ 1 \\ 0 \end{bmatrix}$. For both U and V, addition is adding the corresponding coordi-

nates. Scalar multiplication is multiplying elements of the four tuples by the scalar. Both vector spaces have dimension 2 but the vector space U+V has dimension 3.

Notice that U∩V is the vector space with basis $\begin{bmatrix} 0 \\ 1 \\ 0 \\ 0 \end{bmatrix}$ of dimension 1.   □

Example 5.4 Let P be the vector space with basis $\begin{bmatrix} 1 \\ 0 \\ 0 \\ 0 \end{bmatrix}$ and $\begin{bmatrix} 0 \\ 1 \\ 0 \\ 0 \end{bmatrix}$ and Q be the

vector space with basis $\begin{bmatrix} 0 \\ 0 \\ 1 \\ 0 \end{bmatrix}$ and $\begin{bmatrix} 0 \\ 0 \\ 0 \\ 1 \end{bmatrix}$. Addition and scalar multiplication are as

defined in Example 5.3. Then $\mathbb{R}^4 = P \oplus Q$. This is the vector space consisting of four-dimensional column vectors all of whose entries are real numbers.   □

An important vector space is that generated by the rows or columns of a matrix.
These are referred to as the row space and the column space respectively. These vector spaces will be discussed in Section 6.

5.4   INNER PRODUCT SPACES

In the three-dimensional Euclidean space, the inner product of two vectors $v = (v_1, v_2, v_3)$ and $w = (w_1, w_2, w_3)$ was given by $v \cdot w = v_1 w_1 + v_2 w_2 + v_3 w_3$. The reader can easily verify that

1. $v \cdot w = w \cdot v$.
2. For real numbers a and b,

$$(au + bv) \cdot w = a(u \cdot v) + b(v \cdot w).$$

3. $v \cdot v \geq 0$.
4. $v \cdot v = 0$ if and only if $v = 0$.

An inner product space is a vector space with similar properties. Definition 5.5 is the formal definition of an inner product space.

Definition 5.5 An inner product space is a vector space with an inner product. Let V be a vector space and let $V \times V$ denote the set of ordered pairs of elements of V, $\{v, w\}$. An inner product is a mapping of $V \times V$ into the real numbers[1] that satisfies properties 1–4 below. The real number obtained is denoted by (v, w).

1. The mapping is symmetric. That means

$$(v, w) = (w, v).$$

2. The mapping is bilinear. That means that for real numbers a, b, c, and d,

$$(av + bw, cx + dy) = ac(v, x) + bc(w, x) + ad(v, y) + bd(w, y).$$

3. The inner product of a vector with itself is nonnegative. That can be expressed by the inequality

$$(v, v) \geq 0.$$

4. The inner product of a vector with itself is zero if and only if it is the zero vector. In symbols

$$(v, v) = 0$$

if and only if $v = 0$.
   The norm of v is the square root of the inner product of a vector in a vector space with itself. It is denoted by $\|v\|$. Thus,

$$\|v\| = (v, v)^{1/2}. \qquad \bullet$$

Example 5.5 Consider the vector space with elements of the form $v = (v_1, v_2, \ldots, v_n)$. An inner product would be

$$(v, w) = \sum_{i=1}^{n} v_i w_i. \qquad \square$$

[1] Actually it could be a mapping into any field that contains the scalars in the vector space. However, we are only considering vector spaces over the real numbers.

Example 5.6 The set of real $m \times n$ matrices can be viewed as a vector space of dimension mn. One choice of a basis would be the matrices with one element 1 and the remaining elements zero. An inner product of two matrices A and B would be

$$(A, B) = \text{trace}(AB').$$

For the set of $1 \times n$ matrices, the inner product reduces to

$$(A, B) = \sum_{j=1}^{n} a_{1j} b_{j1},$$

the inner product of two n-dimensional vectors. □

Two well-known and important inequalities for inner product spaces are the Cauchy–Schwarz inequality and the triangle inequality. These are stated and proved in Theorems 5.1 and 5.2.

Theorem 5.1 The Cauchy–Schwarz inequality. Let u and v be vectors in a vector space V. Then

$$(u, v) \leq \|u\| \|v\|. \tag{5.5}$$

If $(u, v) = \|u\| \|v\|$, then u is a scalar multiple of v.

Proof. For any scalar a

$$\|u - av\|^2 \geq 0. \tag{5.6}$$

Now using the defining properties of inner products,

$$\begin{aligned} \|u - av\|^2 &= (u - av, u - av) \\ &= \|u\|^2 - 2a(u, v) + a^2 \|v\|^2. \end{aligned} \tag{5.7}$$

In order for (5.5) to hold true, from the quadratic formula and (5.6),

$$4(u, v)^2 \leq 4 \|u\|^2 \|v\|^2. \tag{5.8}$$

Dividing (5.8) by 4 and taking the square root of both sides yields (5.5).
If equality holds in (5.4), then $(u, v) = \|u\| \|v\|$ and $u - av = 0$ so $u = -av$. ∎

Theorem 5.2 The triangle inequality. For any two vectors u and v,

$$\|u + v\| \leq \|u\| + \|v\|. \tag{5.9}$$

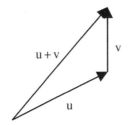

FIGURE 5.1 Sum of two vectors.

Proof. Observe that using properties of the inner product and the Cauchy–Schwarz inequality that

$$\begin{aligned}
\|u + v\|^2 &= (u + v, u + v) \\
&= \|u\|^2 + 2(u, v) + \|v\|^2 \\
&\leq \|u\|^2 + 2\|u\|\|v\| + \|v\|^2 \\
&= (\|u\| + \|v\|)^2.
\end{aligned} \tag{5.10}$$

The result follows by taking the square root of the expressions in (5.10). ∎

Theorem 5.2 is in a sense a statement of the well-known fact that a straight line is the shortest distance between two points. Figure 5.1 illustrates the sum of two vectors in the plane. The magnitudes of the vectors are the lengths of the sides of the triangle. Notice how from the picture

$$\|u + v\| \leq \|u\| + \|v\|.$$

5.5 LINEAR TRANSFORMATIONS

A linear transformation between two vector spaces is a function or mapping that takes linear combinations of vectors in one space into linear combinations of vectors in another space. More formally:

Definition 5.6 Let U and V be two vector spaces and let T be a mapping from U to V. That means that each element of U is mapped to a unique element of V. Let u_1 and u_2 be vectors in U and let a and b be real numbers. Then T is a linear transformation if and only if

$$T(au_1 + bu_2) = aT(u_1) + bT(u_2). \qquad \bullet$$

For example, let A be a matrix and suppose that u_1 and u_2 are column vectors with the same number of entries as A has columns. Consider the vector space U generated by all such column vectors. Let a and b be scalars. Then using the fact that matrix multiplication is distributive,

$$A(au_1 + bu_2) = aAu_1 + bAu_2.$$

Thus, A is a linear transformation from the vector space U to a vector space generated by the column vectors Au.

On the other hand there is a matrix that corresponds to a linear transformation between vector spaces. Let U and V be vector spaces with basis $u_1, u_2, \ldots, u_n$ and $v_1, v_2, \ldots, v_m$. If T is a linear transformation from U to V, there are scalars $a_{i1}, a_{i2}, \ldots, a_{im}$ such that

$$T(u_i) = \sum_{j=1}^{m} a_{ij} v_j, \quad 1 \le i \le m.$$

Then the $n \times m$ matrix

$$A = [a_{ij}]_{1 \le i \le n, \ 1 \le j \le m} = \begin{bmatrix} a_{11} & a_{12} & \cdots & a_{1m} \\ a_{21} & a_{22} & \cdots & a_{2m} \\ \vdots & \vdots & & \vdots \\ a_{n1} & a_{n2} & \cdots & a_{nm} \end{bmatrix}$$

is the matrix associated with the linear transformation T. Let U, V, and w be vector spaces with basis $u_1, u_2, \ldots, u_n$, $v_1, v_2, \ldots, v_m$, and $w_1, w_2, \ldots, w_\ell$.

What linear transformation corresponds to matrix multiplication? Let $S : V \to W$ and $T : U \to V$ be linear transformations with matrices A and B, respectively. Then

$$ST(u_i) = S\left(\sum_{j=1}^{m} a_{ij} v_j \right) = \sum_{j=1}^{m} a_{ij} S(v_j) = \sum_{j=1}^{m} a_{ij} \left(\sum_{k=1}^{\ell} b_{jk} w_k \right) = \sum_{k=1}^{\ell} \left(\sum_{j=1}^{m} a_{ij} b_{jk} \right) w_k, \quad 1 \le i \le n.$$

Thus, the composition of two linear transformations corresponds to multiplication of their matrices.

What about the inverse of a linear transformation? In order for T to have an inverse transformation:

1. It has to be one to one, which means if that for two vectors p and q we have $T(p) = T(q)$, then $p = q$.
2. The transformation T must be onto meaning that for each element $v \in V$ there is an element $u \in U$ such that $T(u) = v$. Then $T(U) = V$.

If 1 and 2 above are satisfied, the matrix of the transformation must be nonsingular. It must be a square matrix and no linear combination of its rows or columns may be zero.

Define the identity transformations $I_U : U \to U$ and $I_V : V \to V$ as those where for all elements $u \in U$ and $v \in V$ $I_U(u) = u$ and $I_V(v) = v$. The matrices of these transformations are the identity matrix. Transformation Q is the inverse of T if and only if $QT = I_U$ and $TQ = I_V$.

The inverse matrix A^{-1} would be the matrix of the inverse of transformation T with matrix A.

Example 5.7 Consider the vector space of column vectors $\begin{bmatrix} a \\ b \\ c \end{bmatrix}$. Two bases of this

vector space are $e_1 = \begin{bmatrix} 1 \\ 0 \\ 0 \end{bmatrix}, e_2 = \begin{bmatrix} 0 \\ 1 \\ 0 \end{bmatrix}, e_3 \begin{bmatrix} 0 \\ 0 \\ 1 \end{bmatrix}$ and $f_1 = \begin{bmatrix} 0 \\ 1 \\ 1 \end{bmatrix}, f_2 = \begin{bmatrix} 1 \\ 0 \\ 1 \end{bmatrix}, f_3 = \begin{bmatrix} 1 \\ 1 \\ 0 \end{bmatrix}$. Let T be

a linear transformation where $T(e_i) = f_i$, $i = 1, 2, 3$. Since

$$f_1 = e_2 + e_3, \; f_2 = e_1 + e_3, \; f_3 = e_1 + e_2,$$

the matrix of the transformation would be

$$M = \begin{bmatrix} 0 & 1 & 1 \\ 1 & 0 & 1 \\ 1 & 1 & 0 \end{bmatrix}.$$

The reader may verify that

$$M^{-1} = \frac{1}{2} \begin{bmatrix} -1 & 1 & 1 \\ 1 & -1 & 1 \\ 1 & 1 & -1 \end{bmatrix}$$

so that

$$e_1 = \frac{1}{2}(-f_1 + f_2 + f_3)$$

$$e_2 = \frac{1}{2}(f_1 - f_2 + f_3)$$

$$e_3 = \frac{1}{2}(f_1 + f_2 - f_3).$$

□

An important type of linear transformation is a projection. Suppose that vector space $W = U \oplus V$. Then any vector $w \in W$ may be written $w = u + v$ where $u \in U$ and $v \in V$. A projection W into U would be the mapping where $T(w) = u$. Likewise a projection of W into V would be the mapping S where $S(w) = v$.

Example 5.8 Consider the Cartesian plane with points (x, y). Draw the vectors from the origin to this point. Define $P_1(x, y) = (x, 0)$ and $P_2(x, y) = (0, y)$. Observe that $R^2 = R^1 \oplus R^1$. Consider the vector (3, 4). Its projections would be (3, 0) and (0, 4) as shown in Figure 5.2. □

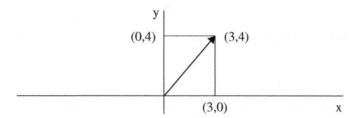

FIGURE 5.2   Projections of a vector.

Example 5.9 Consider the vector space $\mathbb{R}^4 = P \oplus Q$ in Example 5.4. Consider

the projection into P. Then, for example, $\text{Proj}_P \begin{bmatrix} 1 \\ 1 \\ 1 \\ 1 \end{bmatrix} = \begin{bmatrix} 1 \\ 1 \\ 0 \\ 0 \end{bmatrix}$. The matrix for Proj_P would

be $\begin{bmatrix} 1 & 0 & 0 & 0 \\ 0 & 1 & 0 & 0 \\ 0 & 0 & 0 & 0 \\ 0 & 0 & 0 & 0 \end{bmatrix}$. $\qquad\qquad$ □

5.6   SUMMARY

We defined what is meant by a vector space and gave examples of vector spaces. We also defined the notions of linear dependence and independence and gave examples. Inner product spaces were defined. The triangle inequality and the Cauchy–Schwarz inequality were stated and proved. The definition of linear transformations and their relationship with matrices and matrix operations were presented.

EXERCISES

5.1   What are one-dimensional subspaces of the vector space in Example 5.1?

5.2   Show that the E_{ij}, $1 \leq i \leq n$, $1 \leq j \leq m$ in Example 5.2 is a basis for the vector space described there.

5.3   Show that the polynomials of degree at most three form a vector space where the sum of two polynomials is formed by adding coefficients of terms with the same exponent and the scalar multiples are obtained by multiplying all of the coefficients of a polynomial by the same real number. Exhibit a basis of the vector space and show that the elements span the space and are linearly independent. What are the subspaces of this vector space?

5.4 Show that functions of the form $f(x) = c_1 e^x + c_2 x e^x$ form a vector space of dimension 2. What are the one-dimensional subspaces?

5.5 In Definition 5.5, replace the bilinearity with

$$(av + bw, x) = a(v, x) + b(w, x).$$

Use this property together with the other defining properties of an inner product to obtain the result in Property 2.

5.6 Show that the inner products in Examples 5.5 and 5.6 actually do satisfy the defining properties of an inner product.

5.7 Let X and Y be random variables. Define the inner product

$$(X, Y) = \text{cov}(X, Y)$$

and the norm

$$\|X\| = \sqrt{\text{Var}(X)}.$$

Show that the defining properties of an inner product are satisfied. You may freely use the properties of the variance and covariance studied in elementary statistics courses.

5.8 For the random variables in Exercise 5.7, recall that the correlation coefficient is defined by

$$\rho = \frac{\text{cov}(X, Y)}{\sqrt{\text{Var}(X)}\sqrt{\text{Var}(Y)}}.$$

Using the Cauchy–Schwarz inequality, obtain the well-known property that

$$-1 \le \rho \le 1.$$

How do you interpret $\rho = 1$ or $\rho = -1$?

5.9 Establish the result in (5.4) without using (5.3).

5.10 What other ways are there to express $\mathbb{R}^4$ as the direct sum of subspaces in Example 5.4? You can use as many as four subspaces.

5.11 **A.** In Example 5.7 verify that e_1, e_2, e_3 and f_1, f_2, f_3 are indeed basis of the vector space in the problem.

B. Express the column vector $\begin{bmatrix} 1 \\ 2 \\ 4 \end{bmatrix}$ in terms of both sets of basis vectors.

C. Show that $g_1 = \begin{bmatrix} 2 \\ 1 \\ 1 \end{bmatrix}$, $g_2 = \begin{bmatrix} 1 \\ 2 \\ 1 \end{bmatrix}$, $g_3 = \begin{bmatrix} 1 \\ 1 \\ 2 \end{bmatrix}$ is also a set of basis vectors for the

vector space in Example 5.7. Express e_1, e_2, e_3 and f_1, f_2, f_3 in terms of g_1, g_2, g_3. What are the matrices of the transformations that take the e's and f's to the g's and what are the inverse transformation matrices?

5.12 For the vector space $W = U \oplus V$, consider the projection mappings S and T.

A. Show that S and T are linear transformations.

B. Show that $S^2 = S$ and $T^2 = T$.

5.13 In Example 5.9, what is the matrix for Proj_Q?

5.14 Consider the vector space in Exercise 5.3. Let D be the operation of differentiation.

A. Show that this is a linear transformation and find its matrix.

B. Show that the matrix obtained in Part A is nilpotent.

5.15 Let V be the vector space with basis $\cos\theta$ and $\sin\theta$. Again let D be the operation of differentiation. Let G be the operation of antidifferentiation (forget the constant of integration). What are the matrices of these linear transformations? Also show that G and D are inverse matrices.

SECTION 6

THE RANK OF A MATRIX AND SOLUTIONS TO SYSTEMS OF EQUATIONS

6.1 INTRODUCTION

The rank of a matrix, some of its properties, and its importance for the existence of solutions to a system of equations will be discussed in this section.

A very important vector space is the one that is generated by the rows or by the columns of a matrix. The dimension of that vector space is called the row rank or column rank of a matrix. These two ranks will be shown to be the same. Some theorems will be proved about the ranks of the sum or the product of a matrix.

A system of equations $Ax = b$ is consistent if and only if the rank of the matrix of coefficients and the matrix of coefficients with the vector on the right-hand side as an additional column, the augmented matrix, is the same. If A is of full rank, the system has a unique solution; if A is of less than full rank, the system has infinitely many solutions. We will present these results and give examples of them in this section.

6.2 THE RANK OF A MATRIX

The linear combinations of the columns of an $m \times n$ matrix A generate a vector space called the column space that will be denoted by $C(A)$. Likewise, the linear combinations of the rows of a matrix generate a vector space that will be denoted by $R(A)$.

Matrix Algebra for Linear Models, First Edition. Marvin H. J. Gruber.
© 2014 John Wiley & Sons, Inc. Published 2014 by John Wiley & Sons, Inc.

The dimension of C(A) is called the column rank. The dimension of R(A) is called the row rank. The following theorem is true but not altogether obvious.

Theorem 6.1 The row rank of an $m \times n$ matrix is equal to the column rank.

This proof is due to Wardlaw (2005).

Proof. The case where $A = 0$ is trivial because the row rank and the column rank are both zero. Let r be the smallest positive integer where there is an $m \times r$ matrix B and an $r \times n$ matrix C such that $A = BC$. Then the r rows of C are the minimal spanning set of R(A) and the r columns of B are the minimal spanning set of C(B). The row and column ranks are both r. ∎

What still requires proof is the existence of the factorization. A proof of this is available in Harville (2008). We will give a different proof that the row and column ranks of a matrix are the same in Part II using the singular value decomposition.

There are several ways to find the rank of a matrix. One method is to find the largest matrix with a nonzero determinant. Here is the rationale. If an $m \times n$ matrix has rank r, there exists an $r \times r$ submatrix with a non-zero determinant. The r rows of the $m \times n$ matrix that include the rows of the submatrix must be independent. If they were dependent, the rows of the submatrix would be dependent and the determinant would be zero. Likewise any larger submatrix would have determinant zero because the rows would be dependent since the matrix has rank r. Thus, one way to find the rank of a matrix is to determine the largest dimension of a matrix with a nonzero determinant.

Another way to find the rank of a matrix is to reduce the matrix to an upper triangular form and see how many zero rows there are. The rank would be m minus the number of zero rows.

Example 6.1 The matrix

$$H = \begin{bmatrix} 9 & 4 & 3 & 2 \\ 4 & 4 & 0 & 0 \\ 3 & 0 & 3 & 0 \\ 2 & 0 & 0 & 2 \end{bmatrix}$$

has rank three because det $(H) = 0$ but

$$\begin{vmatrix} 4 & 0 & 0 \\ 0 & 3 & 0 \\ 0 & 0 & 2 \end{vmatrix} = 24 \neq 0.$$

Also, by elementary row operations

$$
\begin{bmatrix} 9 & 4 & 3 & 2 \\ 4 & 4 & 0 & 0 \\ 3 & 0 & 3 & 0 \\ 2 & 0 & 0 & 2 \end{bmatrix} \rightarrow \begin{bmatrix} 4 & 4 & 0 & 0 \\ 3 & 0 & 3 & 0 \\ 2 & 0 & 0 & 2 \\ 9 & 4 & 3 & 2 \end{bmatrix} \rightarrow \begin{bmatrix} 4 & 4 & 0 & 0 \\ 3 & 0 & 3 & 0 \\ 2 & 0 & 0 & 2 \\ 0 & 0 & 0 & 0 \end{bmatrix}
$$

$$
\rightarrow \begin{bmatrix} 1 & 1 & 0 & 0 \\ 1 & 0 & 1 & 0 \\ 1 & 0 & 0 & 1 \\ 0 & 0 & 0 & 0 \end{bmatrix} \rightarrow \begin{bmatrix} 1 & 1 & 0 & 0 \\ 0 & -1 & 1 & 0 \\ 0 & 0 & -1 & 1 \\ 0 & 0 & 0 & 0 \end{bmatrix}.
$$

□

Some important properties of the rank of a matrix are given in Theorem 6.2.

Theorem 6.2 1. The rank of a matrix $A + B$ is less than or equal to the rank of A plus the rank of B:

$$
\text{rank}(A + B) \leq (\text{rank}(A) + \text{rank}(B)).
$$

2. The rank of matrix AB is less than or equal to the minimum of the ranks of A and B:

$$
\text{rank}(AB) \leq \min(\text{rank}(A), \text{rank}(B)).
$$

Proof of 1. Indeed

$$
C(A + B) \subseteq C(A) + C(B)
$$

so

$$
\begin{aligned}
\text{rank}(A + B) &= \dim C(A + B) \\
&\leq \dim(C(A) + C(B)) \leq \dim C(A) + \dim C(B). \\
&= \text{rank}(A) + \text{rank}(B).
\end{aligned}
$$

Proof of 2. The rows of AB are linear combinations of the rows of B so the number of linearly independent rows of AB is less than that of B. Thus,

$$
\text{rank}(AB) \leq \text{rank}(B).
$$

Likewise the columns of AB are linear combinations of the columns of A, so

$$
\text{rank}(AB) \leq \text{rank}(A).
$$

It is not hard to show that

$$
\text{rank}(A) = \text{rank}(A').
$$

The row rank of A would be the same as the column rank of A'.   ■

Another useful result is the following.

Theorem 6.3 If M is an $m \times n$ matrix of rank r and A, a nonsingular $m \times m$ matrix, AM has rank r.

Proof. From Theorem 6.2 rank(AM)$\leq$r. If rank(AM)$<$r, then there is a nonzero vector x with at least $n-r$ components equal to zero where AMx=0. Then Mx=A^{-1}AMx=0, meaning that r columns of M are linearly dependent. Thus, the rank of M would have to be less than r. This is clearly a contradiction. ■

The following result will be proved again in Part II when we discuss the singular value decomposition. We will give a proof that is due to Schott (2005).

Theorem 6.4 The following equality holds:

$$\text{rank}(A) = \text{rank}(A'A) = \text{rank}(AA').$$

Proof. Assume rank(A)=r. A full column rank mxr matrix A_1 exists such that (possibly after changing some rows and columns of A)

$$A = [A_1 \ A_1 C] = A_1[I_r \ C]$$

where C is an $r \times (n-r)$ matrix. As a result

$$A'A = \begin{bmatrix} I_r \\ C' \end{bmatrix} A'_1 A_1 [I_r \ C].$$

Let

$$E = \begin{bmatrix} I_r & 0 \\ -C' & I_{n-r} \end{bmatrix}.$$

Then

$$EA'AE' = \begin{bmatrix} A'_1 A_1 & 0 \\ 0 & 0 \end{bmatrix}.$$

The matrix E has determinant 1 so that E is nonsingular and

$$\text{rank}(A'A) = \text{rank}(EA'AE) = \text{rank}(A'_1 A_1).$$

Suppose that $A'_1 A_1$ is of less than full rank. Then there is an $r \times 1$ nonzero vector x such that

$$A'_1 A_1 x = 0.$$

Then

$$x'A_1 A_1 x = 0$$

and

$$A_1 x = 0,$$

contradicting the linear independence of r rows of A_1. Hence,

$$\text{rank}(A'A) = \text{rank}(A'_1 A_1).$$ ■

A useful result about the product of three matrices is as follows.

Theorem 6.5 Let A, B, and C be $p \times m$, $m \times n$, and $n \times q$ matrices, respectively.
Then

$$\text{rank}(ABC) \geq \text{rank}(AB) + \text{rank}(BC) - \text{rank}(B).$$

Proof (due to Schott (2005)). From Exercise 6.6

$$\text{rank}\left(\begin{bmatrix} B & BC \\ AB & 0 \end{bmatrix}\right) \geq \text{rank}(AB) + \text{rank}(BC). \tag{6.1}$$

Observe that

$$\begin{bmatrix} B & BC \\ AB & 0 \end{bmatrix} = \begin{bmatrix} I_m & 0 \\ A & I_p \end{bmatrix} \begin{bmatrix} B & 0 \\ 0 & -ABC \end{bmatrix} \begin{bmatrix} I_n & C \\ 0 & I_q \end{bmatrix}. \tag{6.2}$$

The first and third matrices in (6.2) are nonsingular. Hence,

$$\text{rank}\left(\begin{bmatrix} B & BC \\ AB & 0 \end{bmatrix}\right) = \text{rank}\left(\begin{bmatrix} B & 0 \\ 0 & -ABC \end{bmatrix}\right). \tag{6.3}$$
$$= \text{rank}(B) + \text{rank}(ABC)$$

Combine (6.1) and (6.3) to obtain the desired result.   ■

Example 6.2 Let

$$A = \begin{bmatrix} 1 & 2 & 3 \\ 2 & 4 & 6 \\ 3 & 6 & 9 \end{bmatrix}, B = \begin{bmatrix} 4 & 8 & 12 \\ 1 & 3 & 5 \\ 12 & 24 & 36 \end{bmatrix}.$$

Notice that A is a rank one matrix, while B has rank 2. Now

$$C = A + B = \begin{bmatrix} 5 & 10 & 15 \\ 3 & 7 & 11 \\ 15 & 30 & 45 \end{bmatrix}$$

so C has rank 2, while $R(A) + R(B) = 3$. Also

$$D = AB = \begin{bmatrix} 42 & 86 & 130 \\ 84 & 172 & 260 \\ 126 & 258 & 290 \end{bmatrix}$$

has rank 1, the minimum of the ranks of A and B. □

6.3　SOLVING SYSTEMS OF EQUATIONS WITH COEFFICIENT MATRIX OF LESS THAN FULL RANK

Let A be an $m \times n$ matrix of non-full rank. Consider a system of equations $Ax = b$. Such a system of equations is consistent if and only if the dependence of the rows of A is the same as that of the vector b. In other words for a vector t, $t'A = 0$ implies $t'b = 0$, and vice versa. If $t'A = 0$ but $t'b \neq 0$ for some t, the system of equations is inconsistent. A consistent system of equations has at least one solution. An inconsistent system of equations has no solutions. When A is of nonfull rank, it has infinitely many solutions.

Such systems may be solved by Gauss elimination. We will give some examples. In Part III, we will give explicit formulae for the solutions using generalized inverses.

Example 6.3 Consider the system of equations

$$2x - y - z = 4$$
$$-x + 2y - z = -2$$
$$-x - y + 2z = -2.$$

Adding these equations up gives zero. The rank of the coefficient matrix and the augmented matrix is two. Observe that

$$\begin{bmatrix} 2 & -1 & -1 & | & 4 \\ -1 & 2 & -1 & | & -2 \\ -1 & -1 & 2 & | & -2 \end{bmatrix} \to \begin{bmatrix} -1 & 2 & -1 & | & -2 \\ -1 & -1 & 2 & | & -2 \\ 2 & -1 & -1 & | & 4 \end{bmatrix} \to \begin{bmatrix} -1 & 2 & -1 & | & -2 \\ -1 & -1 & 2 & | & -2 \\ 0 & 0 & 0 & | & 0 \end{bmatrix}$$

$$\to \begin{bmatrix} -1 & 2 & -1 & | & -2 \\ 0 & -3 & 3 & | & 0 \\ 0 & 0 & 0 & | & 0 \end{bmatrix}.$$

Then we have

$$-x + 2y - z = -2$$
$$-3x + 3z = 0.$$

Then $y=z$ and $x=2+z$ where z is arbitrary. For example, if $z=2$, $x=4$ and $y=2$. The set of solutions form a vector space of dimension 1. $\qquad\square$

Example 6.4 An Inconsistent System of Equations. Consider the system of equations

$$\begin{aligned} x - 3y &= -2 \\ -2x + 6y &= 10 \end{aligned}.$$

The matrix

$$A = \begin{bmatrix} 1 & -3 \\ -2 & 6 \end{bmatrix}, b = \begin{bmatrix} -2 \\ 10 \end{bmatrix}.$$

If $t' = (2, 1), t'A = 0$ but $t'b \neq 0$. Observe that

$$\left[\begin{array}{cc|c} 1 & -3 & -2 \\ -2 & 6 & 10 \end{array}\right] \rightarrow \left[\begin{array}{cc|c} 1 & -3 & -2 \\ 0 & 0 & 6 \end{array}\right].$$

Clearly, the equation $0x + 0y = 6$ means $6 = 0$, nonsense! $\qquad\square$

Example 6.5 Consider the linear model in (1.4). The objective is to find the least square estimators. The normal equations are

$$\begin{aligned} 9\mu + 4\alpha_1 + 3\alpha_2 + 2\alpha_3 &= Y.. \\ 4\mu + 4\alpha_1 &= Y_1. \\ 3\mu + 3\alpha_2 &= Y_2. \\ 2\mu + 2\alpha_3 &= Y_3. \end{aligned}$$

where $Y.. = \sum_{i=1}^{3} Y_i.$, $Y_1. = \sum_{j=1}^{4} Y_{1j}$, $Y_2. = \sum_{j=1}^{3} Y_{2j}$, $Y_3. = \sum_{j=1}^{2} Y_{3j}$. In what follows $\bar{Y}.. = \frac{1}{9}Y..$, $\bar{Y}_1. = \frac{1}{4}Y_1.$, $\bar{Y}_2. = \frac{1}{3}Y_2.$, and $\bar{Y}_3. = \frac{1}{2}Y_3.$ To solve the normal equations, observe that

$$\left[\begin{array}{cccc|c} 9 & 4 & 3 & 2 & Y.. \\ 4 & 4 & 0 & 0 & Y_1. \\ 3 & 0 & 3 & 0 & Y_2. \\ 2 & 0 & 0 & 2 & Y_3 \end{array}\right] \rightarrow \left[\begin{array}{cccc|c} 4 & 4 & 0 & 0 & Y_1. \\ 3 & 0 & 3 & 0 & Y_2. \\ 2 & 0 & 0 & 2 & Y_3. \\ 9 & 4 & 3 & 3 & Y.. \end{array}\right] \rightarrow \left[\begin{array}{cccc|c} 4 & 4 & 0 & 0 & Y_1. \\ 3 & 0 & 3 & 0 & Y_2. \\ 2 & 0 & 0 & 2 & Y_3. \\ 0 & 0 & 0 & 0 & Y.. \end{array}\right] \rightarrow \left[\begin{array}{cccc|c} 1 & 1 & 0 & 0 & \bar{Y}_1. \\ 1 & 0 & 1 & 0 & \bar{Y}_2. \\ 1 & 0 & 0 & 1 & \bar{Y}_3. \\ 0 & 0 & 0 & 0 & 0 \end{array}\right]$$

$$\rightarrow \left[\begin{array}{cccc|c} 1 & 1 & 0 & 0 & \bar{Y}_1. \\ 0 & -1 & 1 & 0 & \bar{Y}_2. - \bar{Y}_1. \\ 0 & -1 & 0 & 1 & \bar{Y}_3. - \bar{Y}_1. \\ 0 & 0 & 0 & 0 & 0 \end{array}\right] \rightarrow \left[\begin{array}{cccc|c} 1 & 1 & 0 & 0 & \bar{Y}_1. \\ 0 & -1 & 1 & 0 & \bar{Y}_2. - \bar{Y}_1. \\ 0 & 0 & -1 & 1 & \bar{Y}_3. - \bar{Y}_1. \\ 0 & 0 & 0 & 0 & 0 \end{array}\right].$$

Then

$$\mu + \alpha_1 = \overline{Y}_{1.}$$
$$-\alpha_1 + \alpha_2 = \overline{Y}_{2.} - \overline{Y}_{1.}$$
$$-\alpha_2 + \alpha_3 = \overline{Y}_{3.} - \overline{Y}_{2.}$$

and the estimators of the parameters are

$$\hat{\alpha}_1 = \overline{Y}_{1.} - \mu$$
$$\hat{\alpha}_2 = \overline{Y}_{2.} - \mu$$
$$\hat{\alpha}_3 = \overline{Y}_{3.} - \mu$$

where μ is arbitrary. Possible choices for estimators of μ are $\hat{\mu} = 0$ and $\hat{\mu} = \overline{Y}_{..}$ Then

$$\hat{\mu} = 0$$
$$\hat{\alpha}_1 = \overline{Y}_{1.}$$
$$\hat{\alpha}_2 = \overline{Y}_{2.}$$
$$\hat{\alpha}_3 = \overline{Y}_{3.}$$

and

$$\hat{\mu} = \overline{Y}_{..}$$
$$\hat{\alpha}_1 = \overline{Y}_{1.} - \overline{Y}_{..}$$
$$\hat{\alpha}_2 = \overline{Y}_{2.} - \overline{Y}_{..}$$
$$\hat{\alpha}_3 = \overline{Y}_{3.} - \overline{Y}_{..}$$

Linear combinations of the form $c_1\hat{\alpha}_1 + c_2\hat{\alpha}_2 + c_3\hat{\alpha}_3$ where $c_1 + c_2 + c_3 = 0$ are independent of μ. These linear combinations will be important later. □

The following holds true for a consistent system of m equations in n unknowns with a coefficient matrix of rank r:

1. The rank of the coefficient matrix and the augmented matrix is the same. Otherwise, the system is inconsistent and has no solutions. For a system of equations $Ax = b$, A is the coefficient matrix and $[A,b]$ is the augmented matrix. In other words the column vector b is adjoined.
2. The number of arbitrary variables in the solutions is m–r.
3. If b=0, then an n×n A must be of rank less than n for the system to have a solution other than zero.
4. When the coefficient matrix is nonsingular and n=m, there is one unique solution.

6.4 SUMMARY

We have defined and established the properties of the rank of a matrix. We explained what is meant by a consistent system of equations. We gave conditions for a system of equations to be consistent that depend on the rank of the coefficient matrix and the augmented matrix. We observed that a consistent system of equations with coefficient matrix of full rank has a unique solution and showed how to obtain the vector space of infinitely many solutions when the coefficient matrix is of nonfull rank.

EXERCISES

6.1   Find the rank of the following matrices:

A.
$$\begin{bmatrix} 4a & a & a & a & 0 \\ a & a & 0 & 0 & 0 \\ a & 0 & a & 0 & 0 \\ a & 0 & 0 & a & 0 \\ a & 0 & 0 & 0 & a \end{bmatrix}$$

B.
$$\begin{bmatrix} \frac{3}{4} & -\frac{1}{4} & -\frac{1}{4} & -\frac{1}{4} \\ -\frac{1}{4} & \frac{3}{4} & -\frac{1}{4} & -\frac{1}{4} \\ -\frac{1}{4} & -\frac{1}{4} & \frac{3}{4} & -\frac{1}{4} \\ -\frac{1}{4} & -\frac{1}{4} & -\frac{1}{4} & \frac{3}{4} \end{bmatrix}$$

C.
$$\begin{bmatrix} 1 & 2 & 3 \\ 2 & 4 & 6 \\ 3 & 6 & 9 \end{bmatrix}$$

D.
$$\begin{bmatrix} 1 & 1 & 0 \\ 1 & 1 & 0 \\ 1 & 0 & 1 \\ 1 & 0 & 1 \end{bmatrix}$$

E.
$$\begin{bmatrix} 2 & 0 & 1 & 1 \\ 0 & 2 & 1 & 1 \\ 1 & 1 & 1 & 1 \\ 1 & 1 & 1 & 1 \end{bmatrix}$$

6.2 Find solutions to the following systems of equations that are consistent:

A.
$$3x - y - z - w = 6$$
$$x - 3y + z + w = 2$$
$$x + y - 3z + 3w = 3$$
$$x + y + z - 3w = 1$$

B.
$$x + y + z = 3$$
$$x - y + z = 2$$
$$2x + 2y + 2z = 8$$

C.
$$2x - 2y + z - w = 0$$
$$x + y + z + w = 4$$
$$-4x + 4y - 2z + 2w = 0$$

6.3 Given the linear model

$$\begin{bmatrix} Y_{11} \\ Y_{12} \\ Y_{21} \\ Y_{22} \end{bmatrix} = \begin{bmatrix} 1 & 1 & 0 & 1 & 0 \\ 1 & 1 & 0 & 0 & 1 \\ 1 & 0 & 1 & 1 & 0 \\ 1 & 0 & 1 & 0 & 1 \end{bmatrix} \begin{bmatrix} \mu \\ \alpha_1 \\ \alpha_2 \\ \beta_1 \\ \beta_2 \end{bmatrix} + \varepsilon,$$

find the normal equations and the least square estimators.

6.4 Show that $\text{rank}(A + B) \geq |\text{rank}(A) - \text{rank}(B)|$.

6.5 Show that if $C(A) \cap C(B) = \{0\}$, then $\text{rank}(A + B) = \text{rank}(A) + \text{rank}(B)$.

6.6 Let A, B, and C be any matrices for which the partitioned matrices below are defined. Show that

A. $\text{rank}([A\ B]) \geq \max\{\text{rank}(A), \text{rank}(B)\}$.

B. $\text{rank}\left(\begin{bmatrix} A & 0 \\ 0 & B \end{bmatrix}\right) = \text{rank}\left(\begin{bmatrix} 0 & B \\ A & 0 \end{bmatrix}\right) = \text{rank}(A) + \text{rank}(B)$.

C.
$$\text{rank}\left(\begin{bmatrix} A & 0 \\ C & B \end{bmatrix}\right) = \text{rank}\left(\begin{bmatrix} C & B \\ A & 0 \end{bmatrix}\right) = \text{rank}\left(\begin{bmatrix} B & C \\ 0 & A \end{bmatrix}\right)$$
$$= \text{rank}\left(\begin{bmatrix} 0 & A \\ B & C \end{bmatrix}\right) \geq \text{rank}(A) + \text{rank}(B)$$

6.7 Let $b_1, b_2, \ldots, b_r$ be linearly independent n-dimensional column vectors. Show that

A. Each of the $n \times n$ matrices $b_i b_i'$ has rank one.

B. The matrix $b_1 b_1' + b_2 b_2'$ has rank two.

C. The matrix $\sum_{i=1}^{r} b_i b_i'$ has rank r for $r \leq n$.

D. If $r = n$, then $\sum_{i=1}^{r} b_i b_i'$ is nonsingular.

6.8 What are the ranks of the following matrices?

A. $\begin{bmatrix} 1 & 3 & 5 \\ 3 & 9 & 15 \\ 5 & 15 & 25 \end{bmatrix}$

B. $\begin{bmatrix} 5 & 15 & 25 \\ 15 & 45 & 75 \\ 25 & 75 & 125 \end{bmatrix}$

C. $\begin{bmatrix} 26 & 18 & 10 \\ 18 & 18 & 18 \\ 10 & 18 & 26 \end{bmatrix}$

D. $\begin{bmatrix} 35 & 21 & 25 \\ 21 & 19 & 23 \\ 25 & 23 & 51 \end{bmatrix}$

6.9 Show that the rank of a matrix is not changed by elementary row operations.

6.10 Let

$$A = \begin{bmatrix} 1 & 2 & 6 \\ 3 & 4 & 12 \end{bmatrix}.$$

Find a 2×2 matrix A_1 and a column vector C where

$$A = A_1 \begin{bmatrix} I_2 & C \end{bmatrix}.$$

6.11 If A is an $m \times n$ matrix and B is an $n \times p$ matrix, then

$$\text{rank}(AB) \geq \text{rank}(A) + \text{rank}(B) - n.$$

6.12 Let M be an $m \times n$ matrix and let A be an $n \times n$ nonsingular matrix. Then $\text{rank}(MA) = \text{rank}(M)$.

PART II

EIGENVALUES, THE SINGULAR VALUE DECOMPOSITION, AND PRINCIPAL COMPONENTS

In this section of the book, we will discuss eigenvalues and eigenvectors of matrices. Using these ideas, we will show how to obtain the singular value decomposition of a matrix and then consider some generalizations. As an application, we will consider principal components, an important tool in deciding which factors are important in a multivariate dataset.

In Section 7, we define and show how to find the eigenvalues and the eigenvectors of a matrix. We give a number of illustrative examples for symmetric and nonsymmetric matrices. We define positive semidefinite matrices and present and illustrate their properties. Then using their properties and the Cauchy–Schwarz inequality, we establish some useful inequalities.

The work of Section 7 is extended in Section 8. We consider the properties of eigenvalues and eigenvectors of some of the special matrices studied in Section 4, for example, orthogonal matrices and matrices of the form $aI + bJ$. We establish the Cayley–Hamilton theorem. The relationships between the value of the trace and the determinant of a matrix and its eigenvalues are taken up. We also consider the eigenvalues of the Kronecker product of two matrices. We define and study a partial ordering between positive semidefinite matrices called the Loewner ordering.

Section 9 uses the material from Section 7 to discuss the singular value decomposition. This is a decomposition of a matrix in terms of orthogonal matrices and eigenvalues of the transpose of the matrix multiplied by the matrix. We give a number of illustrative examples.

Matrix Algebra for Linear Models, First Edition. Marvin H. J. Gruber.
© 2014 John Wiley & Sons, Inc. Published 2014 by John Wiley & Sons, Inc.

In Section 10, we consider several applications of the singular value decomposition. These include the reparameterization of a less than rank model to a full rank model and principal components. We explain what multicollinear data is and give some illustrations of such data.

Section 11 considers two generalizations of the singular value decomposition. They are based on the concept of relative eigenvalues and a generalization of orthogonal matrices.

SECTION 7

FINDING THE EIGENVALUES
OF A MATRIX

7.1 INTRODUCTION

We will first tell what eigenvalues and eigenvectors of matrices are and explain
how to find them. We will give several examples. Then we will define positive-
semi-definite (PSD) matrices and establish some of their properties.

7.2 EIGENVALUES AND EIGENVECTORS OF A MATRIX

Let A be a square symmetric $n \times n$ matrix. Let λ be a scalar. The objective is to find
those scalars for which there is a nonzero column vector x such that $Ax = \lambda x$ or
equivalently

$$(A - \lambda I)x = 0. \tag{7.1}$$

The scalars for which this nonzero vector exists are called eigenvalues, and the
corresponding vectors are called eigenvectors. To each distinct eigenvalue λ_j, there
will be a column vector P_j. These column vectors may be normalized so that $P_j' P_j = 1$.
If A is symmetric, then the eigenvectors corresponding to distinct eigenvalues are
orthogonal. That means that if λ_i and λ_j are distinct eigenvalues, the corresponding
eigenvectors P_i and P_j satisfy the relationship $P_i' P_j = 0$. Observe that by the definition
of an eigenvector,

Matrix Algebra for Linear Models, First Edition. Marvin H. J. Gruber.
© 2014 John Wiley & Sons, Inc. Published 2014 by John Wiley & Sons, Inc.

$$AP_i = \lambda_i P_i \quad \text{and} \quad AP_j = \lambda_j P_j.$$

Multiply the first equation on the right by P_j' and multiply the second equation on the right by P_i'. Then

$$P_j' AP_i = \lambda_i P_j' P_i \quad \text{and} \quad P_i' AP_j = \lambda_j P_i' P_j.$$

Since A is symmetric by transposition, the second equation may be rewritten $P_j' AP_i = \lambda_j P_j' P_i$. This will not work for non-symmetric matrices. By subtraction,

$$(\lambda_i - \lambda_j) P_i' P_j = 0 \text{ so that } P_i' P_j = 0.$$

For an $n \times n$ symmetric matrix, the $n \times n$ matrix

$$P = [P_1 \ P_2 \ \cdots \ P_n]$$

formed by from the column vectors corresponding to distinct eigenvectors is a column orthogonal matrix so $P'P = I$ and $P'AP = \Lambda$.

Recall from Exercise 4.24 that an $n \times n$ nonsingular matrix where $P'P = I$ is an orthogonal matrix so that $PP' = I$ and P' is the inverse of P.

Since we can find an orthogonal matrix where $P'AP = \Lambda$ after premultiplication by P and postmultiplication by P', we find that we can write

$$A = P\Lambda P' = \sum_{i=1}^{n} \lambda_i P_i P_i'.$$

This is called the spectral decomposition of A.

The system of equations in 7.1 has a nonzero solution for the eigenvector x for those λ that satisfy the equation

$$\det(A - \lambda I) = 0. \tag{7.2}$$

Equation 7.2 is called the characteristic equation.

Example 7.1 Eigenvalues and Eigenvectors of a Particular 2×2 Matrix

Consider the matrix

$$A = \begin{bmatrix} 6 & 2 \\ 2 & 3 \end{bmatrix}.$$

The characteristic equation is

$$\begin{vmatrix} 6 - \lambda & 2 \\ 2 & 3 - \lambda \end{vmatrix} = 0$$

or

$$(6-\lambda)(3-\lambda)-4=0,$$

which simplifies to

$$\lambda^2 - 9\lambda + 14 = 0.$$

Now

$$(\lambda-7)(\lambda-2)=0$$

and the eigenvalues are $\lambda=7$ and $\lambda=2$. To obtain the eigenvectors, observe that for $\lambda=7$,

$$\begin{bmatrix} -1 & 2 \\ 2 & -4 \end{bmatrix}\begin{bmatrix} x_1 \\ x_2 \end{bmatrix} = \begin{bmatrix} 0 \\ 0 \end{bmatrix}$$

so that

$$-x_1 + 2x_2 = 0$$
$$2x_1 - 4x_2 = 0.$$

A solution to this system is $x_1=2$ and $x_2=1$. Normalizing it, the eigenvector is

$$P_1 = \begin{bmatrix} \dfrac{2}{\sqrt{5}} \\ \dfrac{1}{\sqrt{5}} \end{bmatrix}.$$

For $\lambda=2$,

$$\begin{bmatrix} 4 & 2 \\ 2 & 1 \end{bmatrix}\begin{bmatrix} x_1 \\ x_2 \end{bmatrix} = \begin{bmatrix} 0 \\ 0 \end{bmatrix}$$

or

$$4x_1 + 2x_2 = 0$$
$$2x_1 + x_2 = 0.$$

A solution to this system is $x_1=-1$ and $x_2=2$. Normalizing the second eigenvector is

$$P_2 = \begin{bmatrix} -\dfrac{1}{\sqrt{5}} \\ \dfrac{2}{\sqrt{5}} \end{bmatrix}.$$

The matrix

$$P = \begin{bmatrix} \dfrac{2}{\sqrt{5}} & -\dfrac{1}{\sqrt{5}} \\ \dfrac{1}{\sqrt{5}} & \dfrac{2}{\sqrt{5}} \end{bmatrix}$$

is an orthogonal matrix. Observe that the spectral decomposition of

$$A = \begin{bmatrix} 6 & 2 \\ 2 & 3 \end{bmatrix} = \begin{bmatrix} \dfrac{2}{\sqrt{5}} & \dfrac{1}{\sqrt{5}} \\ -\dfrac{1}{\sqrt{5}} & \dfrac{2}{\sqrt{5}} \end{bmatrix} \begin{bmatrix} 7 & 0 \\ 0 & 2 \end{bmatrix} \begin{bmatrix} \dfrac{2}{\sqrt{5}} & -\dfrac{1}{\sqrt{5}} \\ \dfrac{1}{\sqrt{5}} & \dfrac{2}{\sqrt{5}} \end{bmatrix}.$$

□

Example 7.1 illustrates the singular value decomposition to be studied in Section 9.

When the characteristic equation has a root with multiplicity greater than one, the vector space generated by the eigenvalues has a subspace whose dimension is the same as the multiplicity of that root. Basis vectors of this subspace may be picked that are orthogonal to each other and to the other eigenvectors. This is illustrated in Example 7.2.

Example 7.2 Eigenvalues and Eigenvectors of a 4×4 Patterned Matrix

Consider the matrix

$$B = \begin{bmatrix} 2 & 2 & 1 & 1 \\ 2 & 2 & 1 & 1 \\ 1 & 1 & 2 & 2 \\ 1 & 1 & 2 & 2 \end{bmatrix}.$$

Then the characteristic equation is $\det(B - \lambda I) = 12\lambda^2 - 8\lambda^3 + \lambda^4 = 0$ with roots $\lambda = 6, 2, 0, 0$. The eigenvectors corresponding to $\lambda = 6$ are the nontrivial solutions to the system of equations

$$-4x_1 + 2x_2 + x_3 + x_4 = 0$$
$$2x_1 - 4x_2 + x_2 + x_4 = 0$$
$$x_1 + x_2 - 4x_3 + 2x_4 = 0$$
$$x_1 + x_2 + 2x_3 - 4x_4 = 0.$$

This system reduces to

$$6x_1 - 6x_2 = 0$$
$$6x_3 - 6x_4 = 0.$$

Thus, $x_1 = x_2$ and $x_3 = x_4$. Then a choice of

$$P_1 = \begin{bmatrix} \frac{1}{2} \\ \frac{1}{2} \\ \frac{1}{2} \\ \frac{1}{2} \end{bmatrix}.$$

The systems of equations that correspond to $\lambda = 2$ are

$$2x_2 + x_3 + x_4 = 0$$
$$2x_1 + x_3 + x_4 = 0$$
$$x_1 + x_2 + 2x_4 = 0$$
$$x_1 + x_2 + 2x_3 = 0.$$

Again $x_1 = x_2$ and $x_3 = x_4$ and a normalized matrix orthogonal to P_1 in the eigenspace is

$$P_2 = \begin{bmatrix} \frac{1}{2} \\ \frac{1}{2} \\ -\frac{1}{2} \\ -\frac{1}{2} \end{bmatrix}.$$

Corresponding to $\lambda = 0$, there is a system of two distinct equations in four unknowns:

$$2x_1 + 2x_2 + x_3 + x_4 = 0$$
$$x_1 + x_2 + 2x_3 + 2x_4 = 0.$$

The coefficient matrix has rank two, so two of the variables are arbitrary. A solution takes the form $x_1 = -x_2$ and $x_3 = -x_4$. The eigenspace corresponding to $\lambda = 0$ is of dimension two, so it is spanned by two independent column vectors. Two such column vectors that are orthogonal to each other and also orthogonal to P_1 and P_2 are

$$P_3 = \begin{bmatrix} \frac{1}{\sqrt{2}} \\ -\frac{1}{\sqrt{2}} \\ 0 \\ 0 \end{bmatrix} \quad \text{and} \quad P_4 = \begin{bmatrix} 0 \\ 0 \\ \frac{1}{\sqrt{2}} \\ -\frac{1}{\sqrt{2}} \end{bmatrix}.$$

Then

$$B = P\Lambda P'$$

where

$$P = \begin{bmatrix} \frac{1}{2} & -\frac{1}{2} & \frac{1}{\sqrt{2}} & 0 \\ \frac{1}{2} & -\frac{1}{2} & -\frac{1}{\sqrt{2}} & 0 \\ \frac{1}{2} & \frac{1}{2} & 0 & \frac{1}{\sqrt{2}} \\ \frac{1}{2} & \frac{1}{2} & 0 & -\frac{1}{\sqrt{2}} \end{bmatrix}$$

and

$$\Lambda = \begin{bmatrix} 6 & 0 & 0 & 0 \\ 0 & 2 & 0 & 0 \\ 0 & 0 & 0 & 0 \\ 0 & 0 & 0 & 0 \end{bmatrix}.$$

□

If a matrix is not symmetric, the process of obtaining eigenvalues and eigenvectors is the same. Although the eigenvectors can be normalized so that the sum of the squares of the components is 1, they will not be orthogonal.

Example 7.3 Eigenvalues and Eigenvectors of a Nonsymmetric Matrix

Consider the matrix

$$M = \begin{bmatrix} 3 & 4 \\ 1 & 3 \end{bmatrix}.$$

Then, to find the eigenvalues, we have

$$\begin{vmatrix} 3-\lambda & 4 \\ 1 & 3-\lambda \end{vmatrix} = \lambda^2 - 6\lambda + 5 = (\lambda-5)(\lambda-1) = 0.$$

The eigenvalues are $\lambda = 5$, $\lambda = 1$. For $\lambda = 5$, the eigenvector is the solution to the equation

$$\begin{bmatrix} -2 & 4 \\ 1 & -2 \end{bmatrix} \begin{bmatrix} x_1 \\ x_2 \end{bmatrix} = \begin{bmatrix} 0 \\ 0 \end{bmatrix}$$

or

$$-2x_1 + 4x_2 = 0$$
$$x_1 - 2x_2 = 0$$

so $x_1 = 2$, $x_2 = 1$. Normalizing, we get $x_1 = 2/\sqrt{5}$, $x_2 = 1/\sqrt{5}$. When $\lambda = 1$, we have

$$\begin{bmatrix} 2 & 4 \\ 1 & 2 \end{bmatrix} \begin{bmatrix} x_1 \\ x_2 \end{bmatrix} = \begin{bmatrix} 0 \\ 0 \end{bmatrix}$$

or

$$2x_1 + 4x_2 = 0$$
$$x_1 + 2x_2 = 0$$

so $x_1 = -2$, $x_2 = 1$. Normalizing $x_1 = -2/\sqrt{5}$, $x_1 = 1/\sqrt{5}$. Notice that the eigenvectors corresponding to $\lambda = 5$ and $\lambda = 1$ are not orthogonal. Form the matrices

$$P = \begin{bmatrix} \dfrac{2}{\sqrt{5}} & -\dfrac{2}{\sqrt{5}} \\ \dfrac{1}{\sqrt{5}} & \dfrac{1}{\sqrt{5}} \end{bmatrix}, \quad \Lambda = \begin{bmatrix} 5 & 0 \\ 0 & 1 \end{bmatrix}, \quad \text{and} \quad P^{-1} = \begin{bmatrix} \dfrac{5}{4\sqrt{5}} & \dfrac{5}{2\sqrt{5}} \\ -\dfrac{5}{4\sqrt{5}} & \dfrac{5}{2\sqrt{5}} \end{bmatrix}$$

and we have

$$M = P\Lambda P^{-1}.$$

Although P is not column orthogonal, it is row orthogonal.

However, if we consider the transpose matrix

$$M' = \begin{bmatrix} 3 & 1 \\ 4 & 3 \end{bmatrix},$$

the characteristic equation and the eigenvalues will be the same as that of M, but the eigenvectors will not be. For $\lambda = 5$, we have

$$\begin{bmatrix} -2 & 1 \\ 4 & -2 \end{bmatrix} \begin{bmatrix} x_1 \\ x_2 \end{bmatrix} = \begin{bmatrix} 0 \\ 0 \end{bmatrix}$$

or

$$-2x_1 + x_2 = 0$$
$$4x_1 - 2x_2 = 0$$

and $x_1 = 1$, $x_2 = 2$. Normalizing $x_1 = 1/\sqrt{5}$, $x_2 = 2/\sqrt{5}$. For $\lambda = 1$,

$$\begin{bmatrix} 2 & 1 \\ 4 & 2 \end{bmatrix} \begin{bmatrix} x_1 \\ x_2 \end{bmatrix} = \begin{bmatrix} 0 \\ 0 \end{bmatrix}$$

or

$$2x_1 + x_2 = 0$$
$$4x_1 + 2x_2 = 0$$

and $x_1 = -1, x_2 = 2$. Normalizing $x_1 = -1/\sqrt{5}, x_2 = 2/\sqrt{5}$. Form the matrices

$$Q = \begin{bmatrix} \dfrac{1}{\sqrt{5}} & -\dfrac{1}{\sqrt{5}} \\ \dfrac{2}{\sqrt{5}} & \dfrac{2}{\sqrt{5}} \end{bmatrix}, \ \Lambda = \begin{bmatrix} 5 & 0 \\ 0 & 1 \end{bmatrix} \ \text{and} \ \ Q^{-1} = \begin{bmatrix} \dfrac{5}{2\sqrt{5}} & \dfrac{5}{4\sqrt{5}} \\ -\dfrac{5}{2\sqrt{5}} & \dfrac{5}{4\sqrt{5}} \end{bmatrix}$$

and we have

$$M' = Q\Lambda Q^{-1}.$$

Observe that if P_i and Q_i represent the columns of P and Q, respectively, then if $i \neq j$ $P_i'Q_j = 0$ and $P_j'Q_i = 0$. □

The decomposition illustrated in Example 7.3 holds true in general. Let Λ be the diagonal matrix of eigenvalues of a matrix A and let C be the matrix of eigenvectors. Then $AC = C\Lambda$ and $A = C\Lambda C^{-1}$. Likewise, if E is the matrix of eigenvectors of A', then $A' = E\Lambda E^{-1}$. When A is symmetric, $C = E$ and C is an orthogonal matrix.

Although all of entries of a matrix may be real if it is not symmetric, it may have complex eigenvalues as illustrated in Example 7.4.

Example 7.4 Eigenvalues of a Skew-Symmetric Matrix

Consider a 2×2 skew-symmetric matrix

$$A = \begin{bmatrix} 0 & a_{12} \\ -a_{12} & 0 \end{bmatrix}.$$

Now

$$\det(A - \lambda I) = \begin{vmatrix} -\lambda & a_{12} \\ -a_{12} & -\lambda \end{vmatrix} = \lambda^2 + a_{12}^2 = 0$$

and the resulting eigenvalues are $\lambda = \pm a_{12} i$. □

All of the eigenvalues and eigenvectors of a symmetric matrix are real as shown in Theorem 7.1.

Theorem 7.1 Let A be a symmetric matrix. Then all its eigenvalues are real and it has real eigenvectors.

Proof. (The idea for this proof comes from Rao (1973).)

Let $x + iy$ be a complex eigenvector. Then if $\lambda = a + ib$ is an eigenvalue, we have

$$A(x + iy) = (a + ib)(x + iy) = ax - by + i(bx + ay).$$

Equate real and imaginary parts. Then

$$Ax = ax - by \quad \text{and} \quad Ay = bx + ay.$$

Now

$$y'Ax = ay'x - by'y$$

and

$$x'Ay = bx'x + ax'y.$$

Now if A is symmetric, $x'Ay = (y'A'x)' = (y'Ax)' = y'Ax$. (This would not be true for nonsymmetric matrices.) Then since $x'y = y'x$,

$$0 = x'Ay - y'Ax = b(x'x + y'y).$$

Hence $b = 0$ and y could be chosen to be zero.   ∎

7.3 NONNEGATIVE DEFINITE MATRICES

An $n \times n$ matrix A is PSD or nonnegative definite if and only if (if):

1. It is symmetric ($A' = A$).
2. For every n-dimensional vector, column vector p

$$p'Ap \geq 0. \tag{7.3}$$

If in addition to (7.3) holding true $p'Ap = 0$ if $p = 0$ or equivalently for every nonzero vector p

$$p'Ap > 0 \tag{7.4}$$

then A is said to be positive definite (PD). Expressions of the form $x'Ax$ are called quadratic forms. When the matrix A is PSD or PD, the quadratic form is PSD or PD.

All of the eigenvalues of a nonnegative definite matrix are nonnegative.

Let x be a nonzero eigenvector of a matrix A. Then $x'Ax = \lambda x'x \geq 0$. Now $x'x$ is positive, so $\lambda \geq 0$. For PD matrices, the greater than or equal to zero is replaced by greater than, and the eigenvalues must be positive.

It will be shown in Section 8 that PD matrices have positive determinants. In addition all of the $k \times k$ submatrices containing the elements in the main diagonal have positive determinants. Such submatrices are called principal minors.

Example 7.5 A PD Matrix

Let

$$A = \begin{bmatrix} 1 & 1 & 0 \\ 1 & 3 & 0 \\ 0 & 0 & 6 \end{bmatrix}.$$

The eigenvalues of A are 6, 3.414, and 0.5858. All of the diagonal elements are positive. All of the two-by-two matrices that contain diagonal elements have a positive determinant. The matrix A has determinant 12.   □

There are a few interesting inequalities that are consequences of the Cauchy–Schwarz inequality that involve nonnegative definite matrices. They will prove to be useful in comparing the efficiency of estimators in Part V. They are given in Theorems 7.2 and 7.3.

Theorem 7.2 Let u and v be n-dimensional vectors. Let A be a PD matrix. Then

$$[u'v]^2 \le u'A^{-1}uv'Av. \tag{7.5}$$

Proof. Using the Cauchy–Schwarz inequality,

$$[u'v]^2 = [u'A^{-1/2}A^{1/2}v]^2 \le u'A^{-1}uv'Av. \tag{7.6}$$

∎

Another interesting result due to Farebrother (1976) is as follows.

Theorem 7.3 Let d be a positive scalar and let A be a PD matrix. Then $dA - uu'$ is PSD if $u'A^{-1}u \le d$.

Proof. Observe that $dA - uu'$ is PD if for $v \ne 0$

$$dv'Av > v'uu'v = (v'u)^2 \tag{7.7}$$

or

$$d > \frac{(v'u)^2}{v'Av}. \tag{7.8}$$

Multiply both sides of inequality (7.8) by $1/u'A^{-1}u$ to obtain for all nonzero v

$$\frac{d}{u'A^{-1}u} > \frac{(v'u)^2}{v'Avu'A^{-1}u}. \tag{7.9}$$

From Theorem 7.2, the maximum value of the right-hand side of (7.9) is 1.Since (7.7) holds for all v, it follows that $\dfrac{d}{u'A^{-1}u} \geq 1$ or equivalently that $u'A^{-1}u \leq d$. ∎

The difference of two rank one matrices cannot be PSD. This is shown by the following theorem due to Terasvirta (1980).

Theorem 7.4 Let u and v be two linearly independent m-dimensional column vectors. Then $uu'-vv'$ is indefinite. Its nonzero eigenvalues are

$$\lambda = \frac{-(u'u - v'v) \pm \left[(u'u - v'v)^2 + 4u'uv'v - 4(u'v)^2\right]^{1/2}}{2}. \tag{7.10}$$

Proof. The signs of the two eigenvalues are different because by the Cauchy–Schwarz inequality, $u'uv'v-(u'v)^2>0$ if u and v are linearly independent.

It must be shown that the expressions in (7.10) are indeed the eigenvalues of $uu'-vv'$. With this in mind, let z be an eigenvector of $uu'-vv'$:

$$(vv' - uu')z = \lambda z. \tag{7.11}$$

Multiply (7.11) by u' to obtain

$$(u'vv' - u'uu')z = \lambda u'z \tag{7.12}$$

or equivalently

$$(uu' + \lambda)u'z = u'vv'z. \tag{7.13}$$

Multiply (7.11) by v'. Then

$$(v'v - \lambda)v'z = v'uu'z. \tag{7.14}$$

From (7.13), it follows that

$$v'z = (u'v)^{-1}(u'u + \lambda)u'z. \tag{7.15}$$

Substitution of (7.15) into (7.14) yields

$$(v'v - \lambda)(u'v)^{-1}(u'u + \lambda)u'z = v'uu'z. \qquad (7.16)$$

Since $(u'v)$ and $u'z$ are scalars,

$$(v'v - \lambda)(u'u + \lambda) = (v'u)^2. \qquad (7.17)$$

Solving this equation by the quadratic formula gives the eigenvalues in (7.10). ∎

Example 7.6 Illustration of the Fact That the Difference of Two Rank 1 Matrices of the Form uu′ Is Not PSD

$$u = \begin{bmatrix} 3 \\ 2 \\ 1 \end{bmatrix}, v = \begin{bmatrix} 1 \\ 2 \\ 3 \end{bmatrix}$$

Then

$$W = uu' - vv' = \begin{bmatrix} 3 \\ 2 \\ 1 \end{bmatrix} \begin{bmatrix} 3 & 2 & 1 \end{bmatrix} - \begin{bmatrix} 1 \\ 2 \\ 3 \end{bmatrix} \begin{bmatrix} 1 & 2 & 3 \end{bmatrix}$$

$$= \begin{bmatrix} 9 & 6 & 3 \\ 6 & 4 & 2 \\ 3 & 2 & 1 \end{bmatrix} - \begin{bmatrix} 1 & 2 & 3 \\ 2 & 4 & 6 \\ 3 & 6 & 9 \end{bmatrix} = \begin{bmatrix} 8 & 4 & 0 \\ 4 & 0 & -4 \\ 0 & -4 & -8 \end{bmatrix}$$

$$\det(W - \lambda I) = 96\lambda - \lambda^3 = 0$$

$$\lambda = 0, \pm 4\sqrt{6}. \qquad \square$$

7.4 SUMMARY

We have explained what eigenvalues and eigenvectors of a matrix are. The eigenvalues and eigenvectors of several matrices were calculated. We noted the difference in several important properties of eigenvalues and eigenvectors for symmetric and nonsymmetric matrices.

Positive semi-definite and PD matrices were defined. Several inequalities that are consequences of the Cauchy–Schwarz inequality were established.

We also showed that the difference of two rank one matrices of the form ww′ where w is a column vector cannot be PSD and gave an illustrative example.

EXERCISES

7.1 Find the eigenvalues and the eigenvectors of the matrices and the corresponding spectral decomposition for

A. $\begin{bmatrix} 68 & -24 \\ -24 & 82 \end{bmatrix}$

B. $\begin{bmatrix} 2 & 1 \\ 1 & 2 \end{bmatrix}$

C. $\begin{bmatrix} 5 & 1 & 1 \\ 1 & 5 & 1 \\ 1 & 1 & 5 \end{bmatrix}$

D. $\begin{bmatrix} 1 & \frac{1}{5} & \frac{1}{5} \\ \frac{1}{5} & 1 & \frac{1}{5} \\ \frac{1}{5} & \frac{1}{5} & 1 \end{bmatrix}$

E. $\begin{bmatrix} 9 & 3 & 3 & 3 \\ 3 & 3 & 0 & 0 \\ 3 & 0 & 3 & 0 \\ 3 & 0 & 0 & 3 \end{bmatrix}.$

7.2 Which of the following matrices are (1) PD and (2) PSD? Find the eigenvalues, eigenvectors, and the spectral decomposition or the decomposition using a nonsingular matrix.

A. $\begin{bmatrix} 1 & 3 \\ 3 & 9 \end{bmatrix}$

B. $\begin{bmatrix} 5 & 10 \\ 10 & 5 \end{bmatrix}$

C. $\begin{bmatrix} 1 & -0.5 \\ -0.5 & 1 \end{bmatrix}$

7.3 For what values of the parameter ρ is the matrix $\begin{bmatrix} 1 & \rho & \rho \\ \rho & 1 & \rho \\ \rho & \rho & 1 \end{bmatrix}$ $-1 \le \rho \le 1$ nonnegative definite? Positive definite?

7.4 Suppose that u and v are linearly dependent vectors. Show that $uu' - vv'$ has eigenvalues 0 and $\mathrm{tr}(uu' - vv')$.

7.5   Suppose

$$u = \begin{bmatrix} 2 \\ 3 \end{bmatrix}, \quad v = \begin{bmatrix} 1 \\ 2 \end{bmatrix}.$$

A. Show that u and v are linearly independent.
B. Find the eigenvalues of $uu' - vv'$.

7.6   Suppose

$$u = \begin{bmatrix} 4 \\ 8 \end{bmatrix}, \quad v = \begin{bmatrix} 1 \\ 2 \end{bmatrix}.$$

A. Show that u and v are linearly dependent.
B. Find the eigenvalues of $uu' - vv'$
    1. By using the characteristic equation
    2. By using the result from Exercise 7.4

7.7   Let $p_1, p_2, \ldots, p_n$ be n-dimensional mutually orthogonal column vectors. Show that they are linearly independent.

7.8   Let $q_1, q_2, \ldots, q_n$, be n-dimensional row vectors where for $n \times n$ symmetric matrix A $q_i A = \lambda_i I$, $i = 1, 2, \ldots, n$. Show that the n vectors are mutually orthogonal and thus the matrix

$$Q = \begin{bmatrix} q_1 \\ q_2 \\ \vdots \\ q_n \end{bmatrix}$$

is an orthogonal matrix. The vectors q_i are right eigenvectors.

7.9   Find the eigenvalues of $M = \begin{bmatrix} 0 & 1 & 1 \\ -1 & 0 & 1 \\ -1 & -1 & 0 \end{bmatrix}$ and observe that they are complex numbers.

7.10   For the matrices below, find P and Q where $M = P\Lambda P^{-1}$, $M' = Q\Lambda Q^{-1}$.

A. $\begin{bmatrix} -6 & 5 \\ -30 & 19 \end{bmatrix}$

B. $\begin{bmatrix} \frac{7}{3} & \frac{1}{3} & \frac{1}{3} \\ 0 & 2 & 1 \\ \frac{2}{3} & \frac{2}{3} & \frac{5}{3} \end{bmatrix}$

7.11 Find the eigenvalues and the eigenvectors of

$$M = \begin{bmatrix} \frac{1}{2} & -\frac{1}{2} & \frac{1}{2} \\ -1 & 1 & 1 \\ \frac{1}{2} & -\frac{1}{2} & \frac{3}{2} \end{bmatrix}$$

and find P and Q where $M = P\Lambda P^{-1}$, $M' = Q\Lambda Q^{-1}$.

SECTION 8

THE EIGENVALUES AND EIGENVECTORS OF SPECIAL MATRICES

8.1   INTRODUCTION

We give some important properties of special matrices and their eigenvalues in this section. These include orthogonal, nonsingular, and idempotent matrices and matrices of the form $aI + bJ$. We show that orthogonal matrices have eigenvalues ± 1 and that idempotent matrices have eigenvalues 0 or 1. When T is a nonsingular matrix, we show that A and TAT^{-1} have the same eigenvalues and that the eigenvectors of TAT^{-1} are those obtained by applying the transformation T to the eigenvectors of A. We also show how to obtain the eigenvalues and eigenvectors of equicorrelation matrices.

The relationship between the eigenvalues, the trace, and the determinant of a matrix is established. A result is established that gives the eigenvalues and eigenvectors of the Kronecker product of two matrices in terms of the eigenvalues and eigenvectors of the two matrices. This enables the computation of the trace and the determinant of the Kronecker product.

We state and prove the Cayley–Hamilton theorem that says that a matrix satisfies its characteristic equation and show how to use it to calculate the inverse of a matrix.

We exhibit a partial ordering between matrices, the Loewner ordering, where one matrix is greater than or equal to another if the difference between them is positive semi-definite. One matrix is greater than another if the difference is positive definite.

Matrix Algebra for Linear Models, First Edition. Marvin H. J. Gruber.
© 2014 John Wiley & Sons, Inc. Published 2014 by John Wiley & Sons, Inc.

8.2   ORTHOGONAL, NONSINGULAR, AND IDEMPOTENT MATRICES

First, we give an interesting property of a matrix pre-multiplied by an orthogonal matrix and post-multiplied by its transpose, PAP'.

Theorem 8.1 Let P be an orthogonal matrix. Then PAP' has the same eigenvalues as A. If x is an eigenvector of A, then $y = Px$ is an eigenvector of PAP'.

Proof. Since $PP' = I$, $\det P = \pm 1$. Then

$$
\begin{aligned}
\det(PAP' - \lambda I) &= \det(PAP' - \lambda PP') \\
&= \det(P(A - \lambda I)P') \\
&= \det P \det(A - \lambda I) \det P' \\
&= \det(A - \lambda I),
\end{aligned}
$$

so the characteristic equation is the same.
If x is an eigenvector of A, then for eigenvalue λ

$$(A - \lambda I)x = 0$$

so that

$$(PA - \lambda P)x = 0$$

if $y = Px$ then $x = P'y$ so that

$$(PA - \lambda P)P'y = 0$$

so that since $PP' = I$, it follows that

$$(PAP' - \lambda I)y = 0. \qquad \blacksquare$$

What are the eigenvalues of an orthogonal matrix? They are ± 1 as established in Theorem 8.2.

Theorem 8.2 Orthogonal matrices have eigenvalues ± 1.

Proof. Let λ be an eigenvalue of orthogonal matrix P. By definition, there exists a nonzero vector x where $P'x = \lambda x$. Then since $PP' = I, x'x = xPP'x = \lambda^2 x'x$, so $\lambda^2 = 1$ and the result follows. $\qquad \blacksquare$

A matrix E is idempotent if $E^2 = E$.

Theorem 8.3 Idempotent matrices have eigenvalues 0 or 1.

Proof. For any vector x, $\lambda x'x = x'Ex = x'E^2x = \lambda^2 x'x$. Then when $x \neq 0$, it follows that $\lambda^2 = \lambda$ so that $\lambda = 1$ or 0. $\qquad \blacksquare$

Example 8.1 Two Important Idempotent Matrices

Two idempotent matrices important to regression analysis are

$$X(X'X)^{-1}X' \quad \text{and} \quad I - X(X'X)^{-1}X'.$$

It is easily seen that

$$X(X'X)^{-1}X'X(X'X)^{-1}X' = X(X'X)^{-1}X'$$

and that

$$
\begin{aligned}
(I - X(X'X)^{-1}X')&(I - X(X'X)^{-1}X') \\
&= I - 2X(X'X)^{-1}X' + X(X'X)^{-1}X'X(X'X)^{-1}X' \\
&= I - X(X'X)^{-1}X'.
\end{aligned}
$$

Also

$$
\begin{aligned}
X(X'X)^{-1}X'(I - X(X'X)^{-1}X') &= X(X'X)^{-1}X' - X(X'X)^{-1}X'X(X'X)^{-1}X' \\
&= X(X'X)^{-1}X' - X(X'X)^{-1}X' \\
&= 0.
\end{aligned}
$$

$\square$

We now compare the eigenvalues of a matrix and its transpose.

Theorem 8.4 The eigenvalues of A and A′ are the same.

Proof. If A is symmetric, there is nothing to prove. Observe that

$$\det(A' - \lambda I) = \det(A - \lambda I)' = \det(A - \lambda I).$$

Thus, the characteristic polynomial is the same for both A and A′. ∎

A result similar to Theorem 8.1 holds for TAT^{-1} when T is a nonsingular matrix. We first give an illustration.

Example 8.2 Eigenvalues of TAT^{-1}

Let

$$M = \begin{bmatrix} 7 & 2 \\ 0 & 1 \end{bmatrix}.$$

Then $\det(M - \lambda I) = \begin{vmatrix} 7-\lambda & 2 \\ 0 & 1-\lambda \end{vmatrix} = (7-\lambda)(1-\lambda) = 0$, so the eigenvalues are $\lambda = 7$ and

$\lambda = 1$. Let $T = \begin{bmatrix} 2 & 1 \\ 1 & 1 \end{bmatrix}$. Then

$$M_T = TMT^{-1} = \begin{bmatrix} 2 & 1 \\ 1 & 1 \end{bmatrix}\begin{bmatrix} 7 & 2 \\ 0 & 1 \end{bmatrix}\begin{bmatrix} 1 & -1 \\ -1 & 2 \end{bmatrix} = \begin{bmatrix} 9 & -4 \\ 4 & -1 \end{bmatrix}$$

and $\det(M_T - \lambda I) = \begin{vmatrix} 9-\lambda & -4 \\ 4 & -1-\lambda \end{vmatrix} = \lambda^2 - 10\lambda + 7 = (\lambda - 7)(\lambda - 1) = 0,$ so the eigen-

values are $\lambda = 7$ and $\lambda = 1$.

To obtain the eigenvectors of M, we have for $\lambda = 7$

$$\begin{bmatrix} 0 & 2 \\ 0 & -7 \end{bmatrix} \begin{bmatrix} x_1 \\ x_2 \end{bmatrix} = \begin{bmatrix} 0 \\ 0 \end{bmatrix},$$

so x_1 is arbitrary and $x_2 = 0$. Thus, $\begin{bmatrix} 1 \\ 0 \end{bmatrix}$ is an eigenvector. For $\lambda = 1$,

$$\begin{bmatrix} 6 & 2 \\ 0 & 0 \end{bmatrix} \begin{bmatrix} x_1 \\ x_2 \end{bmatrix} = 0,$$

so an eigenvector is $\begin{bmatrix} 1 \\ -3 \end{bmatrix}$.

For M_T corresponding to $\lambda = 7$,

$$\begin{bmatrix} 2 & -4 \\ 4 & -8 \end{bmatrix} \begin{bmatrix} y_1 \\ y_2 \end{bmatrix} = 0,$$

so an eigenvector is $\begin{bmatrix} 2 \\ 1 \end{bmatrix}$. Corresponding to $\lambda = 1$,

$$\begin{bmatrix} 8 & -4 \\ 4 & -2 \end{bmatrix} \begin{bmatrix} y_1 \\ y_2 \end{bmatrix} = \begin{bmatrix} 0 \\ 0 \end{bmatrix}$$

and an eigenvalue is $\begin{bmatrix} 1 \\ 2 \end{bmatrix}$.

Now

$$H = \begin{bmatrix} 2 & 1 \\ 1 & 1 \end{bmatrix} \begin{bmatrix} 1 & 1 \\ 0 & -3 \end{bmatrix} = \begin{bmatrix} 2 & -1 \\ 1 & -2 \end{bmatrix}.$$

The columns of H are scalar multiples of the eigenvectors of M_T that were obtained above. □

Theorem 8.5 Let T be a nonsingular matrix. Then A and TAT^{-1} have the same eigenvalues. If x is an eigenvector of A, then $y = TX$ is an eigenvector of TAT^{-1}.

Proof. Observe that

$$
\begin{aligned}
\det(TAT^{-1} - \lambda I) &= \det(TAT^{-1} - \lambda TT^{-1}) \\
&= \det(T(A - \lambda I)T^{-1}) \\
&= \det(T)\det(A - \lambda I)\det(T^{-1}) \\
&= \det(T)\det(A - \lambda I)\frac{1}{\det(T)} \\
&= \det(A - \lambda I).
\end{aligned}
$$

Thus, the characteristic equation and the resulting eigenvalues for A and TAT^{-1} are the same.

Also if x is an eigenvector of A, then for eigenvalue λ

$$(A - \lambda I)x = 0$$

so that

$$(TA - \lambda T)x = 0$$

if $y = Tx$ then $x = T^{-1}y$ so that

$$(TA - \lambda T)T^{-1}y = 0$$

so that since $TT^{-1} = I$, it follows that

$$(TAT^{-1} - \lambda I)y = 0.$$

Actually Theorem 8.5 includes Theorem 8.1 as a special case because orthogonal matrices are nonsingular with $P^{-1} = P'$. ∎

8.3 THE CAYLEY–HAMILTON THEOREM

The Cayley–Hamilton theorem states that a square matrix satisfies its characteristic equation. This proof is adapted from that of Rao (1973).

Theorem 8.6 (Cayley–Hamilton) Every square matrix satisfies its own characteristic equation.

Proof. Observe that

$$\det(A - \lambda I)I = (A - \lambda I)\mathrm{adj}(A - \lambda I). \tag{8.1}$$

Now

$$\mathrm{adj}(A - \lambda I) = C_0 + \lambda C_1 + \cdots \lambda^{n-1}C_{n-1} \tag{8.2}$$

for suitably chosen matrices C_i. Now the characteristic polynomial is given by

$$\det(A - \lambda I) = a_0 + a_1\lambda + \cdots a_n\lambda^n. \tag{8.3}$$

Equating coefficients of like powers of the right-hand side of (8.1) to the right-hand side of (8.3),

$$
\begin{aligned}
a_0 I &= AC_0, \\
a_1 I &= -C_0 I + AC_1, \\
a_2 I &= -C_1 I + A^2 C_2, \\
&\ \ \vdots \\
a_i I &= -C_{i-1} + A^i C_i, \\
&\ \ \vdots \\
a_n I &= -C_n.
\end{aligned}
\tag{8.4}
$$

Multiply the first equation in (8.4) by I, the second by A, the (i+1) st by A^i, and the (n+1)st by A^n

$$a_0 I + a_1 A + \cdots a_n A^n = 0.$$

Thus, A satisfies its characteristic equation.   ∎

If A has an inverse, it can be expressed as

$$A^{-1} = -\frac{1}{a_0}\left(a_1 I + a_2 A + \cdots + a_n A^{n-1}\right). \tag{8.5}$$

Example 8.3 Finding an Inverse Using the Cayley–Hamilton Theorem

Consider the matrix

$$A = \begin{bmatrix} 1 & -2 \\ 4 & 7 \end{bmatrix}.$$

The characteristic equation for this matrix is

$$
\begin{aligned}
\det(A - \lambda I) &= \begin{vmatrix} 1-\lambda & -2 \\ 4 & 7-\lambda \end{vmatrix} = (1-\lambda)(7-\lambda) + 8 \\
&= \lambda^2 - 8\lambda + 15.
\end{aligned}
$$

By Cayley–Hamilton theorem,

$$A^2 - 8A + 15I = 0,$$

and using (8.5) or by multiplication by A^{-1} and solving the equation for A^{-1}, we get

$$A^{-1} = -\frac{1}{15}(A - 8I).$$

Then

$$A^{-1} = -\frac{1}{15}\left(\begin{bmatrix} 1 & -2 \\ 4 & 7 \end{bmatrix} - 8\begin{bmatrix} 1 & 0 \\ 0 & 1 \end{bmatrix}\right)$$

$$= -\frac{1}{15}\begin{bmatrix} -7 & -2 \\ 4 & -1 \end{bmatrix} = \begin{bmatrix} \frac{7}{15} & \frac{2}{15} \\ -\frac{4}{15} & \frac{1}{15} \end{bmatrix}.$$

The reader may check that this is indeed the inverse of A by using the methods developed in Section 3.   □

8.4 THE RELATIONSHIP BETWEEN THE TRACE, THE DETERMINANT, AND THE EIGENVALUES OF A MATRIX

Theorem 8.7 summarizes important relationships between the eigenvalues of a matrix and its trace and its determinant.

Theorem 8.7 The following relationships hold between the eigenvalues of a matrix and its trace and determinant:

1. The trace of a matrix is equal to the sum of the eigenvalues each multiplied by their multiplicity.
2. The determinant of a matrix is equal to the product of the eigenvalues each raised to the power of its multiplicity.

Proof. Consider the characteristic polynomial of a matrix

$$\det(A - \lambda I) = a_0 + a_1\lambda + \cdots + a_{n-1}\lambda^{n-1} + \lambda^n$$
$$= (r_1 - \lambda)^{m_1}(r_2 - \lambda)^{m_2}\cdots(r_n - \lambda)^{m_n}.$$

The $r_1, r_2, \ldots, r_n$ are the eigenvalues of matrix A. The coefficient of λ^{n-1} is the negative of the sum of the eigenvalues weighted by their multiplicities:

$$a_{n-1} = -\sum_{i=1}^{n} m_i r_i.$$

The constant term

$$a_0 = \prod_{i=1}^{n} r_i^{m_i}.$$

If you look at the expansion of $\det(A - \lambda I)$ using the definition of a determinant in (2.2) in Section 2, you will notice that the coefficient of λ^{n-1} is

$$-\sum_{i=1}^{m} a_{ii} = -\text{Trace}(A)$$

and the constant term is $\det(A)$. Thus,

$$\text{Trace}(A) = \sum_{i=1}^{n} m_i r_i \tag{8.6}$$

and

$$\det(A) = \prod_{i=1}^{n} r_i^{m_i}. \tag{8.7}$$
∎

A simple consequence of Theorem 8.7 is that positive-definite matrices have positive determinants because the value of a determinant is the product of its eigenvalues.

Example 8.4 Illustration of the Relationships between the Eigenvalues and the Trace and the Determinant of a Matrix

Let the matrix

$$A = \begin{bmatrix} 3 & 0 & 0 & -1 & 0 & -1 \\ 0 & 3 & 0 & -1 & 0 & -1 \\ 0 & 0 & 3 & -1 & 0 & -1 \\ 0 & 0 & 0 & 2 & 0 & -1 \\ 0 & 0 & 0 & 0 & 2 & -1 \\ 0 & 0 & 0 & 0 & 0 & 1 \end{bmatrix}.$$

The eigenvalues are $\lambda = 3$ with multiplicity 3, $\lambda = 2$ with multiplicity 2, and $\lambda = 1$. Thus,

$$\det(A) = 3^3 2^2 1 = 108$$

and

$$\text{trace}(A) = 3(3) + 2(2) + 1(1) = 14. \qquad \square$$

Notice that A is an upper triangular matrix. In general the eigenvalues of an upper triangular matrix are the diagonal elements with multiplicities being determined by the number of elements in a group of equal elements. Indeed

$$T = \begin{bmatrix} t_{11} & t_{12} & \cdots & t_{1n} \\ 0 & t_{22} & \cdots & t_{2n} \\ 0 & 0 & \ddots & \vdots \\ 0 & 0 & 0 & t_{nn} \end{bmatrix}$$

has characteristic equation

$$\begin{vmatrix} t_{11}-\lambda & t_{12} & \cdots & t_{1n} \\ 0 & t_{22}-\lambda & \cdots & t_{2n} \\ 0 & 0 & \ddots & \vdots \\ 0 & 0 & 0 & t_{nn}-\lambda \end{vmatrix} = \prod_{i=1}^{n}(t_{ii}-\lambda) = 0$$

with roots $\lambda = t_{11}, \lambda = t_{22}, \ldots, \lambda = t_{nn}$.

8.5   THE EIGENVALUES AND EIGENVECTORS OF THE KRONECKER PRODUCT OF TWO MATRICES

Let A and B be square matrices with eigenvalues λ and μ and eigenvectors x and y, respectively. Then

$$(A \otimes B)(x \otimes y) = Ax \otimes By = \lambda x \otimes \mu y = \lambda\mu(x \otimes y). \tag{8.8a}$$

Thus, if A is an $m \times m$ matrix with not necessarily distinct eigenvalues $\lambda_1, \lambda_2, \ldots, \lambda_m$ and B is an $n \times n$ matrix with not necessarily distinct eigenvalues $\mu_1, \mu_2, \ldots, \mu_n$, the eigenvalues of $A \otimes B$ are $\lambda_i \mu_j, i = 1, \ldots, m, j = 1, \ldots, n$. The eigenvectors are the Kronecker product of the eigenvectors of A and B. Then

$$\text{Trace}(A \otimes B) = \sum_{i=1}^{m}\sum_{j=1}^{n}\lambda_i\mu_j = \left(\sum_{i=1}^{m}\lambda_i\right)\left(\sum_{j=1}^{n}\mu_j\right) = \text{TraceA TraceB}. \tag{8.8b}$$

Also

$$\det(A \otimes B) = \prod_{1 \le i \le m, 1 \le j \le m, (i,j) \in Z \times Z}\lambda_i\mu_j = \prod_{i=1}^{m}\lambda_i^n\prod_{j=1}^{m}\mu_j^m = (\det A)^n(\det B)^m. \tag{8.8c}$$

Example 8.5 The Eigenvalues, Eigenvectors, Trace, and Determinant of the Kronecker Product of Two Matrices

Let

$$A = \begin{bmatrix} 5 & -1 \\ -1 & 5 \end{bmatrix}, B = \begin{bmatrix} 3 & 2 \\ 2 & 3 \end{bmatrix}.$$

The eigenvalues of A are 6 and 4. The eigenvalues of B are 5 and 1. Then the eigenvalues of $A \otimes B$ are 30, 20, 6, and 4. We have that

$$\text{Trace}(A \otimes B) = 30 + 20 + 6 + 4 = (6+4)(5+1) = \text{Trace}(A)\,\text{Trace}(B) = 60.$$

Also

$$\det(A \otimes B) = [\det A]^2[\det B]^2 = (24)^2(5)^2 = 14,400.$$

The normalized orthogonal eigenvectors of A are

$$\begin{bmatrix} -\frac{1}{\sqrt{2}} \\ \frac{1}{\sqrt{2}} \end{bmatrix} \text{ and } \begin{bmatrix} \frac{1}{\sqrt{2}} \\ \frac{1}{\sqrt{2}} \end{bmatrix}.$$

For B, they are

$$\begin{bmatrix} \frac{1}{\sqrt{2}} \\ \frac{1}{\sqrt{2}} \end{bmatrix} \text{ and } \begin{bmatrix} -\frac{1}{\sqrt{2}} \\ \frac{1}{\sqrt{2}} \end{bmatrix}.$$

Finally, for $A \otimes B$, the eigenvectors are

$$\begin{bmatrix} -\frac{1}{2} \\ -\frac{1}{2} \\ \frac{1}{2} \\ \frac{1}{2} \end{bmatrix}, \begin{bmatrix} \frac{1}{2} \\ -\frac{1}{2} \\ \frac{1}{2} \\ \frac{1}{2} \end{bmatrix}, \begin{bmatrix} \frac{1}{2} \\ \frac{1}{2} \\ \frac{1}{2} \\ \frac{1}{2} \end{bmatrix} \text{ and } \begin{bmatrix} -\frac{1}{2} \\ \frac{1}{2} \\ -\frac{1}{2} \\ \frac{1}{2} \end{bmatrix}.$$

$\square$

8.6   THE EIGENVALUES AND THE EIGENVECTORS OF A MATRIX OF THE FORM aI + bJ

We first compute the determinant of this matrix. This will enable us to obtain the characteristic equation and the eigenvalues. Observe that

$$\begin{vmatrix} a+b & b & \cdots & b \\ b & a+b & \cdots & b \\ \vdots & \vdots & \ddots & \vdots \\ b & b & \cdots & a+b \end{vmatrix} = \begin{vmatrix} a+nb & b & b & b \\ a+nb & a+b & b & b \\ \vdots & \vdots & \ddots & \vdots \\ a+nb & b & \cdots & a+b \end{vmatrix}$$

$$= (a+nb) \begin{vmatrix} 1 & b & \cdots & b \\ 1 & a+b & \cdots & b \\ \vdots & \vdots & \ddots & \vdots \\ 1 & b & \cdots & a+b \end{vmatrix} = (a+nb) \begin{vmatrix} 1 & b & \cdots & b \\ 0 & a & \cdots & 0 \\ \vdots & \vdots & \ddots & \vdots \\ 0 & 0 & \cdots & a \end{vmatrix}$$

$$= (a+nb)a^{n-1}.$$

Then the characteristic equation is

$$\begin{vmatrix} a+b-\lambda & b & \cdots & b \\ b & a+b-\lambda & \cdots & b \\ \vdots & \vdots & \ddots & \vdots \\ b & b & \cdots & a+b-\lambda \end{vmatrix} = (a-\lambda+nb)(a-\lambda)^{n-1} = 0.$$

The eigenvalues are $\lambda = a$ with multiplicity $n - 1$ and $\lambda = a + nb$. When $\lambda = a$, we have

$$bJ_n \begin{bmatrix} x_1 \\ x_2 \\ \vdots \\ x_n \end{bmatrix} = 0 \text{ or n repeated equations } x_1 + x_2 + \cdots x_n = 0.$$

A set of independent and orthogonal eigenvectors would be

$$\begin{bmatrix} \frac{1}{\sqrt{2}} \\ -\frac{1}{\sqrt{2}} \\ 0 \\ \vdots \\ 0 \end{bmatrix}, \begin{bmatrix} \frac{1}{\sqrt{6}} \\ \frac{1}{\sqrt{6}} \\ -\frac{2}{\sqrt{6}} \\ \vdots \\ 0 \end{bmatrix}, \ldots, \begin{bmatrix} \frac{1}{\sqrt{n(n-1)}} \\ \frac{1}{\sqrt{n(n-1)}} \\ \frac{1}{\sqrt{n(n-1)}} \\ \vdots \\ -\frac{n-1}{\sqrt{n(n-1)}} \end{bmatrix}.$$

For $\lambda = a + nb$, we get

$$(-nbI + nbJ)x = 0$$

leading to

$$(1 - n)x_1 + x_2 + \cdots + x_n = 0$$
$$x_1 + (1 - n)x_2 + \cdots + x_n = 0$$
$$\vdots$$
$$x_1 + x_2 + \cdots + (1 - n)x_n = 0.$$

A solution to these equations is 1_n, so we have the normalized eigenvector $\frac{1}{\sqrt{n}} 1_n$

Example 8.6 The Eigenvalues and Eigenvectors of a Particular Correlation Matrix

Let $\Sigma = \begin{bmatrix} 1 & 0.5 & 0.5 & 0.5 \\ 0.5 & 1 & 0.5 & 0.5 \\ 0.5 & 0.5 & 1 & 0.5 \\ 0.5 & 0.5 & 0.5 & 1 \end{bmatrix}$. The eigenvalues would be 0.5 with multiplicity 3 and 2.5. The characteristic equation is

$$(\lambda - 0.5)^3 (\lambda - 2.5) = \lambda^4 - 4\lambda^3 + 4.5\lambda^2 - 2\lambda + 0.3125.$$

By the Cayley–Hamilton theorem and (8.5),

$$\Sigma^{-1} = -\frac{1}{0.3125}(-2I + 4.5\Sigma - 4\Sigma^2 + \Sigma^3).$$

Observe that

$$\Sigma = .5(I + J)$$
$$\Sigma^2 = .25(I + 6J)$$
$$\Sigma^3 = .125(I + 31J).$$

Then

$$\Sigma^{-1} = -3.2(-2I + 2.25(I + J) - (I + 6J) + 0.125(I + 31J))$$
$$= 2I - 0.4J$$
$$= \begin{bmatrix} 1.6 & -0.4 & -0.4 & -0.4 \\ -0.4 & 1.6 & -0.4 & -0.4 \\ -0.4 & -0.4 & 1.6 & -0.4 \\ -0.4 & -0.4 & -0.4 & 1.6 \end{bmatrix}.$$

The eigenvectors would be

$$\begin{bmatrix} \frac{1}{2} \\ \frac{1}{2} \\ \frac{1}{2} \\ \frac{1}{2} \end{bmatrix}, \begin{bmatrix} \frac{1}{\sqrt{2}} \\ -\frac{1}{\sqrt{2}} \\ 0 \\ 0 \end{bmatrix}, \begin{bmatrix} \frac{1}{\sqrt{6}} \\ \frac{1}{\sqrt{6}} \\ -\frac{2}{\sqrt{6}} \\ 0 \end{bmatrix}, \begin{bmatrix} \frac{1}{\sqrt{12}} \\ \frac{1}{\sqrt{12}} \\ \frac{1}{\sqrt{12}} \\ -\frac{3}{\sqrt{12}} \end{bmatrix}.$$

$\square$

8.7   THE LOEWNER ORDERING

The Loewner ordering gives a way to compare the size of two nonnegative-definite matrices. Matrix A will be said to be greater than or equal to matrix B if and only if A − B is positive semi-definite. Matrix A will be said to be greater than matrix B if and only if A − B is positive definite. Two positive-semi-definite matrices need not be greater than or equal to or less than or equal to one another. In other words, the difference of two positive-semi-definite matrices need not be a positive-semi-definite matrix.

The Loewner ordering in the language of set theory is a partial ordering. A set is partially ordered if for any elements a, b, and c (1) a≤a, (2) a≤b and b≤a implies a=b, and (3) a≤b and b≤c implies that a≤c. A set is totally ordered if (1)–(3) above holds and for any two elements a≤b or b≤a. Thus, the Loewner ordering is a partial ordering but not a total ordering.

An interesting and important fact is given in Theorem 8.8.

Theorem 8.8 If A and B are positive-definite matrices and if B − A is positive semi-definite, then $A^{-1} - B^{-1}$ is positive semi-definite.

Proof. The symmetry of the matrix $A^{-1} - B^{-1}$ should be clear. We first prove that if I − A is positive semi-definite, then $A^{-1} - I$ is positive semi-definite. Let $A = P\Delta P'$ be the spectral decomposition of A. Define $A^{\frac{1}{2}} = P\Delta^{\frac{1}{2}}P'$ where $\Delta^{\frac{1}{2}}$ consists of the positive square roots of the elements of Δ. Since I − A is positive-semi-definite, for all vectors p

$$p'(I - A)p = p'A^{\frac{1}{2}}(A^{-1} - I)A^{\frac{1}{2}}p \geq 0.$$

For every q, there exists a p such that $q = A^{\frac{1}{2}}p$ or $A^{-\frac{1}{2}}q = p$ and

$$q'(A^{-1} - I)q \geq 0.$$

Hence $A^{-1} - I$ is positive semi-definite. Since B − A is positive semi-definite, we have for all p that

$$p'(B - A)p = p'B^{\frac{1}{2}}\left(I - B^{-\frac{1}{2}}AB^{-\frac{1}{2}}\right)B^{\frac{1}{2}}p \geq 0.$$

Since for each q there exists a p where $q = B^{\frac{1}{2}}p$ we have that $I - B^{-\frac{1}{2}}AB^{-\frac{1}{2}}$ is positive semi-definite and $B^{\frac{1}{2}}A^{-1}B^{\frac{1}{2}} - I$ is positive semi-definite, so $A^{-1} - B^{-1}$ is positive semi-definite. ∎

The word positive definite may replace positive semi-definite in the statement of the theorem and throughout the proof.

Example 8.7 Loewner Ordering of the Inverse of Two Matrices

Let

$$G = \begin{bmatrix} 4.08 & 1.44 \\ 1.44 & 4.92 \end{bmatrix} \quad \text{and} \quad H = \begin{bmatrix} 2.72 & 0.96 \\ 0.96 & 3.28 \end{bmatrix}.$$

Then

$$G - H = \begin{bmatrix} 1.36 & 0.48 \\ 0.48 & 1.64 \end{bmatrix},$$

a positive-definite matrix. Now

$$G^{-1} = \begin{bmatrix} 0.2733 & -0.08 \\ -0.08 & 0.2267 \end{bmatrix} \text{ and } H^{-1} = \begin{bmatrix} 0.41 & -0.12 \\ -0.12 & 0.34 \end{bmatrix}.$$

We have

$$H^{-1} - G^{-1} = \begin{bmatrix} 0.1367 & -0.04 \\ -0.04 & 0.1133 \end{bmatrix},$$

a positive-definite matrix. □

Example 8.8 Matrix B – A is positive semi-definite does not necessarily imply that $B^2 - A^2$ is positive semi-definite

This example is taken from Gruber (2010). Let

$$A = \begin{bmatrix} 2 & 1 \\ 1 & 1 \end{bmatrix}, B = \begin{bmatrix} 2 & 1 \\ 1 & 2 \end{bmatrix}.$$

Then

$$B - A = \begin{bmatrix} 0 & 0 \\ 0 & 1 \end{bmatrix},$$

a positive semi-definite matrix. However,

$$B^2 - A^2 = \begin{bmatrix} 5 & 4 \\ 4 & 5 \end{bmatrix} - \begin{bmatrix} 5 & 3 \\ 3 & 2 \end{bmatrix}$$

$$= \begin{bmatrix} 0 & 1 \\ 1 & 3 \end{bmatrix},$$

a matrix with determinant −1 and thus not positive semi-definite. □

8.8 SUMMARY

Important relationships between the eigenvalues of matrices and the eigenvalues of matrices obtained from them by different operations and multiplications are studied. We saw that the eigenvalues of A and TAT^{-1} were the same and that the eigenvectors of TAT^{-1} were the images of the eigenvectors of A under the transformation T. We noted that for the Kronecker product of two matrices, the eigenvalues were the products of the eigenvalues of the matrix in the product and the eigenvectors were the Kronecker product of the eigenvectors of the individual matrices. We obtained a formula for eigenvalues of matrices of the form $aI + bJ$ and hence for equicorrelation matrices.

The trace of a matrix was shown to be equal to the sum of the eigenvalues counting their multiplicities. Likewise the determinant of the matrix is the product of the eigenvalues raised to the powers indicated by their multiplicities.

The Cayley–Hamilton theorem that shows that a square matrix satisfies its characteristic equation was established, and its use for obtaining the inverse of a matrix was shown.

The Loewner ordering was defined where one matrix was greater than or equal to another if the difference between it and the smaller matrix was positive semi-definite. It was noted that this is a partial ordering between matrices and not a total ordering. We showed that for the inverse of two matrices, the ordering is reversed. We gave a counterexample to establish that if A is larger than B, it may not follow that that relationship holds for the squares of the matrices.

EXERCISES

8.1 A. Show that the matrices

$$A = \begin{bmatrix} 4 & 2 \\ 1 & 3 \end{bmatrix}, B = \begin{bmatrix} 4 & \sqrt{2} \\ \sqrt{2} & 3 \end{bmatrix}$$

have the same characteristic equation and eigenvalues but different eigenvectors.

B. Show that the eigenvectors of A and A′ are different. Thus, although a matrix and its transpose have the same eigenvalues, they may not have the same eigenvectors.

C. Show that if H is a symmetric matrix, then the eigenvalues of H′H are the squares of the eigenvalues of H.

D. Show that C is not generally true for a matrix that is not symmetric. (Hint: Consider matrix A in part A.)

8.2 Let

$$A = \begin{bmatrix} 7 & 2 & 2 \\ 2 & 7 & 2 \\ 2 & 2 & 7 \end{bmatrix}, B = \begin{bmatrix} 7 & 1 \\ 1 & 7 \end{bmatrix}.$$

Find without using Equations (8.8b) and (8.8c) directly

A. The eigenvalues and eigenvectors of A and B

B. The eigenvalues and eigenvectors of $A \otimes B$

C. $\det(A \otimes B)$ and $\text{trace}(A \otimes B)$

Check your answers without using the results of Subsection 8.4.

8.3 Let

$$A = \begin{bmatrix} 9 & 2 \\ 2 & 6 \end{bmatrix}, B = \begin{bmatrix} 5 & 2 \\ 2 & 3 \end{bmatrix}, C = \begin{bmatrix} 9 & 2 \\ 2 & 3 \end{bmatrix}, D = \begin{bmatrix} 9 & 3 \\ 3 & 1 \end{bmatrix}$$

    A. Show that A, B, C, and D are positive semi-definite. Which ones are positive definite?

    B. Show that $A > B$ and that $A \geq C$.

    C. Is $C \geq D$ or $D \geq C$? Justify your answer.

8.4 Let A be a positive-semi-definite matrix. Show that for any matrix M such that the product MAM' exists.

    A. The matrix MAM' is positive semi-definite.

    B. If $A \leq B$ then $MAM' \leq MBM'$.

8.5 **A.** Find the eigenvalues and the eigenvectors of the equicorrelation matrix $M = (1-\rho)I + \rho J$.

    B. Give a range of values of ρ where M is positive semi-definite.

    C. Show that $\Sigma^{-1} = \dfrac{1}{1-\rho}\left(I - \dfrac{\rho}{1+(n-1)\rho}J\right)$.

8.6 Let

$$\Sigma = \begin{bmatrix} 1 & 0.9 & 0.9 \\ 0.9 & 1 & 0.9 \\ 0.9 & 0.9 & 1 \end{bmatrix}.$$

    A. Find the eigenvalues and eigenvectors.

    B. Use the Cayley–Hamilton theorem to obtain the inverse. Check your answer using the formula in Exercise 8.5C.

8.7 Show that if E is an idempotent matrix, then $I - E$ is idempotent and that $E(I - E) = 0$.

8.8 Show that $\dfrac{1}{n}J$ and $I - \dfrac{1}{n}J$ are idempotent matrices.

8.9 For Example 8.5, find the Kronecker product and obtain the eigenvalues and eigenvectors without using (8.8a).

8.10 Show that the eigenvalues of a lower triangular matrix are the elements in the main diagonal.

8.11 (Part A of this exercise comes from Gruber (2010).) Let

$$A = \begin{bmatrix} 2 & 1 \\ 1 & 1 \end{bmatrix}, B = \begin{bmatrix} 2 & 1 \\ 1 & 1+\varepsilon \end{bmatrix}, \varepsilon > 0.$$

Show that

    A. $B - A$ is positive semi-definite but $B^2 - A^2$ is not.

    B. $A^{-1} - B^{-1}$ is positive semi-definite.

SECTION 9

THE SINGULAR VALUE DECOMPOSITION (SVD)

9.1 INTRODUCTION

The decomposition of any matrix into the product of an orthogonal matrix, a diagonal matrix with nonzero diagonal elements ranked from highest to lowest, and another orthogonal matrix is called the singular value decomposition (SVD). The diagonal elements are called the singular values. If A is a real matrix to be decomposed, the singular values are the positive square roots of the eigenvalues of $A'A$. When A is symmetric, the singular values are the eigenvalues of A. This is not true if A is not symmetric. The first orthogonal matrix consists of orthonormalized eigenvectors of AA'. The second orthogonal matrix consists of orthonormalized eigenvectors of $A'A$. Since the eigenvectors are not unique, the SVD is not unique. However, the singular values are always the same.

The SVD will be of particular importance in the study of generalized inverses in Part III and quadratic forms in Part IV.

In Subsection 9.2 we prove that for any real matrix, the SVD exists. In Subsection 9.3 we use the SVD to establish that the row rank and the column rank of a matrix are equal. We also give a number of illustrative examples of SVDs of specific matrices.

Matrix Algebra for Linear Models, First Edition. Marvin H. J. Gruber.
© 2014 John Wiley & Sons, Inc. Published 2014 by John Wiley & Sons, Inc.

9.2   THE EXISTENCE OF THE SVD

From the work done in Section 7, it is clear that for any square symmetric matrix A, there exists an orthogonal matrix P such that $P'AP = \Lambda$ where Λ is the diagonal matrix of eigenvalues of A and the matrix P consists of the eigenvectors. Then the matrix A may be expressed in the form $A = P\Lambda P'$. Is there a similar decomposition for any n × m matrix X? The answer is yes. The existence of a similar decomposition, the SVD, is established by Theorem 9.2. In order to establish Theorem 9.2, the following fact is needed.

Theorem 9.1 The nonzero eigenvalues of $X'X$ and XX' are identical.

Proof. The scalar λ is an eigenvalue of XX' if there is a nonzero column vector x such that

$$(XX' - \lambda I)x = 0. \tag{9.1}$$

Multiply (9.1) by X'. Then it follows that

$$(X'X - \lambda I)X'x = 0.$$

Let $y = X'x$. The column vector y is nonzero because if $y = 0$, $XX'x = 0$. But x cannot be an eigenvector corresponding to 0 and nonzero λ at the same time. Then

$$(X'X - \lambda I)y = 0,$$

and λ is an eigenvalue of $X'X$.

Likewise if λ is an eigenvalue of $X'X$, there is a nonzero column vector z such that

$$(X'X - \lambda I)z = 0. \tag{9.2}$$

Then, after multiplication of (9.2) by X,

$$(XX' - \lambda I)Xz = 0,$$

and $v = Xz$ is an eigenvector of XX'.   ∎

In Sections 7 and 8, the matrix Λ represented the diagonal matrix with all of the eigenvalues in the main diagonal. From now on, Λ will represent the nonzero eigenvalues. Since we will generally be talking about positive-semi-definite matrices, these nonzero eigenvalues will be positive numbers.

We now establish the existence of the SVD. The proof is adapted from that of Stewart (1963).

Theorem 9.2 Let X be an n × m of rank s ≤ m. Then X may be expressed in the form

$$X = L' \begin{bmatrix} \Lambda^{1/2} & 0 \\ 0 & 0 \end{bmatrix} M' \tag{9.3}$$

where L and M are orthogonal matrices and Λ is the diagonal matrix of positive eigenvalues of X'X and XX' ranked from highest to lowest. The matrix $\Lambda^{1/2}$ consists of the positive square roots of the positive elements of Λ and

$$LXX'L' = \begin{bmatrix} \Lambda & 0 \\ 0 & 0 \end{bmatrix} \text{ an } n \times n \text{ matrix} \quad \text{and} \quad M'X'XM = \begin{bmatrix} \Lambda & 0 \\ 0 & 0 \end{bmatrix} \text{ an } m \times m \text{ matrix.}$$

Proof. Partition

$$L = \begin{bmatrix} S \\ T \end{bmatrix} \quad \text{and} \quad M = \begin{bmatrix} U & V \end{bmatrix}$$

where S is s×n, T is (n−s)×n, U is m×s, and V is (m−s)×m. The matrix S of eigenvectors of the nonzero eigenvectors of XX' is row orthogonal (SS' = I). For a given L, let

$$U = X'S'\Lambda^{-1/2}.$$

Observe that

$$U'U = \Lambda^{-1/2} SXX'S'\Lambda^{-1/2} = I.$$

Let V be such that M is orthogonal. Then

$$L' \begin{bmatrix} \Lambda^{1/2} & 0 \\ 0 & 0 \end{bmatrix} M' = S'\Lambda^{1/2}U' = S'\Lambda^{1/2} \Lambda^{-1/2} SX = S'SX$$

$$= (I - T'T)X = X$$

because TXX'T' = 0 implies TX = 0. Also

$$X'X = M \begin{bmatrix} \Lambda & 0 \\ 0 & 0 \end{bmatrix} M' = U\Lambda U'.$$

Thus,

$$U'X'XU = \Lambda$$

and

$$MX'XM' = \Lambda.$$

This proves the existence of the SVD.   ∎

9.3 USES AND EXAMPLES OF THE SVD

We can now fill the gap in the proof of Theorem 6.1 about the equality of the row rank and the column rank of a matrix. In Theorem 6.1 if A has rank r and the SVD of $A = S'\Lambda^{1/2}U'$, then one way to do the factorization is $B = S'$ and $C = \Lambda^{1/2}U'$. However, this factorization is not unique. For example, $B = S'\Lambda^{1/2}$ and $C = U'$ would also work. The problem with the proof of Theorem 6.1 was whether the factorization existed. We have now established that it does by using the SVD.

From (9.3) we have

$$X = S'\Lambda^{1/2}U'. \tag{9.4}$$

Another way to write (9.4) is

$$X = \sum_{i=1}^{s} \lambda_i^{1/2} s_i' u_i' \tag{9.5}$$

where s_i' are the columns of S' and u_i are the columns of U.

As promised in Section 6, we show how to use the SVD to establish that for an $n \times m$ matrix, the row and column ranks are equal. This is done in Theorem 9.3.

Theorem 9.3 The row and column rank of X is s, the number of nonzero singular values.

Proof. First, we show that $C(X') = C(U)$. Suppose that m-dimensional vector $p \in C(X')$. Then $p = X't$ where t is an n-dimensional vector. Now $p = U\Lambda^{1/2}St = Ud$ where s-dimensional vector $d = \Lambda^{1/2}St$. Thus, $p \in C(U)$ and $C(X') \subset C(U)$. On the other hand, suppose $p \in C(U)$. Then there is an s-dimensional vector e where $p = Ue$. But then $p = U\Lambda^{1/2}SS'\Lambda^{-1/2}e = X'S'\Lambda^{-1/2}e = X'f$ where n-dimensional vector $f = S'\Lambda^{-1/2}e$.

Thus, $p \in C(X')$ so $C(U) \subset C(X')$. It then follows that $C(X') = C(U)$. Thus, X' and U have the same column rank. Now U has column rank s. Let $v \in C(U)$ have representation

$$v = \sum_{i=1}^{s} c_i u_{(i)}$$

where $u_{(i)}$ are column vectors of U. Suppose that there exist scalars $d_1, d_2, \ldots, d_s$ such that $\sum_{i=1}^{s} d_i u_{(i)} = 0$. Then $u_{(j)}' \sum_{i=1}^{s} d_i u_{(i)} = d_j = 0$, $j = 1, 2, \ldots, s$, so the s column vectors are linearly independent and U has rank s. Then X' has column rank s and X has row rank s. By a similar type of argument, it can be shown that $C(S') = C(X)$ and that the dimension of $C(S')$ is s. Thus, X has column rank s and X has rank s.

We now consider the rank of Kronecker product of two matrices and give another derivation of the result in (8.4) that uses the SVD. ∎

Theorem 9.4 Let A be an $m \times n$ matrix and B be a $p \times q$ matrix. Then

$$\text{rank}(A \otimes B) = \text{rank}(A)\, \text{rank}(B).$$

Proof. Suppose that the SVDs of A and B are

$$A = S'_A \Lambda_A^{1/2} U'_A$$

and

$$B = S'_B \Lambda_B^{1/2} U'_B.$$

Then

$$A \otimes B = \left(S'_A \otimes S'_B \right) \left(\Lambda_A^{1/2} \otimes \Lambda_B^{1/2} \right) \left(U'_A \otimes U'_B \right).$$

Then the number of elements of the matrix of singular values $\Lambda_A^{1/2} \otimes \Lambda_B^{1/2}$ is the product of the number of elements in $\Lambda_A^{1/2}$ and $\Lambda_B^{1/2}$, the rank of the matrix of A and B. Note that $\Lambda_A^{1/2}$ and $\Lambda_B^{1/2}$ only contain the nonzero elements. The result follows. ∎

Example 9.1 The SVD of a Specific Matrix

Let

$$A = \begin{bmatrix} 1 & 1 & 0 & 0 \\ 1 & 0 & 1 & 0 \\ 1 & 0 & 0 & 1 \end{bmatrix}.$$

Then

$$A'A = \begin{bmatrix} 3 & 1 & 1 & 1 \\ 1 & 1 & 0 & 0 \\ 1 & 0 & 1 & 0 \\ 1 & 0 & 0 & 1 \end{bmatrix}$$

and

$$AA' = \begin{bmatrix} 2 & 1 & 1 \\ 1 & 2 & 1 \\ 1 & 1 & 2 \end{bmatrix}.$$

Observe that

$$
\begin{vmatrix} 2-\lambda & 1 & 1 \\ 1 & 2-\lambda & 1 \\ 1 & 1 & 2-\lambda \end{vmatrix} = \begin{vmatrix} (1-\lambda) & -(1-\lambda) & 0 \\ 0 & (1-\lambda) & -(1-\lambda) \\ 1 & 1 & 2-\lambda \end{vmatrix}
$$

$$
= (1-\lambda)^2 \begin{vmatrix} 1 & -1 & 0 \\ 0 & 1 & -1 \\ 1 & 1 & 2-\lambda \end{vmatrix}
$$

$$
= (1-\lambda)^2 \begin{vmatrix} 1 & -1 & 0 \\ 0 & 1 & -1 \\ 0 & 2 & 2-\lambda \end{vmatrix}
$$

$$
= (1-\lambda)^2 (4-\lambda) = 0.
$$

The eigenvalues of AA' and the nonzero eigenvalues of $A'A$ are 4,1,1. For $\lambda=4$, the eigenvectors of AA' satisfy the system of equations

$$
\begin{aligned}
-2x_1 + x_2 + x_3 &= 0 \\
x_1 - 2x_2 + x_3 &= 0 \\
x_1 + x_2 - 2x_3 &= 0.
\end{aligned}
$$

A normalized solution to this system is $x_1 = x_2 = x_3 = 1/\sqrt{3}$. For $\lambda=1$, the eigenvectors satisfy

$$
x_1 + x_2 + x_3 = 0.
$$

Two orthogonal normalized solutions that are also orthogonal to the eigenvectors for $\lambda=4$ are

$$
x_1 = \frac{1}{\sqrt{2}}, \quad x_2 = \frac{-1}{\sqrt{2}}, \quad x_3 = 0
$$

and

$$
x_1 = \frac{1}{\sqrt{6}}, \quad x_2 = \frac{1}{\sqrt{6}}, \quad x_3 = \frac{-2}{\sqrt{6}}.
$$

Since AA' is 3×3 and of full rank, $T=0$ and

$$S' = \begin{bmatrix} \dfrac{1}{\sqrt{3}} & \dfrac{1}{\sqrt{2}} & \dfrac{1}{\sqrt{6}} \\[2ex] \dfrac{1}{\sqrt{3}} & -\dfrac{1}{\sqrt{2}} & \dfrac{1}{\sqrt{6}} \\[2ex] \dfrac{1}{\sqrt{3}} & 0 & -\dfrac{2}{\sqrt{6}} \end{bmatrix}.$$

Now from the proof of Theorem 9.2,

$$U = A'S'\Lambda^{-1/2} = \begin{bmatrix} 1 & 1 & 1 \\ 1 & 0 & 0 \\ 0 & 1 & 0 \\ 0 & 0 & 1 \end{bmatrix} \begin{bmatrix} \dfrac{1}{\sqrt{3}} & \dfrac{1}{\sqrt{2}} & \dfrac{1}{\sqrt{6}} \\[2ex] \dfrac{1}{\sqrt{3}} & -\dfrac{1}{\sqrt{2}} & \dfrac{1}{\sqrt{6}} \\[2ex] \dfrac{1}{\sqrt{3}} & 0 & -\dfrac{2}{\sqrt{6}} \end{bmatrix} \begin{bmatrix} \dfrac{1}{2} & 0 & 0 \\[1ex] 0 & 1 & 0 \\ 0 & 0 & 1 \end{bmatrix}$$

$$= \begin{bmatrix} \dfrac{3}{2\sqrt{3}} & 0 & 0 \\[2ex] \dfrac{1}{2\sqrt{3}} & \dfrac{1}{\sqrt{2}} & \dfrac{1}{\sqrt{6}} \\[2ex] \dfrac{1}{2\sqrt{3}} & -\dfrac{1}{\sqrt{2}} & \dfrac{1}{\sqrt{6}} \\[2ex] \dfrac{1}{2\sqrt{3}} & 0 & -\dfrac{2}{\sqrt{6}} \end{bmatrix}.$$

Then a SVD of

$$A = \begin{bmatrix} \dfrac{1}{\sqrt{3}} & \dfrac{1}{\sqrt{2}} & \dfrac{1}{\sqrt{6}} \\[2ex] \dfrac{1}{\sqrt{3}} & -\dfrac{1}{\sqrt{2}} & \dfrac{1}{\sqrt{6}} \\[2ex] \dfrac{1}{\sqrt{3}} & 0 & -\dfrac{2}{\sqrt{6}} \end{bmatrix} \begin{bmatrix} 2 & 0 & 0 \\ 0 & 1 & 0 \\ 0 & 0 & 1 \end{bmatrix} \begin{bmatrix} \dfrac{3}{2\sqrt{3}} & \dfrac{1}{2\sqrt{3}} & \dfrac{1}{2\sqrt{3}} & \dfrac{1}{2\sqrt{3}} \\[2ex] 0 & \dfrac{1}{\sqrt{2}} & -\dfrac{1}{\sqrt{2}} & 0 \\[2ex] 0 & \dfrac{1}{\sqrt{6}} & \dfrac{1}{\sqrt{6}} & -\dfrac{2}{\sqrt{6}} \end{bmatrix}.$$

$\square$

If a square matrix A is not symmetric, its singular values are not its eigenvalues. This is illustrated by Example 9.2.

Example 9.2 The Eigenvalues and the Singular Values of a Nonsymmetric Square Matrix Need Not Be the Same

Let

$$B = \begin{bmatrix} 7 & -3 \\ 0 & 4 \end{bmatrix}.$$

The eigenvalues are $\lambda = 7$ and $\lambda = 4$. Then

$$B'B = \begin{bmatrix} 49 & -21 \\ -21 & 25 \end{bmatrix}$$

has eigenvalues $37 + 3\sqrt{65}$ and $37 - 3\sqrt{65}$. Then B has singular values $\sqrt{37 + 3\sqrt{65}} = 7.822$ and $\sqrt{37 - 3\sqrt{65}} = 3.57956.$ □

The SVD is not unique because it depends on the choice of eigenvectors of $X'X$ or XX'. Consider Example 9.3.

Example 9.3 The Nonuniqueness of the SVD

Observe that the SVD of

$$M = \begin{bmatrix} 4 & 2 & 2 \\ 2 & 2 & 0 \\ 2 & 0 & 2 \end{bmatrix} = \begin{bmatrix} \frac{2}{\sqrt{6}} & 0 & -\frac{1}{\sqrt{3}} \\ \frac{1}{\sqrt{6}} & -\frac{1}{\sqrt{2}} & \frac{1}{\sqrt{3}} \\ \frac{1}{\sqrt{6}} & \frac{1}{\sqrt{2}} & \frac{1}{\sqrt{3}} \end{bmatrix} \begin{bmatrix} 6 & 0 & 0 \\ 0 & 2 & 0 \\ 0 & 0 & 0 \end{bmatrix} \begin{bmatrix} \frac{2}{\sqrt{6}} & \frac{1}{\sqrt{6}} & \frac{1}{\sqrt{6}} \\ 0 & -\frac{1}{\sqrt{2}} & \frac{1}{\sqrt{2}} \\ -\frac{1}{\sqrt{3}} & \frac{1}{\sqrt{3}} & \frac{1}{\sqrt{3}} \end{bmatrix}$$

$$= \begin{bmatrix} \frac{2}{\sqrt{6}} & 0 & \frac{1}{\sqrt{3}} \\ \frac{1}{\sqrt{6}} & \frac{1}{\sqrt{2}} & -\frac{1}{\sqrt{3}} \\ \frac{1}{\sqrt{6}} & -\frac{1}{\sqrt{2}} & -\frac{1}{\sqrt{3}} \end{bmatrix} \begin{bmatrix} 6 & 0 & 0 \\ 0 & 2 & 0 \\ 0 & 0 & 0 \end{bmatrix} \begin{bmatrix} \frac{2}{\sqrt{6}} & \frac{1}{\sqrt{6}} & -\frac{1}{\sqrt{6}} \\ 0 & \frac{1}{\sqrt{2}} & -\frac{1}{2} \\ \frac{1}{\sqrt{3}} & -\frac{1}{\sqrt{3}} & -\frac{1}{\sqrt{3}} \end{bmatrix}.$$

Thus, it is not unique. □

$A'A$. Then find the matrices U, V of eigenvectors. Obtain $S = \Lambda^{-1/2}U'A'$ as in Example 9.4.

Example 9.4 Another Example of a SVD

Recall the X matrix for the data in Table 1.1:

$$X = \begin{bmatrix} 1015.5 & 831.8 & 78678 \\ 1102.7 & 894.0 & 79367 \\ 1212.8 & 981.6 & 82153 \\ 1359.3 & 1101.7 & 86064 \\ 1472.8 & 1210.1 & 86794 \end{bmatrix}$$

Using Mathematica, we find that for the SVD of X,

$$S' = \begin{bmatrix} -0.425553 & 0.605395 & 0.542521 \\ -0.42929 & 0.331829 & -0.180058 \\ -0.444368 & 0.0777208 & -0.413666 \\ -0.465535 & -0.255498 & -0.469924 \\ -0.469499 & -0.672359 & 0.530379 \end{bmatrix},$$

$$\Lambda = \begin{bmatrix} 184909 & 0 & 0 \\ 0 & 341.739 & 0 \\ 0 & 0 & 10.8487 \end{bmatrix},$$

and

$$U' = \begin{bmatrix} -0.0149735 & -0.012195 & -0.999814 \\ -0.768431 & -0.639642 & 0.0193101 \\ -0.639758 & 0.768576 & 0.000206638 \end{bmatrix}.$$

□

The eigenvalues and the singular values of a nonnegative-definite matrix are the same. However, (see Exercise 9.7) for a square matrix that is not symmetric, although the product of the eigenvalues and the product of the singular values are both equal to the determinant of the matrix, the eigenvalues and the singular values generally are not equal.

Assume that an $n \times n$ nonsingular matrix A is not symmetric. Then its SVD would be

$$A = P\Delta Q$$

where P and Q are orthogonal matrices and Δ is the diagonal matrix of singular values, which may be different from the eigenvalues of A. Then

$$\det(A) = \det(P)\det(\Delta)\det(Q) = \det \Delta = \prod_{i=1}^{n} \delta_i.$$

Thus, the determinant is the product of the singular values, and the product of the singular values is equal to the product of the eigenvalues. However, the sum of the singular values may be different from the sum of the eigenvalues.

It can be shown that any square matrix is the sum of a symmetric matrix and a skew-symmetric matrix. A matrix A is skew symmetric when $A' = -A$; in other words, its transpose is the negative of the matrix. Observe that

$$A = \frac{1}{2}(A + A') + \frac{1}{2}(A - A').$$

By taking transposes and using $A'' = A$, it is easy to see that the first matrix in the above decomposition is symmetric, while the second matrix is skew symmetric.

The reader may verify that the trace of A is the same as the trace of the symmetric portion because the skew-symmetric portion has trace zero. Even though the sum of the diagonal elements and the sum of the eigenvalues are the same, the eigenvalues themselves for the symmetric matrix

$$S = \frac{1}{2}(A + A')$$

are generally different from those of A.

The following result will be very useful for comparison of estimators in Part V. The result was first obtained by Theobald (1974).

Theorem 9.5 A symmetric $n \times n$ matrix A is nonnegative definite if and only if $\mathrm{tr}AB \geq 0$ for all nonnegative-definite matrices B.

Proof. Suppose $\mathrm{tr}AB \geq 0$ for all nonnegative matrices B. For any n-dimensional column vector p, let $B = pp'$. Then

$$\mathrm{tr}App' = p'Ap \geq 0.$$

Thus, A is nonnegative definite.

Now suppose that A is nonnegative definite. The SVD of

$$B = P\Lambda P' = \sum_{i=1}^{n} \lambda_i P_i P_i'$$

where λ_i are the eigenvalues of B. The column vectors P_i are the orthogonal eigenvectors of B. The $\lambda_i \geq 0$ because B is positive semi-definite. Since A is nonnegative definite, it follows that

$$\mathrm{tr}(AB) = \sum_{i=1}^{n} \lambda_i P_i' A P_i \geq 0. \qquad \blacksquare$$

By using the SVD, some bounds may be found on the ratio of quadratic forms.

Theorem 9.6 Suppose that A is a positive-semi-definite $m \times m$ matrix with nonzero eigenvalues $\lambda_1 \geq \lambda_2 \geq \ldots \geq \lambda_s$. Then

$$\lambda_s \leq \frac{x'Ax}{x'x} \leq \lambda_1$$

if $s = m$. If $s < m$,

$$0 \leq \frac{x'Ax}{x'x} \leq \lambda_1$$

Proof. Observe that if $s < m$,

$$0 \leq \frac{x'Ax}{x'x} = \frac{\sum_{i=1}^{s} \lambda_i x'U_i U_i' x}{\sum_{i=1}^{s} x'U_i U_i' x + x'VV'x} \leq \lambda_1.$$

If $s = m$, then $VV' = 0$, so

$$\lambda_s \leq \frac{x'Ax}{x'x} = \frac{\sum_{i=1}^{s} \lambda_i x'U_i U_i' x}{\sum_{i=1}^{s} x'U_i U_i' x} \leq \lambda_1.$$ ∎

9.4   SUMMARY

We explained what the SVD is and gave a number of examples of it. We established its existence and nonuniqueness. We observed that for square matrices, singular values and eigenvalues are not always equal. We established some inequalities based on the SVD.

EXERCISES

9.1   In the above proof of Theorem 9.2, start out with the orthogonal eigenvectors of X'X. Define $S = \Lambda^{-1/2}U'X'$. Choose T so that L is orthogonal.
   A. Show that $SS' = I$.
   B. Show that $XV = 0$.
   C. Show that

$$L'\begin{bmatrix} \Lambda^{1/2} & 0 \\ 0 & 0 \end{bmatrix} M' = X.$$

9.2   Show by the method of proof of Theorem 9.3 that $C(S') = C(X)$.

9.3 Find the SVD of

A. $A = \begin{bmatrix} 1 & 1 & 0 \\ 1 & 1 & 0 \\ 1 & 0 & 1 \\ 1 & 0 & 1 \end{bmatrix}$.

B. $B = \begin{bmatrix} 1 & 1 & -5 & -5 \\ 1 & 1 & -5 & -5 \\ -5 & -5 & 1 & 1 \\ -5 & -5 & 1 & 1 \end{bmatrix}$.

C. $C = \begin{bmatrix} 1 & 0 & 1 & 0 \\ 1 & 0 & 0 & 1 \\ 0 & 1 & 1 & 0 \\ 0 & 1 & 0 & 1 \end{bmatrix}$.

D. $D = \begin{bmatrix} \frac{7}{6} & \frac{1}{6} & \frac{2}{3} \\ \frac{1}{6} & \frac{7}{6} & \frac{2}{3} \\ \frac{2}{3} & \frac{2}{3} & \frac{2}{3} \end{bmatrix}$.

9.4 Let A be a nonnegative-definite $m \times m$ matrix of rank r. Using the SVD, show that there is an $m \times r$ matrix B where $A = BB'$.

9.5 For each of the matrices in 9.3, find the factorization of the matrices discussed in conjunction with the proof of Theorem 6.1.

9.6 **A.** Let A and B be positive-definite matrices such that $A \le B$ with SVDs $A = P\Lambda P'$ and $B = P\Delta P'$. Show that $\Lambda \le \Delta$ and for the individual eigenvalues $\lambda_i \le \delta_i$ for all i.
B. Extend the result in part A to the case where A and B are positive semi-definite.

9.7 Let $A = PD_1P'$ and $B = PD_2P'$ where P is an orthogonal matrix and D_1 and D_2 are diagonal matrices. Show that for such matrices $AB = BA$.

9.8 **A.** Show that in fact $\frac{1}{2}(A + A')$ is symmetric and $\frac{1}{2}(A - A')$ is skew-symmetric.
B. Show that the trace of the symmetric matrix is the same as that of A and the trace of the skew-symmetric matrix is zero.

9.9 **A.** Decompose the matrix

$$B = \begin{bmatrix} 7 & 2 \\ 1 & 7 \end{bmatrix}$$

into the sum of a symmetric matrix and a skew-symmetric matrix.

   B. Find the eigenvalues and the singular values of

   1. The matrix B
   2. The symmetric matrix
   3. The skew-symmetric matrix

   C. In each case compare the eigenvalues and the singular values. What do you notice about the skew-symmetric matrix?

9.10 Let A and B be two nonnegative-definite matrices. Show that $A \le B$ if and only if $tr(AC) \le tr(BC)$ for all nonnegative-definite matrices C.

9.11 Let A and B be positive-definite $m \times m$ matrices. The matrix A has eigenvalues $\lambda_1 \ge \lambda_2 \ge \ldots \ge \lambda_m$. The matrix B has eigenvalues $\nu_1 \ge \nu_2 \ge \ldots \ge \nu_m$. Show that

$$\frac{\lambda_m}{\nu_1} \le \frac{x'Ax}{x'Bx} \le \frac{\lambda_1}{\nu_m}.$$

SECTION 10

APPLICATIONS OF THE SINGULAR VALUE DECOMPOSITION

10.1 INTRODUCTION

This section will present several applications of eigenvalues, eigenvectors, and the singular value decomposition. These will include the reparameterization of a non-full-rank model to a full-rank model, principal components, the multicollinearity problem, and principal component regression.

10.2 REPARAMETERIZATION OF A NON-FULL-RANK MODEL TO A FULL-RANK MODEL

The reparameterization is accomplished by making transformations on the parameters to be estimated and on the model matrix X. The model (1.1) may be rewritten as

$$\begin{aligned} Y &= XUU'\beta + \varepsilon \\ &= XU\gamma + \varepsilon, \end{aligned} \tag{10.1}$$

where $\gamma = U'\beta$. The reparameterized model is

$$Y = XU\gamma + \varepsilon. \tag{10.2}$$

Matrix Algebra for Linear Models, First Edition. Marvin H. J. Gruber.
© 2014 John Wiley & Sons, Inc. Published 2014 by John Wiley & Sons, Inc.

With respect to the reparameterized model (10.2), the least square estimator is

$$\begin{aligned}
g &= (U'X'XU')^{-1} U'X'Y \\
&= \Lambda^{-1} U'X'Y \\
&= \Lambda^{-1/2} Z,
\end{aligned} \tag{10.3}$$

where $Z = SY$.

Coordinatewise

$$g_i = \frac{z_i}{\lambda_i^{1/2}}. \tag{10.4}$$

This reparameterization is also possible for the full-rank case. This is the principal component regression.

Example 10.1 A Reparameterization

Consider the linear model

$$Y = \begin{bmatrix} 1_2 & 1_2 & 0 & 0 \\ 1_2 & 0 & 1_2 & 0 \\ 1_2 & 0 & 0 & 1_2 \end{bmatrix} \begin{bmatrix} \mu \\ \beta_1 \\ \beta_2 \\ \beta_3 \end{bmatrix} + \varepsilon.$$

We need to find the matrices U and S mentioned above. From the singular value decomposition, we have

$$S = \begin{bmatrix} \frac{1}{\sqrt{6}} & \frac{1}{\sqrt{6}} & \frac{1}{\sqrt{6}} & \frac{1}{\sqrt{6}} & \frac{1}{\sqrt{6}} & \frac{1}{\sqrt{6}} \\ -\frac{1}{2} & -\frac{1}{2} & 0 & 0 & \frac{1}{2} & \frac{1}{2} \\ -\frac{1}{2\sqrt{3}} & -\frac{1}{2\sqrt{3}} & \frac{1}{\sqrt{3}} & \frac{1}{\sqrt{3}} & -\frac{1}{2\sqrt{3}} & -\frac{1}{2\sqrt{3}} \end{bmatrix},$$

$$U = \begin{bmatrix} \frac{\sqrt{3}}{2} & 0 & 0 \\ \frac{1}{2\sqrt{3}} & -\frac{1}{\sqrt{2}} & -\frac{1}{\sqrt{6}} \\ \frac{1}{2\sqrt{3}} & 0 & \frac{\sqrt{2}}{\sqrt{3}} \\ \frac{1}{2\sqrt{3}} & \frac{1}{\sqrt{2}} & -\frac{1}{\sqrt{6}} \end{bmatrix},$$

and

$$\Lambda^{1/2} = \begin{bmatrix} \sqrt{8} & 0 & 0 \\ 0 & \sqrt{2} & 0 \\ 0 & 2 & \sqrt{2} \end{bmatrix}.$$

After finding XU, we get the reparameterized model

$$Y = XU\gamma + \varepsilon$$

$$= \begin{bmatrix} \frac{2}{\sqrt{3}} & -\frac{1}{\sqrt{2}} & -\frac{1}{\sqrt{6}} \\ \frac{2}{\sqrt{3}} & -\frac{1}{\sqrt{2}} & -\frac{1}{\sqrt{6}} \\ \frac{2}{\sqrt{3}} & 0 & \frac{2}{\sqrt{6}} \\ \frac{2}{\sqrt{3}} & 0 & \frac{2}{\sqrt{6}} \\ \frac{2}{\sqrt{3}} & \frac{1}{\sqrt{2}} & -\frac{1}{\sqrt{6}} \\ \frac{2}{\sqrt{3}} & \frac{1}{\sqrt{2}} & -\frac{1}{\sqrt{6}} \end{bmatrix} \begin{bmatrix} \gamma_1 \\ \gamma_2 \\ \gamma_3 \end{bmatrix} + \varepsilon,$$

where $\gamma = U'\beta$ as given below. We have

$$\gamma_1 = \frac{1}{2\sqrt{3}}(3\mu + \beta_1 + \beta_2 + \beta_3)$$

$$\gamma_2 = \frac{1}{\sqrt{2}}(-\beta_1 + \beta_3)$$

$$\gamma_3 = \frac{1}{\sqrt{6}}(-\beta_1 + 2\beta_2 - \beta).$$

Now $Y' = [y_{11} \quad y_{12} \quad y_{21} \quad y_{22} \quad y_{31} \quad y_{31}]$. Since $Z = SY$, we have

$$Z_1 = \frac{1}{\sqrt{6}} Y_{..}$$

$$Z_2 = \frac{1}{2}(-Y_{1.} + Y_{3.})$$

$$Z_3 = \frac{1}{2\sqrt{3}}(-Y_{1.} + 2Y_{2.} - Y_{3.}),$$

where the subscripted dots mean summation with respect to that index. Also

$$Z = \begin{bmatrix} \sqrt{8} & 0 & 0 \\ 0 & \sqrt{2} & 0 \\ 0 & 0 & \sqrt{2} \end{bmatrix} \begin{bmatrix} \gamma_1 \\ \gamma_2 \\ \gamma_3 \end{bmatrix} + \varepsilon,$$

and the least square estimators with respect to this model are

$$g_1 = \frac{1}{\sqrt{8}} z_1, \quad g_2 = \frac{1}{\sqrt{2}} z_2, \quad g_3 = \frac{1}{\sqrt{2}} z_3.$$

□

Example 10.2 A Principal Component Regression

The data below is taken from the Economic Report of the President 2010. The growth rates in real gross domestic product from 2001 to 2008 are given for five countries the United States, Germany, France, Italy, and Spain. A regression will be performed where the United States is the response or dependent variable and the European countries are the prediction or independent variable.

Country	2001	2002	2003	2004	2005	2006	2007	2008
United States	1.1	1.8	2.5	3.6	3.1	2.7	2.1	2.8
Germany	1.2	0	−0.2	1.2	0.7	3.2	2.5	1.2
France	1.8	1.1	1.1	2.3	1.9	2.4	2.3	0.3
Italy	1.8	0.5	0	1.5	0.7	2.0	1.6	−1.0
Spain	3.6	2.7	3.1	3.3	3.6	4.0	3.6	0.9

For the regression model

$$Y = X\beta + \varepsilon,$$

we have

$$Y = \begin{bmatrix} 1.1 \\ 1.8 \\ 2.5 \\ 3.6 \\ 3.1 \\ 2.7 \\ 2.1 \\ 2.8 \end{bmatrix}, \; X = \begin{bmatrix} 1 & 1.2 & 1.8 & 1.8 & 3.6 \\ 1 & 0 & 1.1 & 0.5 & 2.7 \\ 1 & -0.2 & 1.1 & 0 & 3.1 \\ 1 & 1.2 & 2.3 & 1.5 & 3.3 \\ 1 & 0.7 & 1.9 & 0.7 & 3.6 \\ 1 & 3.2 & 2.4 & 2.0 & 4.0 \\ 1 & 2.5 & 2.3 & 1.6 & 3.6 \\ 1 & 1.2 & 0.3 & -1.0 & 0.9 \end{bmatrix}.$$

The least square estimator

$$b = \begin{bmatrix} 1.08781 \\ -0.0963699 \\ 2.8152 \\ -1.48692 \\ -0.591194 \end{bmatrix}.$$

The matrix

$$U = \begin{bmatrix} -0.22446 & -0.204651 & 0.367647 & -0.0289083 & -0.878485 \\ -0.332293 & 0.807523 & 0.461543 & -0.124984 & 0.0940529 \\ -0.425223 & -0.0301302 & -0.120575 & 0.896253 & 0.021675 \\ -0.274166 & 0.344599 & -0.79706 & -0.240769 & -0.335873 \\ -0.763687 & -0.431705 & 0.0444016 & -0.34972 & 0.325788 \end{bmatrix}.$$

The singular values of X are

$$\lambda_1 = 11.856, \quad \lambda_2 = 2.78331, \quad \lambda_3 = 1.89106, \quad \lambda_4 = 0.528956, \quad \lambda_5 = 0.377542.$$

The least square estimator with respect to the reparameterized model is

$$g = U'b = \begin{bmatrix} -0.550085 \\ -0.47279 \\ 1.17492 \\ 3.06849 \\ -0.596858 \end{bmatrix}.$$

$\square$

10.3 PRINCIPAL COMPONENTS

Suppose you have a matrix X of n observations of p > 1 factors. Consider an estimator S of the covariance matrix of X, Σ. Let X_i be the column vectors of X that represents the observations of each of the p factors. Let A be a matrix and let

$$Y = XA. \tag{10.5}$$

Then the columns of Y have the representation

$$Y_i = \sum_{j=1}^{p} a_{ji} X_j, \quad 1 \le i \le p. \tag{10.6}$$

It can be shown that if A is the column orthogonal matrix whose columns are the eigenvectors of S, then the sample variance of the Y_i is maximized. The singular value decomposition of

$$S = A\Lambda A' = \sum_{i=1}^{p} \lambda_i A_i A_i', \tag{10.7}$$

where the eigenvalues $\lambda_1 \ge \lambda_2 \ge \ldots \ge \lambda_p$. The eigenvalues λ_i are the relative maximum values of the sample variance of the Y_i. The representations in (10.6) are called the principal components. Their sample variances are the summands in (10.7). Thus, in a practical problem one can see which linear combination of the factors is the most important in the analysis.

Example 10.3 A Principal Component Analysis

The data below concerns the components of the gross national product from 1995 to 2007. Three components of personal consumption expenditures are money spent on durable goods, non-durable goods, and services.

	Year	Durable Goods	Non-durable Goods	Services
1	1995	612	1485	2879
2	1996	653	1556	3049
3	1997	693	1619	3236
4	1998	750	1684	3446
5	1999	818	1805	3660
6	2000	863	1947	3929
7	2001	884	2017	4154
8	2002	924	2080	4347
9	2003	943	2190	4571
10	2004	984	2344	4868
11	2005	1024	2516	5168
12	2006	1049	2688	5488
13	2007	1078	2833	5823

The principal component analysis consists of the eigenvalues and eigenvectors of the covariance matrix. Using Minitab, we find that the covariance matrix is

$$M = \begin{bmatrix} 23541.6 & 65055.6 & 142454.6 \\ 65055.6 & 1914939 & 413696.6 \\ 142454.6 & 413696.6 & 897529.8 \end{bmatrix}.$$

The eigenvalues of M are

$$\lambda_1 = 2067760, \quad \lambda_2 = 767342, \quad \lambda_3 = 908.251.$$

The eigenvectors of M are

$$\begin{bmatrix} -0.0534923 \\ -0.939406 \\ -0.338608 \end{bmatrix}, \begin{bmatrix} 0.147697 \\ -0.342807 \\ 0.927723 \end{bmatrix}, \quad \text{and} \quad \begin{bmatrix} 0.987585 \\ 0.00038527 \\ -0.157085 \end{bmatrix},$$

representing durable goods by DG, nondurable goods by NDG, and services by S. The principal components are the vectors using the components of the eigenvectors. Thus, rounding off to four decimal places, the principal components are

$$PC1 = 0.0535DG - 0.9394NDG - 0.3386S$$
$$PC2 = 0.1477DG - 0.3428NDG + 0.9277S$$
$$PC3 = 0.9876DG - 0.0004NDG - 0.1571S.$$

To find the proportion of the variability corresponding to a principal component, divide the eigenvalue by the trace of the correlation matrix or the sum of the eigenvalues. The percentage accounted for is obtained by multiplying this proportion by 100. Thus, in this case, the first principal component accounts for 72.9% of the variability, the second 27.1%, and the third 0.03%. (There is a slight error due to roundoff.)   □

10.4   THE MULTICOLLINEARITY PROBLEM

Consider a linear regression model from (1.1). Sometimes in practice there may be a dependence between the independent variables. This is called multicollinearity. Two cases are of interest. They include:

1. Exact multicollinearity where the X matrix in the regression model is of less than full rank and the X'X matrix is singular;
2. Near multicollinearity where the X'X is nonsingular but has at least one very small eigenvalue.

The case of exact multicollinearity may be dealt with by solving the normal equation

$$X'Xb = X'Y \qquad (10.8)$$

making use of a generalized inverse so that

$$b = GX'Y. \qquad (10.9)$$

How to find generalized inverses and their definition and properties will be taken up in Part III.

For the case of near multicollinearity, the least square estimators may be very imprecise. Observe that for the least square estimator in (10.4),

$$\text{Var}(g_i) = \frac{\sigma^2}{\lambda_i}. \qquad (10.10)$$

This variance could be quite large for a small eigenvalue, and g_i as an estimator of γ_i could be very imprecise. Also in studying the effect of the regression on the variation, there might be ambiguity as to which variable the variation is coming from.

In multicollinear data since two or more of the variables might have a strong linear relationship, the correlation coefficients between those variables would be expected to have absolute values close to one. When there are high correlations between variables, one should suspect multicollinearity.

Two methods of dealing with this are principal component regression where the estimators corresponding to very small eigenvalues are neglected and the use of a ridge regression estimator of the form

$$\hat{\beta}_i = (X'X + kI)^{-1} X'Y. \qquad (10.11)$$

These estimators are discussed quite extensively in Gruber (1998) and Gruber (2010). Their derivation will be taken up in Part V.

Example 10.4 An Example of Multicollinear Data

The following data on price indexes of food, apparel, housing, transportation, and medical care from 2000 to 2009 is taken from the 2009 Economic Report of the President.

Year	Food	Apparel	Housing	Transportation	Medical care
2000	167.8	129.6	169.6	153.3	260.8
2001	173.1	127.3	176.4	154.3	272.8
2002	176.2	124.0	180.3	152.9	285.6
2003	180.0	120.9	184.8	157.6	297.1
2004	186.2	120.4	189.5	163.1	310.1
2005	190.7	119.5	195.7	173.9	323.2
2006	195.2	119.5	203.2	180.9	336.2
2007	202.9	119.0	209.6	184.7	351.1
2008	214.1	118.9	216.3	195.5	364.1
2019	218.0	120.1	217.1	179.3	375.6

A regression will be performed where food is the response matrix and apparel, housing, transportation, and medical care are the predictor variables. The table below gives the correlations between the predictor variables.

	Apparel	Housing	Transportation
Housing	−0.889		
Transportation	−0.769	0.962	
Medical Care	−0.913	0.997	0.953

Notice the high correlations between housing and transportation, housing and medical care, and transportation and medical care. The eigenvalues of the $X'X$ matrix are $\lambda_1 = 1.37287 \times 10^6$, $\lambda_2 = 1645.32$, $\lambda_3 = 73.3135$, $\lambda_4 = 2.89265$, $\lambda_5 = 0.000278545$.

The large ratio between the largest and the smallest eigenvalue would indicate severe multicollinearity. For the reparameterized model, the regression coefficients are

$$g_1 = -0.444956, \quad g_2 = -0.186834, \quad g_3 = -0.278113,$$
$$g_4 = 0.405328, \quad g_5 = 29.5366. \qquad \Box$$

10.5 SUMMARY

We have considered the applications of eigenvalues and the singular value decomposition that were presented in Sections 7–9. These included the reparameterization of a non-full-rank model to a full-rank model, principal components, and principal component regression and how these ideas can be used in studying multicollinear data.

EXERCISES

10.1 Consider the following linear models:

$$\mathbf{A.}\ \ Y = \begin{bmatrix} 1 & 1 & 0 \\ 1 & 0 & 1 \end{bmatrix} \begin{bmatrix} \mu \\ \beta_1 \\ \beta_2 \end{bmatrix} + \varepsilon.$$

$$\mathbf{B.}\ \ Y = \begin{bmatrix} 1 & 1 & 0 & 1 & 0 \\ 1 & 1 & 0 & 0 & 1 \\ 1 & 0 & 1 & 1 & 0 \\ 1 & 0 & 1 & 0 & 1 \end{bmatrix} \begin{bmatrix} \mu \\ \tau_1 \\ \tau_2 \\ \theta_1 \\ \theta_2 \end{bmatrix} + \varepsilon.$$

Reparameterize each of these models to one of full rank and find the least square estimators with respect to the reparameterized models.

10.2 Using Minitab or some other statistical software package, do a principal component analysis for the data in Example 10.2.

10.3 **A.** For the data of Example 10.4, obtain the least square estimators $b = (X'X)^{-1}X'Y$ and determine which variables are statistically significant.

  B. Do a principal component analysis of the predictor variables in Example 10.4 and tell which of the principal components are important and which can be omitted from the analysis.

SECTION 11

RELATIVE EIGENVALUES AND GENERALIZATIONS OF THE SINGULAR VALUE DECOMPOSITION

11.1 INTRODUCTION

This section gives two generalizations of the singular value decomposition. The first generalization is a decomposition of two matrices A and B into the product of an orthogonal matrix, a diagonal matrix with nonnegative elements, and a nonsingular matrix that is the same for both matrices. The second generalization requires the definition of a matrix M being T orthogonal where T is a positive-definite matrix. Given positive-definite matrices, the generalization of the singular value decomposition consists of obtaining a matrix as the product of an S orthogonal matrix, a diagonal matrix, and a T orthogonal matrix. Relative eigenvalues will be useful for these generalizations of the singular value decomposition. For two matrices A and B with A symmetric and B positive definite, the relative eigenvalues and eigenvectors are those of AB^{-1}. We first take up relative eigenvalues and eigenvectors and then take up each of the two generalizations of the singular value decomposition.

11.2 RELATIVE EIGENVALUES AND EIGENVECTORS

Let A be an $n \times n$ symmetric matrix and let B be an $n \times n$ positive-definite matrix. Then the scalars λ and vectors x such that

$$(A - \lambda B)x = 0 \tag{11.1}$$

Matrix Algebra for Linear Models, First Edition. Marvin H. J. Gruber.
© 2014 John Wiley & Sons, Inc. Published 2014 by John Wiley & Sons, Inc.

are the relative eigenvalues and eigenvectors of A with respect to the positive-definite matrix B. In order that there exists nonzero x such that (11.1) holds true, λ must satisfy

$$\det(A - \lambda B) = 0. \qquad (11.2)$$

This is equivalent to obtaining the eigenvalues of AB^{-1} or $B^{-1}A$ when both matrices are nonsingular. When B is the identity matrix, the relative eigenvalues are the usual eigenvalues studied previously.

Example 11.1 Obtaining Relative Eigenvalues and Eigenvectors

Let $A = \begin{bmatrix} 1 & 2 \\ 2 & 4 \end{bmatrix}$, $B = \begin{bmatrix} 3 & 1 \\ 1 & 3 \end{bmatrix}$. Then

$$
\begin{aligned}
\det(A - \lambda B) &= \begin{vmatrix} 1-3\lambda & 2-\lambda \\ 2-\lambda & 4-3\lambda \end{vmatrix} \\
&= (1-3\lambda)(4-3\lambda) - (2-\lambda)^2 \\
&= 11\lambda - 8\lambda^2 = 0
\end{aligned}
$$

for $\lambda = 11/8, 0$.

Now

$$Ax = 0 \text{ when } x_1 + 2x_2 = 0 \text{ or } x_1 = -2,\ x_2 = 1.$$

Furthermore,

$$
\begin{bmatrix} 1 & 2 \\ 2 & 4 \end{bmatrix} - \frac{11}{8}\begin{bmatrix} 3 & 1 \\ 1 & 3 \end{bmatrix} = \begin{bmatrix} -\dfrac{25}{8} & \dfrac{5}{8} \\ \dfrac{5}{8} & -\dfrac{1}{8} \end{bmatrix}
$$

and thus $-5x_1 + x_2 = 0$, which holds true when $x_1 = 1$, $x_2 = 5$. The normalized eigenvectors are

$$
v_1 = \begin{bmatrix} \dfrac{1}{\sqrt{26}} \\ \dfrac{5}{\sqrt{26}} \end{bmatrix} \quad \text{and} \quad v_2 = \begin{bmatrix} -\dfrac{2}{\sqrt{5}} \\ \dfrac{1}{\sqrt{5}} \end{bmatrix}.
$$

Notice that these eigenvectors are not orthogonal.   □

Example 11.2 Relative Eigenvalues and Eigenvectors: Another Example

Let

$$
H = \begin{bmatrix} 1 & 2 & 3 \\ 2 & 4 & 6 \\ 3 & 6 & 9 \end{bmatrix}, K = \begin{bmatrix} 5 & 2 & 2 \\ 2 & 5 & 2 \\ 2 & 2 & 5 \end{bmatrix}.
$$

Then

$$\det(H - \lambda K) = \begin{vmatrix} 1-5\lambda & 2-2\lambda & 3-2\lambda \\ 2-2\lambda & 4-5\lambda & 6-2\lambda \\ 3-2\lambda & 6-2\lambda & 9-5\lambda \end{vmatrix} = 162\lambda^2 - 81\lambda^3 = 0.$$

The relative eigenvalues are $\lambda = 2$, $\lambda = 0,0$. For $\lambda = 2$, the eigenvectors are obtained by solving

$$\begin{bmatrix} -9 & -2 & -1 \\ -2 & -6 & 2 \\ -1 & 2 & -1 \end{bmatrix} \begin{bmatrix} y_1 \\ y_2 \\ y_2 \end{bmatrix} = \begin{bmatrix} 0 \\ 0 \\ 0 \end{bmatrix}$$

or

$$-9y_1 - 2y_2 - y_3 = 0$$
$$-2y_1 - 6y_2 + 2y_3 = 0.$$
$$-y_1 + 2y_2 - y_3 = 0$$

Adding the first and third equations, we have that $y_3 = -5y_1$. From substitution into the first equation, we get $y_2 = -2y_1$. Thus, a relative eigenvector is $[1 \quad -2 \quad -5]'$. For $\lambda = 0$, we must solve $Ay = 0$ or

$$\begin{bmatrix} 1 & 2 & 3 \\ 2 & 4 & 6 \\ 3 & 6 & 9 \end{bmatrix} \begin{bmatrix} y_1 \\ y_2 \\ y_3 \end{bmatrix} = \begin{bmatrix} 0 \\ 0 \\ 0 \end{bmatrix}$$

or

$$y_1 + 2y_2 + 3y_3 = 0.$$

Since H is a rank one matrix, there are two arbitrary variables. Two relative eigenvectors are $[-2 \quad 1 \quad 0]'$ and $[-3 \quad 0 \quad 1]'$. □

When a matrix A is symmetric and B is positive definite, the matrices can be simultaneously diagonalized by a nonsingular matrix. The details are given in Theorem 11.1.

Theorem 11.1 Let A be an $n \times n$ symmetric matrix and let B be an $n \times n$ positive-definite matrix. Then there exists an $n \times n$ nonsingular matrix Q such that $B = QQ'$ and $A = QDQ'$ where D is a diagonal matrix.

Proof. Let R be any nonsingular matrix such that $B^{-1} = R'R$. Let $C = RAR'$. Let P be the orthogonal matrix that diagonalizes C so that $C = PDP'$ and $P'CP = D$ where D is a diagonal matrix (the matrix whose diagonal elements are the eigenvalues of C). Let $Q = R^{-1}P$. It will be shown that this is the desired Q. Observe that

$$QQ' = R^{-1}PP'(R')^{-1} = R^{-1}(R')^{-1} = B \tag{11.3}$$

and that

$$QDQ' = R^{-1}PP'CPP'(R')^{-1} = R^{-1}C(R')^{-1} = R^{-1}RAR'(R')^{-1} = A. \tag{11.4}$$

∎

The following result uses Theorem 11.1 to give another proof of Theorem 8.8, a result that will be useful for comparison of the efficiency of estimators in Part V.

Theorem 11.2 Let A and B be two positive-definite matrices such that $A \leq B$. Then $B^{-1} \leq A^{-1}$.

Proof. Use the decomposition in Theorem 11.1. Since $A \leq B$, it follows that $D \leq I$. Then $I \leq D^{-1}$. Now $B^{-1} = Q'^{-1}IQ^{-1} \leq Q'^{-1}D^{-1}Q^{-1} = A^{-1}$.

The following result follows from Theorem 9.6 for positive-definite $m \times m$ matrices A and B. ∎

Theorem 11.3 Suppose the relative eigenvalues of A and B are

$\theta_1 \geq \theta_2 \geq \ldots \geq \theta_m$. Then

$$\theta_m \leq \frac{x'Ax}{x'Bx} \leq \theta_1. \tag{11.5}$$

Proof. Let Θ be the diagonal matrix of relative eigenvalues. From Theorem 11.1,

$$\frac{x'Ax}{x'Bx} = \frac{x'T\Theta T'x}{x'TT'x} = \frac{y'\Theta y}{y'y}. \tag{11.6}$$

From Theorem 9.6,

$$\theta_m \leq \frac{y'\Theta y}{y'y} \leq \theta_1. \tag{11.7}$$

Putting (11.6) and (11.7) together, the result (11.5) follows. ∎

Example 11.3 Generalized Singular Value Decomposition

Let A and B be matrices as in Example 11.1. Then

$$B^{-1} = \begin{bmatrix} \dfrac{3}{8} & -\dfrac{1}{8} \\[2mm] -\dfrac{1}{8} & \dfrac{3}{8} \end{bmatrix}.$$

The singular value decomposition of

$$B^{-1} = \begin{bmatrix} -\dfrac{1}{\sqrt{2}} & \dfrac{1}{\sqrt{2}} \\ \dfrac{1}{\sqrt{2}} & \dfrac{1}{\sqrt{2}} \end{bmatrix} \begin{bmatrix} \dfrac{1}{2} & 0 \\ 0 & \dfrac{1}{4} \end{bmatrix} \begin{bmatrix} -\dfrac{1}{\sqrt{2}} & \dfrac{1}{\sqrt{2}} \\ \dfrac{1}{\sqrt{2}} & \dfrac{1}{\sqrt{2}} \end{bmatrix}.$$

Let

$$R = \begin{bmatrix} \dfrac{1}{\sqrt{2}} & 0 \\ 0 & \dfrac{1}{\sqrt{2}} \end{bmatrix} \begin{bmatrix} -\dfrac{1}{\sqrt{2}} & \dfrac{1}{\sqrt{2}} \\ \dfrac{1}{\sqrt{2}} & \dfrac{1}{\sqrt{2}} \end{bmatrix} = \begin{bmatrix} -\dfrac{1}{2} & \dfrac{1}{2} \\ \dfrac{1}{2\sqrt{2}} & \dfrac{1}{2\sqrt{2}} \end{bmatrix}.$$

Let

$$C = RAR' = \begin{bmatrix} -\dfrac{1}{2} & \dfrac{1}{2} \\ \dfrac{1}{2\sqrt{2}} & \dfrac{1}{2\sqrt{2}} \end{bmatrix} \begin{bmatrix} 1 & 2 \\ 2 & 4 \end{bmatrix} \begin{bmatrix} -\dfrac{1}{2} & \dfrac{1}{2\sqrt{2}} \\ \dfrac{1}{2} & \dfrac{1}{2\sqrt{2}} \end{bmatrix}$$

$$= \begin{bmatrix} \dfrac{1}{4} & \dfrac{3}{4\sqrt{2}} \\ \dfrac{3}{4\sqrt{2}} & \dfrac{9}{8} \end{bmatrix}.$$

The singular value decomposition of

$$C = \begin{bmatrix} \sqrt{\dfrac{2}{11}} & -\dfrac{3}{\sqrt{11}} \\ \dfrac{3}{\sqrt{11}} & \sqrt{\dfrac{2}{11}} \end{bmatrix} \begin{bmatrix} \dfrac{11}{8} & 0 \\ 0 & 0 \end{bmatrix} \begin{bmatrix} \sqrt{\dfrac{2}{11}} & \dfrac{3}{\sqrt{11}} \\ -\dfrac{3}{\sqrt{11}} & \sqrt{\dfrac{2}{11}} \end{bmatrix}.$$

Thus,

$$P = \begin{bmatrix} \sqrt{\dfrac{2}{11}} & -\dfrac{3}{\sqrt{11}} \\ \dfrac{3}{\sqrt{11}} & \sqrt{\dfrac{2}{11}} \end{bmatrix}$$

and

$$D = \begin{bmatrix} \dfrac{11}{8} & 0 \\ 0 & 0 \end{bmatrix}.$$

Take

$$Q = R^{-1}P = \begin{bmatrix} -1 & \sqrt{2} \\ 1 & \sqrt{2} \end{bmatrix} \begin{bmatrix} \sqrt{\dfrac{2}{11}} & -\dfrac{3}{\sqrt{11}} \\ \dfrac{3}{\sqrt{11}} & \sqrt{\dfrac{2}{11}} \end{bmatrix} = \begin{bmatrix} 2\sqrt{\dfrac{2}{11}} & \dfrac{5}{\sqrt{11}} \\ 4\sqrt{\dfrac{2}{11}} & -\dfrac{1}{\sqrt{11}} \end{bmatrix}.$$

The reader may verify that $B = QQ'$ and $A = QDQ'$ □

11.3 GENERALIZATIONS OF THE SINGULAR VALUE DECOMPOSITION: OVERVIEW

Two generalizations of the singular value decomposition will be presented in Subsections 11.4 and 11.5 together with some examples and statistical applications. These generalizations are described in Van Loan (1976).

First a generalization of singular values, the B singular values, must be defined. The B singular values of a matrix A are the square roots of the relative eigenvalues of $A'A$ and $B'B$. A more formal definition is given in Definition 11.1.

Definition 11.1 The B singular values of a matrix A are the elements of the set $\mu(A, B)$ defined by

$$\mu(A, B) = \{\mu \mid \mu \geq 0, \det(A'A - \mu^2 B'B) = 0\}$$

where the dimension of A is $m_a \times n$ and the dimension of B is $m_b \times n$ with $m_a \geq n$.

Note that if B is the identity matrix, then the singular values would be those in the singular value decomposition that we have already presented.   ●

Example 11.4 Singular Values of One Matrix Relative to Another

Let H and K be as in Example 11.2. Let us find the K singular values of the matrix H. Now

$$H'H = \begin{bmatrix} 14 & 28 & 42 \\ 28 & 56 & 84 \\ 42 & 84 & 126 \end{bmatrix}, \quad K'K = \begin{bmatrix} 33 & 24 & 24 \\ 24 & 33 & 24 \\ 24 & 24 & 33 \end{bmatrix}$$

and

$$\det(H'H - \mu^2 K'K) = \begin{vmatrix} 14 - 33\mu^2 & 28 - 24\mu^2 & 42 - 24\mu^2 \\ 28 - 24\mu^2 & 56 - 33\mu^2 & 84 - 24\mu^2 \\ 42 - 24\mu^2 & 84 - 24\mu^2 & 126 - 33\mu^2 \end{vmatrix}$$

$$= 34{,}020\mu^4 - 6561\mu^6$$

$$= \mu^4(34{,}020 - 656\mu^2) = 0.$$

The K singular values of H are $\mu = 0$ and $\mu = \sqrt{140/27} = 2\sqrt{35}/3\sqrt{3}$. □

11.4 THE FIRST GENERALIZATION

The B singular value decomposition is a decomposition of two matrices A and B into the product of an orthogonal matrix, a diagonal matrix with nonnegative elements, and a nonsingular matrix that is the same for both of the matrices. When B is the identity matrix, the B singular value decomposition of A reduces to the singular value decomposition that has already been discussed. Theorem 11.4 defines and proves the existence of the B singular value decomposition. The statement and proof of the theorem are based on that of Van Loan (1976).

Up to now diagonal matrices have always been square matrices. In Theorem 11.4 the diagonal matrices may be rectangular diagonal matrices where all elements not in the ith row and column are zero. There may or not be zero diagonal elements. Two examples of rectangular diagonal matrices are

$$\begin{bmatrix} 1 & 0 & 0 \\ 0 & 2 & 0 \\ 0 & 0 & 3 \\ 0 & 0 & 0 \end{bmatrix} \quad \text{and} \quad \begin{bmatrix} 4 & 0 & 0 & 0 \\ 0 & 3 & 0 & 0 \\ 0 & 0 & 2 & 0 \end{bmatrix}.$$

Theorem 11.4 Let A and B be real matrices with dimensions $m_a \times n$ ($m_a \geq n$) and $m_b \times n$, respectively. There exist orthogonal matrices U of dimension $m_a \times m_a$ and V of dimension $m_b \times m_b$, a nonsingular matrix X of dimension $n \times n$, and diagonal matrices D_A and D_B with nonnegative elements such that

$$U'AX = D_A = \text{diag}(\alpha_1, \ldots, \alpha_n), \quad \alpha_i \geq 0 \tag{11.8a}$$

or

$$A = UD_A X^{-1} \tag{11.8b}$$

and

$$V'BX = D_B = \text{diag}(\beta_1, \ldots, \beta_q), \quad \beta_i \geq 0 \tag{11.9a}$$

or

$$B = VD_B X^{-1}.$$ (11.9b)

The matrix D_A has dimension $m_a \times n$. The matrix D_B has dimension $m_b \times q$ where $q = \min\{m_b, n\}$. Thus, if $m_a \neq n$ or $m_b \neq q$, D_A and D_B would be rectangular diagonal matrices. Matrix B has rank r, and as a result for the diagonal elements of D_B, $\beta_1 \geq \beta_2 \geq \ldots \geq \beta_r > \beta_{r+1} = \ldots \beta_q = 0$. If $\alpha_j = 0$ for any j, $r+1 \leq j \leq n$, then $\mu(A, B) = \{\mu | \mu \geq 0\}$. Otherwise, $\mu(A, B) = \{\alpha_i / \beta_i | i = 1, 2, \ldots, r\}$.

Proof. For the singular value decomposition of $\begin{bmatrix} A \\ B \end{bmatrix}$, we have

$$Q' \begin{bmatrix} A \\ B \end{bmatrix} Z_1 = \Gamma = \mathrm{diag}(\gamma_1, \ldots, \gamma_n)$$ (11.10)

where the diagonal elements of Γ are such that $\gamma_1 \geq \gamma_2 \geq \cdots \gamma_k > \gamma_{k+1} = \cdots \gamma_n = 0$ where the rank of $\begin{bmatrix} A \\ B \end{bmatrix} = k$. Partition the orthogonal $n \times n$ matrix Z_1 into an $n \times k$ matrix Z_{11} and an $n \times (n-k)$ dimensional matrix Z_{12}; that is,

$$Z_1 = [Z_{11} \quad Z_{12}].$$ (11.11)

Let $A_1 = A Z_{11} D^{-1}$ where D is the submatrix of Γ with positive diagonal elements. Let $B_1 = B Z_{11} D^{-1}$. The reader is invited to show in Exercise 11.4 that $A_1' A_1 + B_1' B_1 = I_k$. The singular value decomposition of

$$B_1 = V \begin{bmatrix} \beta_1 & 0 & 0 \\ 0 & \ddots & 0 \\ 0 & 0 & \beta_p \end{bmatrix} Z_2, p = \min\{m_b, k\}$$ (11.12)

where the first r elements are positive and the remaining elements are zero. It follows that

$$V' B Z_1 \begin{bmatrix} D^{-1} Z_2 & 0 \\ 0 & I_{n-k} \end{bmatrix} = \begin{bmatrix} \beta_1 & 0 & 0 \\ 0 & \ddots & 0 \\ 0 & 0 & \beta_q \end{bmatrix} = D_B$$ (11.13)

where $q = \min\{m_b, n\}$ and elements $\beta_{p+1} = \ldots \beta_q = 0$ (Exercise 11.4). Now since $A_1' A_1 + B_1' B_1 = I_k$, it follows that

$$(A_1 Z_2)'(A_1 Z_2) = Z_2'(I - B_1' B_1) Z_2 = \begin{bmatrix} 1 - \beta_1^2 & 0 & 0 \\ 0 & \ddots & 0 \\ 0 & 0 & 1 - \beta_k^2 \end{bmatrix}$$ (11.14)

and that the columns of $A_1 Z_2$ are mutually orthogonal. Let $\alpha_i = \sqrt{1 - \beta_i^2}$.
The reader may show in Exercise (11.5) that

$$U = A_1 Z_2 \begin{bmatrix} \dfrac{1}{\alpha_1} & 0 & 0 \\ 0 & \ddots & 0 \\ 0 & 0 & \dfrac{1}{\alpha_1} \end{bmatrix} \qquad (11.15)$$

is an orthogonal matrix. Define $\alpha_i = 0$, for $i = k+1, \ldots, n$. It follows that

$$U'AZ_1 \begin{bmatrix} D^{-1}Z_2 & 0 \\ 0 & I_{n-k} \end{bmatrix} = D_A = \begin{bmatrix} \alpha_1 & 0 & 0 & 0 \\ 0 & \ddots & 0 & 0 \\ 0 & 0 & \alpha_k & 0 \\ 0 & 0 & 0 & 0 \end{bmatrix}. \qquad (11.16)$$

The result of the theorem follows for

$$X = Z_1 \begin{bmatrix} D^{-1}Z_2 & 0 \\ 0 & I_{n-k} \end{bmatrix} \qquad (11.17)$$

in (11.8) and (11.10). ∎

The following corollary is easy to establish (Exercise 11.6).

Corollary 11.1 Let A and B and notations be as in Theorem 11.1. Then

$$\begin{aligned} X'A'AX = D_A' D_A \quad &\text{or} \quad A'A = X'^{-1} D_A' D_A X^{-1} \\ X'B'BX = D_B' D_B \quad &\text{or} \quad B'B = X'^{-1} D_B' D_B X^{-1}. \end{aligned} \qquad (11.18)$$

Thus, the matrix X constructed in the proof of Theorem 11.4 simultaneously diagonalizes $A'A$ and $B'B$. ∎

Example 11.5 Generalized Singular Value Decomposition of Two Specific Matrices

Notations in this example are the same as in the proof of Theorem 11.4.

Let $A = \begin{bmatrix} 2 & 1 \\ 1 & 2 \end{bmatrix}$, $B = \begin{bmatrix} 3 & 1 \\ 1 & 3 \end{bmatrix}$. Then, after obtaining the singular value decomposition of $\begin{bmatrix} A \\ B \end{bmatrix}$, we find that

$$Z_1 = \begin{bmatrix} \dfrac{1}{\sqrt{2}} & -\dfrac{1}{\sqrt{2}} \\ \dfrac{1}{\sqrt{2}} & \dfrac{1}{\sqrt{2}} \end{bmatrix}$$

and

$$D = \begin{bmatrix} 5 & 0 \\ 0 & \sqrt{5} \end{bmatrix}.$$

The rank of Z_1 is 2 so $Z_1 = Z_{11}$.
  The matrix

$$A_1 = AZ_{11}D^{-1} = \begin{bmatrix} \dfrac{3}{5\sqrt{2}} & -\dfrac{1}{\sqrt{10}} \\ \dfrac{3}{5\sqrt{2}} & \dfrac{1}{\sqrt{10}} \end{bmatrix}$$

and

$$B_1 = BZ_{11}D^{-1} = \begin{bmatrix} \dfrac{2\sqrt{2}}{5} & -\sqrt{\dfrac{2}{5}} \\ \dfrac{2\sqrt{2}}{5} & \sqrt{\dfrac{2}{5}} \end{bmatrix}.$$

In the singular value decomposition of B_1

$$V = \begin{bmatrix} -\dfrac{1}{\sqrt{2}} & \dfrac{1}{\sqrt{2}} \\ \dfrac{1}{\sqrt{2}} & \dfrac{1}{\sqrt{2}} \end{bmatrix},$$

the diagonal matrix is

$$D_B = \begin{bmatrix} \dfrac{2}{\sqrt{5}} & 0 \\ 0 & \dfrac{4}{5} \end{bmatrix},$$

and

$$Z_2 = \begin{bmatrix} 0 & 1 \\ 1 & 0 \end{bmatrix}.$$

Also

$$D_A = \begin{bmatrix} \dfrac{1}{\sqrt{5}} & 0 \\ 0 & \dfrac{3}{5} \end{bmatrix}.$$

The diagonal matrix above is equal to D_B because the matrices are of full rank. Also

$$A_1 Z_2 = \begin{bmatrix} -\dfrac{1}{\sqrt{10}} & \dfrac{3}{5\sqrt{2}} \\[2ex] \dfrac{1}{\sqrt{10}} & \dfrac{3}{5\sqrt{2}} \end{bmatrix}$$

and

$$U = \begin{bmatrix} -\dfrac{1}{\sqrt{2}} & \dfrac{1}{\sqrt{2}} \\[2ex] \dfrac{1}{\sqrt{2}} & \dfrac{1}{\sqrt{2}} \end{bmatrix}.$$

Now

$$X = \begin{bmatrix} -\dfrac{1}{\sqrt{10}} & \dfrac{1}{5\sqrt{2}} \\[2ex] \dfrac{1}{\sqrt{10}} & \dfrac{1}{5\sqrt{2}} \end{bmatrix}.$$

Then the generalized singular value decomposition of

$$A = \begin{bmatrix} 2 & 1 \\ 1 & 2 \end{bmatrix} = \begin{bmatrix} -\dfrac{1}{\sqrt{2}} & \dfrac{1}{\sqrt{2}} \\[2ex] \dfrac{1}{\sqrt{2}} & \dfrac{1}{\sqrt{2}} \end{bmatrix} \begin{bmatrix} \dfrac{1}{\sqrt{5}} & 0 \\[2ex] 0 & \dfrac{3}{5} \end{bmatrix} \begin{bmatrix} -\sqrt{\dfrac{5}{2}} & \sqrt{\dfrac{5}{2}} \\[2ex] \sqrt{\dfrac{5}{2}} & \sqrt{\dfrac{5}{2}} \end{bmatrix}$$

and

$$B = \begin{bmatrix} 3 & 1 \\ 1 & 3 \end{bmatrix} = \begin{bmatrix} -\dfrac{1}{\sqrt{2}} & \dfrac{1}{\sqrt{2}} \\[2ex] \dfrac{1}{\sqrt{2}} & \dfrac{1}{\sqrt{2}} \end{bmatrix} \begin{bmatrix} \dfrac{2}{\sqrt{5}} & 0 \\[2ex] 0 & \dfrac{4}{5} \end{bmatrix} \begin{bmatrix} -\sqrt{\dfrac{5}{2}} & \sqrt{\dfrac{5}{2}} \\[2ex] \sqrt{\dfrac{5}{2}} & \sqrt{\dfrac{5}{2}} \end{bmatrix}.$$

Furthermore,

$$A'A = \begin{bmatrix} 5 & 4 \\ 4 & 5 \end{bmatrix} = \begin{bmatrix} -\sqrt{\dfrac{5}{2}} & \dfrac{5}{\sqrt{2}} \\[2ex] \sqrt{\dfrac{5}{2}} & \dfrac{5}{\sqrt{2}} \end{bmatrix} \begin{bmatrix} \dfrac{1}{5} & 0 \\[2ex] 0 & \dfrac{9}{25} \end{bmatrix} \begin{bmatrix} -\sqrt{\dfrac{5}{2}} & \dfrac{5}{\sqrt{2}} \\[2ex] \sqrt{\dfrac{5}{2}} & \dfrac{5}{\sqrt{2}} \end{bmatrix}$$

and

$$B'B = \begin{bmatrix} 10 & 6 \\ 6 & 10 \end{bmatrix} = \begin{bmatrix} -\dfrac{1}{\sqrt{2}} & \dfrac{1}{\sqrt{2}} \\ \dfrac{1}{\sqrt{2}} & \dfrac{1}{\sqrt{2}} \end{bmatrix} \begin{bmatrix} \frac{4}{5} & 0 \\ 0 & \dfrac{16}{25} \end{bmatrix} \begin{bmatrix} -\sqrt{\dfrac{5}{2}} & \dfrac{5}{\sqrt{2}} \\ \dfrac{5}{\sqrt{2}} & \dfrac{5}{\sqrt{2}} \end{bmatrix}.$$

□

11.5   THE SECOND GENERALIZATION

This subsection will take up another generalization of the singular value decomposition. In order to do this, some preliminaries will be necessary.

First, we generalize the idea of orthogonal matrix to one that is orthogonal with respect to another matrix. This generalization is given by Definition 11.2.

Definition 11.2 Let P be a positive-definite $n \times n$ matrix. An $n \times n$ matrix Q is P orthogonal if $Q'PQ = I$.   ●

Clearly, an orthogonal matrix is a special case where P is the identity matrix.

Example 11.6 One Matrix That Is Orthogonal with Respect to Another Matrix

Let

$$P = \begin{bmatrix} 5 & 2 \\ 2 & 7 \end{bmatrix}, \quad Q = \begin{bmatrix} \dfrac{\sqrt{5}}{5} & -\dfrac{2\sqrt{155}}{155} \\ 0 & \dfrac{\sqrt{155}}{31} \end{bmatrix}.$$

The reader may verify that $Q'PQ = I$. Thus, matrix Q is P orthogonal.   □

A non-negative-definite matrix may be written as a product of another matrix and its transpose. One way to do this is the Cholesky decomposition. The matrix in this product has all zeros above the main diagonal. In symbols $A = LL'$ where L is a lower triangular matrix, that is a matrix with all zeros above the main diagonal. For example,

$$\begin{bmatrix} 4 & 1 \\ 1 & 4 \end{bmatrix} = \begin{bmatrix} 2 & 0 \\ \dfrac{1}{2} & \dfrac{\sqrt{15}}{2} \end{bmatrix} \begin{bmatrix} 2 & \dfrac{1}{2} \\ 0 & \dfrac{\sqrt{15}}{2} \end{bmatrix}.$$

There are other ways to write a matrix in the form $A = LL'$ besides the Cholesky decomposition. For example, if the singular value decomposition of $A = U\Lambda U'$, then $A = LL'$ where $L = U\Lambda^{1/2}$. However, this L would generally not be lower triangular.

The machinery is now available to present another generalization of the singular value decomposition. This is contained in Theorem 11.5.

Theorem 11.5 Let A be an $m \times n$ matrix, S a positive-definite $m \times m$ matrix, and T a positive-definite $n \times n$ matrix. Assume that $m \geq n$. There exists an S orthogonal $m \times m$ matrix U and a T orthogonal $n \times n$ matrix V such that

$$A = UDV^{-1} \tag{11.19}$$

where the diagonal matrix D consists of the square roots of the relative eigenvalues of A'SA and T. The decomposition in (11.19) is the S, T singular value decomposition of A.

Proof. Let $S = LL'$ and $T = KK'$ be the Cholesky factorizations of S and T.

Let $C = L'A(K^{-1})'$. The singular value decomposition of

$$C = QDZ' \tag{11.20}$$

Define $U = (L^{-1})'Q$ and $V = (K^{-1})'Z$. The reader is invited to show that U is S orthogonal and V is T orthogonal. Now

$$U^{-1}AV = Q'L'A(K^{-1})Z = Q'CZ = D, \tag{11.21}$$

and as a result

$$A = UDV^{-1}. \tag{11.22}$$

Let d_i^2 be the eigenvalues of C'C. Since

$$\begin{aligned}
\det(A'SA - d_i^2 T) &= \det(A'LL'A - d_i^2 KK') \\
&= \det K \ \det(K^{-1}A'LL'A(K')^{-1} - d_i^2 I)\det K' \\
&= \det K \ \det(C'C - d_i^2 I)\det K',
\end{aligned} \tag{11.23}$$

and $\det K \neq 0$, these eigenvalues are also relative eigenvalues of A'SA and T. ∎

Example 11.7 An S, T Singular Value Decomposition

Let

$$A = \begin{bmatrix} 7 & 1 \\ 1 & 7 \end{bmatrix}.$$

We will find the S, T singular value decomposition of A where

$$S = \begin{bmatrix} 2 & 1 \\ 1 & 2 \end{bmatrix}$$

and

$$T = \begin{bmatrix} 3 & 1 \\ 1 & 3 \end{bmatrix}.$$

For the Cholesky decomposition of S,

$$L = \begin{bmatrix} \sqrt{2} & 0 \\ \dfrac{1}{\sqrt{2}} & \sqrt{\dfrac{3}{2}} \end{bmatrix}$$

and for the Cholesky decomposition of T,

$$K = \begin{bmatrix} \sqrt{3} & 0 \\ \dfrac{1}{\sqrt{3}} & 2\sqrt{\dfrac{2}{3}} \end{bmatrix}.$$

The matrix

$$C = L'A(K')^{-1} = \begin{bmatrix} 5\sqrt{\dfrac{3}{2}} & \sqrt{3} \\ \dfrac{1}{\sqrt{2}} & 5 \end{bmatrix}.$$

The singular value decomposition of

$$C = QDZ' = \begin{bmatrix} \dfrac{\sqrt{3}}{2} & -\dfrac{1}{2} \\ \dfrac{1}{2} & \dfrac{\sqrt{3}}{2} \end{bmatrix} \begin{bmatrix} 4\sqrt{3} & 0 \\ 0 & 3\sqrt{2} \end{bmatrix} \begin{bmatrix} \sqrt{\dfrac{2}{3}} & \dfrac{1}{\sqrt{3}} \\ -\dfrac{1}{\sqrt{3}} & \sqrt{\dfrac{2}{3}} \end{bmatrix}.$$

Now by direct computation

$$U = (L')^{-1}Q = \begin{bmatrix} \dfrac{1}{\sqrt{6}} & -\dfrac{1}{\sqrt{2}} \\ \dfrac{1}{\sqrt{6}} & \dfrac{1}{\sqrt{2}} \end{bmatrix}$$

may be shown to be S orthogonal and

$$V = (K')^{-1}Z = \begin{bmatrix} \dfrac{1}{2\sqrt{2}} & -\dfrac{1}{2} \\ \dfrac{1}{2\sqrt{2}} & \dfrac{1}{2} \end{bmatrix}$$

may be shown to be T orthogonal. Then the S, T singular value decomposition of

$$
A = \begin{bmatrix} \dfrac{1}{\sqrt{6}} & -\dfrac{1}{\sqrt{2}} \\[2mm] \dfrac{1}{\sqrt{6}} & \dfrac{1}{\sqrt{2}} \end{bmatrix} \begin{bmatrix} 4\sqrt{3} & 0 \\ 0 & 3\sqrt{2} \end{bmatrix} \begin{bmatrix} \sqrt{2} & \sqrt{2} \\ -1 & 1 \end{bmatrix}.
$$

$\square$

11.6 SUMMARY

We have defined and given examples of relative eigenvalues and eigenvectors. Two generalizations of the singular value decomposition were presented together with examples.

EXERCISES

11.1 Let

$$
A = \begin{bmatrix} 4 & 1 \\ 1 & 4 \end{bmatrix}, \; B = \begin{bmatrix} 5 & 2 \\ 2 & 5 \end{bmatrix}.
$$

    A. Find Q and D in the decomposition of Theorem 11.4.
    B. Find the generalized SVD of Theorem 11.4.

11.2 For the matrices in 11.1
    A. Find the B singular values of A.
    B. Find the A singular values of B.

11.3 Show that if A and B are positive-definite matrices, the A singular values of B are the reciprocals of the B singular values of A.

11.4 Show that in the proof of Theorem 11.4, $A_1'A_1 + B_1'B_1 = I_K$.

11.5 Verify Equation 11.10.

11.6 Verify Equation 11.12.

11.7 Prove Corollary 11.1.

11.8 Let P be a positive-definite matrix with singular value decomposition $U\Lambda U'$. Let Q be a P orthogonal matrix. Then the matrix $\Lambda^{1/2}U'Q$ is column orthogonal.

11.9 Suppose the pth diagonal element of a positive-semi-definite matrix A is zero. Show by using the Cholesky decomposition that all of the elements of the pth row and the pth column must be zero.

11.10 Find the Cholesky decomposition with positive entries:

A. $M = \begin{bmatrix} 16 & 4 \\ 4 & 9 \end{bmatrix}$

B. $B = \begin{bmatrix} 4 & 1 & 1 \\ 1 & 4 & 1 \\ 1 & 1 & 4 \end{bmatrix}$.

11.11 Suppose the Cholesky decomposition of $A = LL'$. Show that $(L')^{-1}$ is A orthogonal.

11.12 Establish that U and V in Theorem 11.5 are S and T orthogonal, respectively.

PART III

GENERALIZED INVERSES

We have already discussed systems of equations $Ax = b$ where the matrix A is of less than full rank and where the number of equations and the number of unknowns are not equal. When these equations are consistent, they have infinitely many solutions. These solutions may be expressed in the form $x = Gy$ where G is a generalized inverse. This part of the book consisting of Sections 12–16 discusses generalized inverses and their properties with a number of examples that illustrate these properties.

Section 12 tells what a generalized inverse is and gives a number of examples of them. A less than full rank matrix has infinitely many generalized inverses. The infinite number of solutions to systems of equations may be characterized in terms of these generalized inverses. An important generalized inverse is the Moore–Penrose inverse. A less than full rank matrix has infinitely many generalized inverses but only one Moore–Penrose inverse. We define the Moore–Penrose inverse and give a number of illustrations of it.

In Section 13, we show how generalized inverses may be expressed in terms of the Moore–Penrose inverse. We show that the Moore–Penrose inverse is unique. We also show how the generalized inverses can be expressed in terms of the orthogonal matrices and the singular values in the singular value decomposition of the matrix whose generalized inverse is being studied. In addition, least square and minimum norm inverses are introduced.

Section 14 takes up the properties of least square generalized inverses and minimum norm generalized inverses. Least square generalized inverses are the

Matrix Algebra for Linear Models, First Edition. Marvin H. J. Gruber.
© 2014 John Wiley & Sons, Inc. Published 2014 by John Wiley & Sons, Inc.

generalized inverse of a matrix A when $(y-Ax)'(y-Ax)$ is minimized. Minimum norm generalized inverses are generalized inverses that minimize the norm of Ax. The norm is the square root of the inner product of Ax with itself. Relationships between least square generalized inverses, minimum norm generalized inverses, and Moore–Penrose inverses are studied.

Characterizations of generalized inverses using nonsingular matrices and elementary matrices instead of orthogonal matrices are studied in Section 15. We also give and illustrate a formula for the generalized inverse of a partitioned matrix. Recall that for the full rank case the inverse of a partitioned matrix was studied in Subsection 3.5. The formula in Section 15 is the same with generalized inverses replacing ordinary inverses.

Least square estimators for the nonfull rank model are studied in Section 16. Since these estimators are given in terms of generalized inverses they are not unique.

However, certain linear combinations of these estimators are unique; these linear combinations are the estimable parametric functions. We give the different characterizations of the estimable functions with examples. These estimable functions play an important role in the study of analysis of variance in Part IV.

SECTION 12

BASIC IDEAS ABOUT GENERALIZED INVERSES

12.1 INTRODUCTION

This section explains what a generalized inverse is and shows how it is useful for obtaining solutions of systems of linear equations of less than full rank. Two equivalent definitions of a generalized inverse are given. A method of obtaining the generalized inverse of a matrix is presented. Reflexive generalized inverses and Moore–Penrose inverses are introduced. Reflexive generalized inverses are characterized by the fact that the generalized inverse of the generalized inverse is the original matrix. Not all generalized inverses have this property. In addition to being reflexive for any matrix A with Moore–Penrose inverse M, AM and MA are symmetric matrices. An important property of the Moore–Penrose inverse is its uniqueness. This is established in Section 13.

12.2 WHAT IS A GENERALIZED INVERSE AND HOW IS ONE OBTAINED?

We first give an example to illustrate the need for generalized inverses.

Matrix Algebra for Linear Models, First Edition. Marvin H. J. Gruber.
© 2014 John Wiley & Sons, Inc. Published 2014 by John Wiley & Sons, Inc.

Example 12.1 Illustration of a Generalized Inverse

Consider the system of equations

$$x + y + z = 3$$
$$x - y + z = 1.$$
$$2x + 2y + 2z = 6$$

In matrix form, the system is

$$\begin{bmatrix} 1 & 1 & 1 \\ 1 & -1 & 1 \\ 2 & 2 & 2 \end{bmatrix} \begin{bmatrix} x \\ y \\ z \end{bmatrix} = \begin{bmatrix} 3 \\ 1 \\ 6 \end{bmatrix}.$$

The reader may verify that the coefficient matrix is of rank 2, less than full rank. Thus, the coefficient matrix does not have an inverse. The above system of equations has infinitely many solutions. Two of these solutions are

$$\begin{bmatrix} x \\ y \\ z \end{bmatrix} = \begin{bmatrix} \frac{1}{2} & \frac{1}{2} & 0 \\ \frac{1}{2} & -\frac{1}{2} & 0 \\ 0 & 0 & 0 \end{bmatrix} \begin{bmatrix} 3 \\ 1 \\ 6 \end{bmatrix} = \begin{bmatrix} 2 \\ 1 \\ 0 \end{bmatrix}$$

and

$$\begin{bmatrix} x \\ y \\ z \end{bmatrix} = \begin{bmatrix} 0 & 0 & 0 \\ 0 & -\frac{1}{2} & \frac{1}{4} \\ 0 & \frac{1}{2} & \frac{1}{4} \end{bmatrix} \begin{bmatrix} 3 \\ 1 \\ 6 \end{bmatrix} = \begin{bmatrix} 0 \\ 1 \\ 2 \end{bmatrix}.$$

The matrices

$$\begin{bmatrix} \frac{1}{2} & \frac{1}{2} & 0 \\ \frac{1}{2} & -\frac{1}{2} & 0 \\ 0 & 0 & 0 \end{bmatrix} \text{ and } \begin{bmatrix} 0 & 0 & 0 \\ 0 & -\frac{1}{2} & \frac{1}{4} \\ 0 & \frac{1}{2} & \frac{1}{4} \end{bmatrix}$$

are both examples of generalized inverses that will be defined below. □

Given a system of equations $Ax = y$, a matrix G such that $x = Gy$ is a solution is called a generalized inverse. To find the above solutions to the systems of equations in Example 12.1, generalized inverses were obtained.

One method of obtaining a generalized inverse is to:

1. Find a square submatrix of rank equal to the rank of the matrix and obtain its inverse. There may be more than one such submatrix with rank the same as that of the rank

of the matrix. Different choices of such submatrices will yield different generalized inverses. Generalized inverses are not unique. A matrix may have infinitely generalized inverses.

2. Replace that submatrix with the transpose of the inverse and put zeros for the rest of the entries.

3. Transpose the resulting matrix.

That technique was used to obtain both of the generalized inverses in Example 12.1.

Example 12.2 Illustration of Above Technique for Obtaining a Generalized Inverse

Consider the matrix

$$M = \begin{bmatrix} 1 & 1 & 1 \\ 1 & -1 & 1 \\ 2 & 2 & 2 \end{bmatrix}.$$

We see that M has rank 2. The submatrix

$$M_{11} = \begin{bmatrix} 1 & 1 \\ 1 & -1 \end{bmatrix}$$

has inverse

$$M_{11}^{-1} = \begin{bmatrix} \frac{1}{2} & \frac{1}{2} \\ \frac{1}{2} & -\frac{1}{2} \end{bmatrix}.$$

Since M_{11}^{-1} is symmetric, it is equal to its transpose. Replace the submatrix M_{11} by the transpose of its inverse and replace the other elements by zeros to obtain

$$M_1^- = \begin{bmatrix} \frac{1}{2} & \frac{1}{2} & 0 \\ \frac{1}{2} & -\frac{1}{2} & 0 \\ 0 & 0 & 0 \end{bmatrix}.$$

The generalized inverse of a matrix M will be denoted by M^-. The generalized inverse

$$M_2^- = \begin{bmatrix} 0 & 0 & 0 \\ 0 & -\frac{1}{2} & \frac{1}{4} \\ 0 & \frac{1}{2} & \frac{1}{4} \end{bmatrix}$$

was obtained by inverting the 2×2 submatrix in the lower right-hand corner and following the same procedure used to find the first generalized inverse. In fact for matrix M, we could use any 2×2 submatrix that with rows taken from rows one and two or two and three of M. We cannot use the 2×2 s submatrices that use rows one and three because they are of rank one. $\square$

The reader may show that the generalized inverses obtained above satisfy the matrix identity MGM=M. This identity is another way of defining a generalized inverse. This leads to the following equivalent definitions of a generalized inverse and Theorem 12.1 whose proof establishes that two definitions are indeed equivalent.

Definition 12.1 Let A be an $m \times n$ matrix. Let x be an n-dimensional vector and let y be an m-dimensional vector. An $n \times m$ matrix G is a generalized inverse if and only if $x = Gy$ is a solution to the system of equations $Ax = y$.   ●

Definition 12.2 A matrix G is a generalized inverse of A if and only if the matrix identity AGA=A is satisfied.   ●

Clearly, the generalized inverse of a nonsingular matrix is its ordinary unique inverse. We now establish the equivalence of the two definitions of generalized inverse.

Theorem 12.1 Definitions 12.1 and 12.2 are equivalent definitions of a generalized inverse.

Proof. Definition 12.1 implies Definition 12.2. Let M(A) be the totality of column vectors Ax, the column space of A. Assume that $y \in M(A)$. Then $AGy = y$. Let $y = a_j \in M(A)$, where a_j are the columns of A. Then $AGa_i = a_i$ with $1 \le i \le n$. As a result, $AGA = A$.

Definition 12.2 implies Definition 12.1. Assume that AGA=A and that $y \in M(A)$. Let $x = Gy$. It must be shown that it satisfies the equation $Ax = y$. Since $y \in M(A)$, it follows that $AGy = AGAx = y$ and that Gy satisfies $Ax = y$.

One solution to a system of equations is $x = Gy$. If this is true, then for an arbitrary vector z, $x = Gy + (I - GA)z$ is a solution. This is easy to establish because $Ax = AGy + (A - AGA)z$.   ■

Example 12.3 System of Equations with Infinitely Many Solutions

The solutions to the system of equations

$$\begin{bmatrix} 1 & 1 & 1 \\ 1 & -1 & 1 \\ 2 & 2 & 2 \end{bmatrix} \begin{bmatrix} x \\ y \\ z \end{bmatrix} = \begin{bmatrix} 3 \\ 1 \\ 6 \end{bmatrix}$$

include all vectors of the form

$$\begin{bmatrix} x \\ y \\ z \end{bmatrix} = \begin{bmatrix} \frac{1}{2} & \frac{1}{2} & 0 \\ \frac{1}{2} & -\frac{1}{2} & 0 \\ 0 & 0 & 0 \end{bmatrix}\begin{bmatrix} 3 \\ 1 \\ 6 \end{bmatrix} + \left(\begin{bmatrix} 1 & 0 & 0 \\ 0 & 1 & 0 \\ 0 & 0 & 1 \end{bmatrix} - \begin{bmatrix} 1 & 0 & 1 \\ 0 & 1 & 0 \\ 0 & 0 & 0 \end{bmatrix} \right)\begin{bmatrix} w_1 \\ w_2 \\ w_3 \end{bmatrix}.$$

$$= \begin{bmatrix} 2 \\ 1 \\ 0 \end{bmatrix} + \begin{bmatrix} -w_3 \\ 0 \\ w_3 \end{bmatrix}$$

Hence solutions include $x = 2 - w_3$, $y = 1$, and $z = w_3$ for all choices of w_3. For example, if $w_3 = 2$, $x = 0$, $y = 1$, and $z = 2$.

If $x = Gy + (I - GA)z$ is a solution to $Ax = y$ for arbitrary z, it follows that for the particular choice if $z = y$, $x = (G + I - GA)y$ is a solution to $Ax = y$ and $H = G + I - GA$ is also a generalized inverse of A. $\qquad\square$

Example 12.4 Another Generalized Inverse of the Coefficient Matrix in Example 12.3

Let G be as in Example 12.3.
   Then

$$H = \begin{bmatrix} \frac{1}{2} & \frac{1}{2} & 0 \\ \frac{1}{2} & -\frac{1}{2} & 0 \\ 0 & 0 & 0 \end{bmatrix} + \begin{bmatrix} 1 & 0 & 0 \\ 0 & 1 & 0 \\ 0 & 0 & 1 \end{bmatrix} - \begin{bmatrix} 1 & 0 & 1 \\ 0 & 1 & 0 \\ 0 & 0 & 0 \end{bmatrix}.$$

$$= \begin{bmatrix} \frac{1}{2} & \frac{1}{2} & -1 \\ \frac{1}{2} & -\frac{1}{2} & 0 \\ 0 & 0 & 1 \end{bmatrix}$$

The generalized inverse

$$G = \begin{bmatrix} \frac{1}{2} & \frac{1}{2} & 0 \\ \frac{1}{2} & -\frac{1}{2} & 0 \\ 0 & 0 & 0 \end{bmatrix} \text{ of } A = \begin{bmatrix} 1 & 1 & 1 \\ 1 & -1 & 1 \\ 2 & 2 & 2 \end{bmatrix}$$

satisfies the matrix equality $GAG = G$. A generalized inverse that satisfies this matrix identity is a reflexive generalized inverse. Another reflexive generalized inverse is

$$H = \begin{bmatrix} 0 & 0 & 0 \\ \frac{1}{2} & -\frac{1}{2} & 0 \\ \frac{1}{2} & \frac{1}{2} & 0 \end{bmatrix}.$$

$\qquad\square$

A formal definition of a reflexive generalized inverse is given by Definition 12.3.

Definition 12.3 A generalized inverse is reflexive if and only if $GAG = G$. $\qquad\bullet$

12.3 THE MOORE–PENROSE INVERSE

The Moore–Penrose inverse is a generalized inverse that is always unique. It is a reflexive generalized inverse where AG and GA are always symmetric. The formal statement of these facts is given by Definition 12.4.

Definition 12.4 A matrix M is the Moore–Penrose inverse of A if:

1. M is a generalized inverse of A.
2. M is reflexive.
3. MA is symmetric.
4. AM is symmetric.   ●

Generalized inverses will be denoted by A⁻. The Moore–Penrose inverses will be denoted by A⁺. The generalized inverse and the Moore–Penrose inverse of a non-singular matrix are its ordinary inverse.

The generalized inverses in Example 12.2 are reflexive but do not satisfy properties 3 and 4 for a Moore–Penrose inverse.

Example 12.5 The Moore–Penrose Inverse of a Specific Matrix

The Moore–Penrose inverse of

$$M = \begin{bmatrix} 1 & 1 & 1 \\ 1 & -1 & 1 \\ 2 & 2 & 2 \end{bmatrix}$$

is given by

$$M^+ = \frac{1}{20} \begin{bmatrix} 1 & 5 & 2 \\ 2 & -10 & 4 \\ 1 & 5 & 2 \end{bmatrix}.$$

Indeed

$$MM^+ = \frac{1}{20} \begin{bmatrix} 4 & 0 & 8 \\ 2 & 20 & 0 \\ 8 & 0 & 16 \end{bmatrix}$$

is a symmetric matrix. Furthermore,

$$M^+M = \frac{1}{20} \begin{bmatrix} 10 & 0 & 10 \\ 0 & 20 & 0 \\ 10 & 0 & 10 \end{bmatrix},$$

also a symmetric matrix.

Then

$$MM^+M = \frac{1}{20}\begin{bmatrix} 4 & 0 & 8 \\ 0 & 20 & 0 \\ 8 & 0 & 16 \end{bmatrix}\begin{bmatrix} 1 & 1 & 1 \\ 1 & -1 & 1 \\ 2 & 2 & 2 \end{bmatrix} = \begin{bmatrix} 1 & 1 & 1 \\ 1 & -1 & 1 \\ 2 & 2 & 2 \end{bmatrix} = M$$

and

$$M^+MM^+ = \frac{1}{20}\begin{bmatrix} 1 & 5 & 2 \\ 2 & -10 & 4 \\ 1 & 5 & 2 \end{bmatrix}\frac{1}{20}\begin{bmatrix} 4 & 0 & 8 \\ 0 & 20 & 0 \\ 8 & 0 & 16 \end{bmatrix} = \frac{1}{20}\begin{bmatrix} 1 & 5 & 2 \\ 2 & -10 & 4 \\ 1 & 5 & 2 \end{bmatrix} = M^+.$$

Thus, the matrix M satisfies the defining properties of the Moore–Penrose inverse. Also for the system of equations in Example 12.3,

$$\begin{bmatrix} x \\ y \\ z \end{bmatrix} = \frac{1}{20}\begin{bmatrix} 1 & 5 & 2 \\ 2 & -10 & 4 \\ 1 & 5 & 2 \end{bmatrix}\begin{bmatrix} 3 \\ 1 \\ 6 \end{bmatrix} = \begin{bmatrix} 1 \\ 1 \\ 1 \end{bmatrix}.$$

□

Consider the linear model

$$y = X\beta + \varepsilon \tag{12.1}$$

where X is of less than full rank. Then if G is a generalized inverse of $X'X$, the least square estimator is

$$b = GX'y. \tag{12.2}$$

Then there are as many least square estimators as there are generalized inverses. However, certain linear combinations of the least square parameters are unique. The characterization of these linear combinations will be presented in Section 16. An illustration of this will be given in Example 12.6.

Example 12.6 Least Square Estimator in a Less-than-Full-Rank Linear Model

In the linear model, (12.1), let $\beta' = \begin{bmatrix} \mu & \tau_1 & \tau_2 & \tau_2 & \tau_4 \end{bmatrix}'$

$$X = \begin{bmatrix} 1_5 & 1_5 & 0 & 0 & 0 \\ 1_5 & 0 & 1_5 & 0 & 0 \\ 1_5 & 0 & 0 & 1_5 & 0 \\ 1_5 & 0 & 0 & 0 & 1_5 \end{bmatrix}$$

$$X'X = \begin{bmatrix} 20 & 5 & 5 & 5 & 5 \\ 5 & 5 & 0 & 0 & 0 \\ 5 & 0 & 5 & 0 & 0 \\ 5 & 0 & 0 & 5 & 0 \\ 5 & 0 & 0 & 0 & 5 \end{bmatrix}.$$

One generalized inverse is

$$G = \begin{bmatrix} 0 & 0 & 0 & 0 & 0 \\ 0 & \frac{1}{5} & 0 & 0 & 0 \\ 0 & 0 & \frac{1}{5} & 0 & 0 \\ 0 & 0 & 0 & \frac{1}{5} & 0 \\ 0 & 0 & 0 & 0 & \frac{1}{5} \end{bmatrix}.$$

The least square estimator is

$$\hat{\mu} = \bar{Y}.., \quad \hat{\tau}_i = \bar{Y}_{i.}, \quad i = 1, 2, 3, 4.$$

The Moore–Penrose inverse is

$$M = \frac{1}{125} \begin{bmatrix} 4 & 1 & 1 & 1 & 1 \\ 1 & 19 & -6 & -6 & -6 \\ 1 & -6 & 19 & -6 & -6 \\ 1 & -6 & -6 & 19 & -6 \\ 1 & -6 & -6 & -6 & 19 \end{bmatrix}.$$

The least square estimator is

$$\hat{\mu} = \frac{1}{125}\left(4Y.. + \sum_{i=1}^{4} Y_{i.} \right) = \frac{5}{4}\bar{Y}..$$

$$\hat{\tau}_i = \frac{1}{125}\left(Y.. + 19Y_{i.} - 6\sum_{j \neq i} Y_{j.} \right)$$

$$= -\frac{5}{4}\bar{Y}.. + \bar{Y}_{i.}, \quad i = 1, 2, 3, 4$$

While the least square estimators are different, we have that for both least square estimators,

$$\hat{\theta}_i = \hat{\mu} + \hat{\tau}_i = \bar{Y}_{i.}, \quad i = 1, 2, 3, 4. \qquad \square$$

Thus, the linear combinations of the $\hat{\theta}_i$ are the same for both least square estimators.

We will see how to find a Moore–Penrose inverse using the singular value decomposition in Section 13.

12.4 SUMMARY

We have motivated the need for generalized inverses to solve systems of equations of less than full rank. Two equivalent definitions of generalized inverses were given, and their equivalence was established. We introduced reflexive generalized inverses and Moore–Penrose inverses and gave examples of them.

EXERCISES

12.1   **A.** Find two generalized inverses of the matrix $B = \begin{bmatrix} 1 & 1 & 1 \\ 1 & 2 & 3 \\ 2 & 4 & 6 \end{bmatrix}$.

   B. Determine whether the matrices you obtained in A are reflexive.

   C. Show by direct verification of the Penrose axioms that the Moore–Penrose inverse is

$$B^+ = \frac{1}{30} \begin{bmatrix} 40 & -3 & -6 \\ 10 & 0 & 0 \\ -20 & 3 & 6 \end{bmatrix}.$$

12.2   Show that for the matrices

$$G = \begin{bmatrix} \frac{1}{2} & \frac{1}{2} & 0 \\ \frac{1}{2} & -\frac{1}{2} & 0 \\ 0 & 0 & 0 \end{bmatrix} \quad \text{and} \quad A = \begin{bmatrix} 1 & 1 & 1 \\ 1 & -1 & 1 \\ 2 & 2 & 2 \end{bmatrix} \quad \text{that although } G \text{ is a reflexive}$$

generalized inverse of A, GA and AG are not symmetric matrices. Hence G is not the Moore–Penrose inverse of A. Show by direct verification of the defining properties that the Moore–Penrose inverse of A is

$$A^+ = \frac{1}{20} \begin{bmatrix} 1 & 5 & 2 \\ 2 & -10 & 4 \\ 1 & 5 & 2 \end{bmatrix}.$$

12.3   Consider the matrix

$$C = \begin{bmatrix} 1 & -1 \\ -1 & 1 \end{bmatrix}.$$

   A. Show that the matrices $\begin{bmatrix} 1 & 0 \\ 0 & 0 \end{bmatrix}$ and $\begin{bmatrix} 0 & 0 \\ 0 & 1 \end{bmatrix}$ are reflexive generalized inverses of C.

B. Show by direct verification of the defining properties of a Moore–Penrose inverse that

$$C^+ = \begin{bmatrix} \frac{1}{4} & -\frac{1}{4} \\ -\frac{1}{4} & \frac{1}{4} \end{bmatrix}$$

is the Moore–Penrose inverse.

12.4 For Example 12.6

A. Show that M is the Moore–Penrose inverse of $X'X$.

B. Show that still another generalized inverse of $X'X$ is

$$H = \begin{bmatrix} \frac{1}{20} & 0 & 0 & 0 & 0 \\ -\frac{1}{20} & \frac{1}{5} & 0 & 0 & 0 \\ -\frac{1}{20} & 0 & \frac{1}{5} & 0 & 0 \\ -\frac{1}{20} & 0 & 0 & \frac{1}{5} & 0 \\ -\frac{1}{20} & 0 & 0 & 0 & \frac{1}{5} \end{bmatrix}.$$

Show that H is not a reflexive generalized inverse.

C. Find the least square estimator corresponding to this generalized inverse H and verify that $\widehat{\theta}_i = \widehat{\mu} + \widehat{\tau}_i = \overline{Y}_{i\cdot}$, $i = 1, 2, 3, 4$.

12.5 Let α be any scalar. Show that if G is a generalized inverse of A, then $H = G + \alpha \cdot (I - GA)$ is a generalized inverse of A.

SECTION 13

CHARACTERIZATIONS OF GENERALIZED INVERSES USING THE SINGULAR VALUE DECOMPOSITION

13.1 INTRODUCTION

The generalized inverse of any matrix can be displayed in terms of the inverse of the diagonal matrix containing the singular values and the orthogonal matrices of its singular value decomposition. This technique enables the expression of any generalized inverse in terms of the Moore–Penrose inverse. Thus, solutions to systems of linear equations can be expressed in terms of the solution obtained using the Moore–Penrose inverse. Parametric functions of the least square estimator that are linear functions of the dependent variable may be written uniquely using the least square estimator obtained using the Moore–Penrose inverse. In this section we will consider expressions for the Moore–Penrose inverse based on the singular value decomposition. We will also show how to use the matrices in the singular value decomposition to find reflexive, least square, and minimum norm generalized inverses. Most of the material in this section is based on results presented in Gruber (2010).

13.2 CHARACTERIZATION OF THE MOORE–PENROSE INVERSE

In Section 12, there were a number of examples and exercises to verify that a given matrix was a Moore–Penrose inverse of another matrix by verifying the four defining properties. In Theorem 13.1, we give an explicit formula for the Moore–Penrose inverse.

Matrix Algebra for Linear Models, First Edition. Marvin H. J. Gruber.
© 2014 John Wiley & Sons, Inc. Published 2014 by John Wiley & Sons, Inc.

Theorem 13.1 Let X be an $n \times m$ matrix of rank $s \leq m$. Recall that the singular value decomposition of $X = S'\Lambda^{1/2}U'$. Then $X^+ = U\Lambda^{-1/2}S$.

Proof. The proof consists of showing that X^+ satisfies the defining properties for a Moore-Penrose generalized inverse. First, establish that X^+ is a generalized inverse of X. This holds true because

$$XX^+X = S'\Lambda^{1/2}U'U\Lambda^{-1/2}SS'\Lambda^{1/2}U' = S'\Lambda^{1/2}\Lambda^{-1/2}\Lambda^{1/2}U' = S'\Lambda^{1/2}U' = X.$$

Thus, X^+ is a generalized inverse of X. Second, establish the reflexive property. Observe that

$$X^+XX^+ = U\Lambda^{-1/2}SS'\Lambda^{1/2}U'U\Lambda^{-1/2}S = U\Lambda^{-1/2}\Lambda^{1/2}\Lambda^{-1/2}S = U\Lambda^{-1/2}S = X^+.$$

Third, establish that X^+X and XX^+ are symmetric matrices. Observe that

$$X^+X = U\Lambda^{-1/2}SS'\Lambda^{1/2}U' = UU',$$
$$XX^+ = S'\Lambda^{1/2}U'U\Lambda^{-1/2}S = S'S.$$

The matrices UU' and $S'S$ are both symmetric. ∎

The above theorem establishes the existence of the Moore–Penrose inverse by giving an explicit formula for it.

Example 13.1 Using Theorem 13.1 to Obtain the Moore–Penrose Inverse

In Example 9.1 the singular value decomposition of

$$A = \begin{bmatrix} 1 & 1 & 0 & 0 \\ 1 & 0 & 1 & 0 \\ 1 & 0 & 0 & 1 \end{bmatrix}$$

was

$$A = \begin{bmatrix} \dfrac{1}{\sqrt{3}} & \dfrac{1}{\sqrt{2}} & \dfrac{1}{\sqrt{6}} \\ \dfrac{1}{\sqrt{3}} & -\dfrac{1}{\sqrt{2}} & \dfrac{1}{\sqrt{6}} \\ \dfrac{1}{\sqrt{3}} & 0 & -\dfrac{2}{\sqrt{6}} \end{bmatrix} \begin{bmatrix} 2 & 0 & 0 \\ 0 & 1 & 0 \\ 0 & 0 & 1 \end{bmatrix} \begin{bmatrix} \dfrac{3}{2\sqrt{3}} & \dfrac{1}{2\sqrt{3}} & \dfrac{1}{2\sqrt{3}} & \dfrac{1}{2\sqrt{3}} \\ 0 & \dfrac{1}{\sqrt{2}} & -\dfrac{1}{\sqrt{2}} & 0 \\ 0 & \dfrac{1}{\sqrt{6}} & \dfrac{1}{\sqrt{6}} & -\dfrac{2}{\sqrt{6}} \end{bmatrix}.$$

Thus,

$$
A^+ = \begin{bmatrix} \frac{3}{2\sqrt{3}} & 0 & 0 \\ \frac{1}{2\sqrt{3}} & \frac{1}{\sqrt{2}} & \frac{1}{\sqrt{6}} \\ \frac{1}{2\sqrt{3}} & -\frac{1}{\sqrt{2}} & \frac{1}{\sqrt{6}} \\ \frac{1}{2\sqrt{3}} & 0 & -\frac{2}{\sqrt{6}} \end{bmatrix} \begin{bmatrix} \frac{1}{2} & 0 & 0 \\ 0 & 1 & 0 \\ 0 & 0 & 1 \end{bmatrix} \begin{bmatrix} \frac{1}{\sqrt{3}} & \frac{1}{\sqrt{3}} & \frac{1}{\sqrt{3}} \\ \frac{1}{\sqrt{2}} & -\frac{1}{\sqrt{2}} & 0 \\ \frac{1}{\sqrt{6}} & \frac{1}{\sqrt{6}} & -\frac{2}{\sqrt{6}} \end{bmatrix}
$$

$$
= \frac{1}{4} \begin{bmatrix} 1 & 1 & 1 \\ 3 & -1 & -1 \\ -1 & 3 & -1 \\ -1 & -1 & 3 \end{bmatrix}
$$

$\square$

It has already been pointed out that a matrix has only one Moore–Penrose inverse. This uniqueness is now established in Theorem 13.2.

Theorem 13.2 Let A be an $n \times m$ matrix. Then A has only one Moore–Penrose inverse.

Proof. Assume that B and C are Moore–Penrose inverses of A. Then

$$AB = (AB)' = B'A' = B'A'C'A' = (AB)'(AC)' = ABAC = AC.$$

Likewise

$$BA = (BA)' = A'B' = A'C'A'B' = (CA)'(BA)' = CABA = CA.$$

Then

$$B = BAB = BAC = CAC = C.$$

This establishes the uniqueness of the Moore–Penrose inverse.

∎

13.3   GENERALIZED INVERSES IN TERMS OF THE MOORE–PENROSE INVERSE

The material in this subsection is based on results presented in Chapter 2 of Gruber (2010).

Any generalized inverse may be represented in terms of the Moore–Penrose inverse. The representation is given in Theorem 13.3.

Theorem 13.3 A representation of X^- is given by

$$
X^- = \begin{bmatrix} U & V \end{bmatrix} \begin{bmatrix} \Lambda^{-1/2} & A_1 \\ A_2 & A_3 \end{bmatrix} \begin{bmatrix} S \\ T \end{bmatrix},
$$
$$
= X^+ + VA_2 S + UA_1 T + VA_3 T
$$

(13.1)

with $A_1 = U'X^-T'$, $A_2 = V'X^-S'$, and $A_3 = V'X^-T'$.

Proof. The proof requires the following lemma that is interesting in its own right.

Lemma 13.1 For every generalized inverse of X,

$$U'X^-S' = \Lambda^{-1/2}. \tag{13.2}$$

Furthermore,

$$UU'X^-S'S = X^+. \tag{13.3}$$

Proof of the Lemma. Since $XX^-X = X$, it follows that

$$S'\Lambda^{1/2}U'X^-S'\Lambda^{1/2}U' = S'\Lambda^{1/2}U'. \tag{13.4}$$

Multiply the left-hand side of Equation (13.4) by $\Lambda^{-1/2}S$ and the right-hand side by $U\Lambda^{-1/2}$. The result in (13.2) then follows. Multiplication on the left of (13.2) by U and the right by S yields (13.3).

Proof of the Theorem. Applying the lemma, observe that

$$\begin{aligned} X^- &= (UU' + VV')X^-(S'S + T'T) \\ &= UU'X^-S'S + VV'X^-S'S + UU'X^-T'T + VV'X^-T'T \\ &= X^+ + VA_2S + UA_1T + VA_3T, \end{aligned}$$

where the terms in the last expression are as in the statement of the theorem. ∎

Theorem 13.4 gives a method of construction of generalized inverses that satisfy some but not all of the different Penrose axioms.

Theorem 13.4 Let X be a matrix with singular value decomposition

$$X = [S' \quad T'] \begin{bmatrix} \Lambda^{1/2} & 0 \\ 0 & 0 \end{bmatrix} \begin{bmatrix} U' \\ V' \end{bmatrix}. \tag{13.5}$$

Then:

1. The matrix

$$M = [U \quad V] \begin{bmatrix} \Lambda^{-1/2} & C_1 \\ C_2 & C_3 \end{bmatrix} \begin{bmatrix} S \\ T \end{bmatrix} \tag{13.6}$$

   with C_1, C_2, and C_3 arbitrary is a generalized inverse of X.
2. If $C_3 = C_2\Lambda^{1/2}C_1$, the matrix M is a reflexive generalized inverse of X.

3. If $C_1 = 0$, then XM is a symmetric matrix. A generalized inverse that satisfies this property is called a least square G inverse.

4. If $C_2 = 0$, then MX is a symmetric matrix. A generalized inverse that satisfies this condition is called a minimum norm G inverse.

5. If conditions 2, 3, and 4 are all satisfied, then M is the Moore–Penrose inverse of X.

Proof.

1. Straightforward matrix multiplication may be used to verify that the matrix M given in (13.6) is a generalized inverse of X given by (13.5). From (13.6) using the facts that $U'U = I$, $U'V = 0$, $SS' = I$, and $TS' = 0$, it follows that

$$XMX = S'\Lambda^{1/2}U'(U\Lambda^{-1/2}S + VC_2S + UC_1T + VC_3T)S'\Lambda^{1/2}U'$$
$$= S'\Lambda^{1/2}U' = X'.$$

2. Notice that $XM = S'S + S\Lambda^{1/2}C_1T$ and as a result

$$MXM = U\Lambda^{-1/2}S + UC_1T + VC_2S + VC_2\Lambda^{1/2}C_1T = M$$

   if and only if condition 2 holds.

3. Observe that $XM = S'S + S'\Lambda^{1/2}C_1T$. Then $XM = S'S$ if and only if

$$C_1 = 0.$$

4. Notice that $MX = UU' + VC_2\Lambda^{1/2}U'$. Then $MX = UU'$ if and only if

$$C_2 = 0.$$

5. Condition 5 is clear by the formula for the Moore–Penrose inverse. It follows from (13.5) that

$$X'X = [U \quad V]\begin{bmatrix} \Lambda & 0 \\ 0 & 0 \end{bmatrix}\begin{bmatrix} U' \\ V' \end{bmatrix} = U\Lambda U'. \qquad (13.7)$$

∎

The following corollary is a consequence of Theorem 13.4.

Corollary 13.1 1. The Moore–Penrose inverse of $X'X$ is

$$(X'X)^+ = U\Lambda^{-1}U'. \qquad (13.8)$$

2. For every generalized inverse of $X'X$,

$$U'(X'X)^-U = \Lambda^{-1} \qquad (13.9)$$

and consequently

$$UU'(X'X)^- UU' = (X'X)^+. \tag{13.10}$$

3. A representation of $(X'X)^-$ is

$$(X'X)^- = (X'X)^+ + UB_1 V' + VB_1' U' + VB_2 V' \tag{13.11}$$

where $B_1 = U'(X'X)^- V$ and $B_2 = V'(X'X)^- V$.

4. The matrix

$$H = [U \quad V] \begin{bmatrix} \Lambda^{-1} & B_1 \\ B_1' & B_2 \end{bmatrix} \begin{bmatrix} U' \\ V' \end{bmatrix} \tag{13.12}$$

is a generalized inverse of $X'X$. The matrices $HX'X$ and $X'XH$ are symmetric if $B_1 = 0$. The matrix H is a reflexive generalized inverse when $B_2 = B_1 \Lambda B_1'$. ∎

The following facts are useful and easy to verify:

1. If N is a nonsingular matrix, then

$$(UNU')^+ = UN^{-1}U'. \tag{13.13}$$

2. Let P be an orthogonal matrix. Then

$$(PAP')^+ = PA^+P'. \tag{13.14}$$

3. For a matrix of the form

$$M = \begin{bmatrix} A & 0 \\ 0 & B \end{bmatrix}$$

where A and B are square matrices and the 0s are zero matrices of appropriate dimensions

$$M^+ = \begin{bmatrix} A^+ & 0 \\ 0 & B^+ \end{bmatrix}. \tag{13.15}$$

A number of examples illustrating the application of Theorem 13.3 and its corollary will now be given.

Example 13.2 Finding a Moore–Penrose Inverse and a Reflexive Generalized Inverse of a Matrix by Using Its Singular Value Decomposition

Consider the matrix

$$C = \begin{bmatrix} -1 & 1 \\ 1 & -1 \end{bmatrix}.$$

Its singular value decomposition is

$$C = \begin{bmatrix} -1 & 1 \\ 1 & -1 \end{bmatrix} = \begin{bmatrix} \frac{1}{\sqrt{2}} & \frac{1}{\sqrt{2}} \\ -\frac{1}{\sqrt{2}} & \frac{1}{\sqrt{2}} \end{bmatrix} \begin{bmatrix} 2 & 0 \\ 0 & 0 \end{bmatrix} \begin{bmatrix} -\frac{1}{\sqrt{2}} & \frac{1}{\sqrt{2}} \\ \frac{1}{\sqrt{2}} & \frac{1}{\sqrt{2}} \end{bmatrix}.$$

Then the Moore–Penrose inverse is

$$C^+ = \begin{bmatrix} -\frac{1}{\sqrt{2}} & \frac{1}{\sqrt{2}} \\ \frac{1}{\sqrt{2}} & \frac{1}{\sqrt{2}} \end{bmatrix} \begin{bmatrix} \frac{1}{2} & 0 \\ 0 & 0 \end{bmatrix} \begin{bmatrix} \frac{1}{\sqrt{2}} & -\frac{1}{\sqrt{2}} \\ \frac{1}{\sqrt{2}} & \frac{1}{\sqrt{2}} \end{bmatrix} = \begin{bmatrix} -\frac{1}{4} & \frac{1}{4} \\ \frac{1}{4} & -\frac{1}{4} \end{bmatrix}.$$

When you find a Moore–Penrose inverse using the singular value decomposition, you can always check your answer by verifying that your answer satisfies the defining properties.

To construct a reflexive generalized inverse according to Theorem 13.3, we will replace the elements 0 in the first row and second column and the second row first column by 1 in the diagonal matrix of singular values. According to Theorem 13.3, the element in the second row second column must be 2. Then we have

$$C^{-R} = \begin{bmatrix} -\frac{1}{\sqrt{2}} & \frac{1}{\sqrt{2}} \\ \frac{1}{\sqrt{2}} & \frac{1}{\sqrt{2}} \end{bmatrix} \begin{bmatrix} \frac{1}{2} & 1 \\ 1 & 2 \end{bmatrix} \begin{bmatrix} \frac{1}{\sqrt{2}} & -\frac{1}{\sqrt{2}} \\ \frac{1}{\sqrt{2}} & \frac{1}{\sqrt{2}} \end{bmatrix} = \begin{bmatrix} \frac{3}{4} & \frac{9}{4} \\ \frac{1}{4} & \frac{3}{4} \end{bmatrix}.$$

This reflexive generalized inverse does not satisfy the other two axioms for a Moore–Penrose inverse. To get a matrix where CC^- is symmetric, we can replace the 1 in the second column of the first row by zero. We can put anything we want in the first and second columns of the second row. One possibility is

$$C^{-3} = \begin{bmatrix} -\frac{1}{\sqrt{2}} & \frac{1}{\sqrt{2}} \\ \frac{1}{\sqrt{2}} & \frac{1}{\sqrt{2}} \end{bmatrix} \begin{bmatrix} \frac{1}{2} & 0 \\ 1 & 2 \end{bmatrix} \begin{bmatrix} \frac{1}{\sqrt{2}} & -\frac{1}{\sqrt{2}} \\ \frac{1}{\sqrt{2}} & \frac{1}{\sqrt{2}} \end{bmatrix} = \begin{bmatrix} \frac{5}{4} & \frac{3}{4} \\ \frac{7}{4} & \frac{1}{4} \end{bmatrix}.$$

This is a least square generalized inverse that is not reflexive.

To get a generalized inverse where C^-C is symmetric, we write

$$C^{-4} = \begin{bmatrix} -\frac{1}{\sqrt{2}} & \frac{1}{\sqrt{2}} \\ \frac{1}{\sqrt{2}} & \frac{1}{\sqrt{2}} \end{bmatrix} \begin{bmatrix} \frac{1}{2} & 1 \\ 0 & 2 \end{bmatrix} \begin{bmatrix} \frac{1}{\sqrt{2}} & -\frac{1}{\sqrt{2}} \\ \frac{1}{\sqrt{2}} & \frac{1}{\sqrt{2}} \end{bmatrix} = \begin{bmatrix} \frac{1}{4} & \frac{3}{4} \\ 1 & \frac{1}{2} \end{bmatrix}.$$

The matrix C^{-4} is a nonreflexive minimum norm generalized inverse. A generalized inverse that is not reflexive but satisfies Penrose axioms 3 and 4 (which is both a least square and minimum norm generalized inverse) is

$$C^{-43} = \begin{bmatrix} -\frac{1}{\sqrt{2}} & \frac{1}{\sqrt{2}} \\ \frac{1}{\sqrt{2}} & \frac{1}{\sqrt{2}} \end{bmatrix} \begin{bmatrix} \frac{1}{2} & 0 \\ 0 & 2 \end{bmatrix} \begin{bmatrix} \frac{1}{\sqrt{2}} & -\frac{1}{\sqrt{2}} \\ \frac{1}{\sqrt{2}} & \frac{1}{\sqrt{2}} \end{bmatrix} = \begin{bmatrix} \frac{3}{4} & \frac{5}{4} \\ \frac{5}{4} & \frac{3}{4} \end{bmatrix}.$$

□

Example 13.3 Another Computation of a Moore–Penrose Inverse and a Reflexive Generalized Inverse

Consider the matrix and its singular value decomposition

$$\begin{bmatrix} 2 & 1 & 1 \\ 1 & 1 & 0 \\ 1 & 0 & 1 \end{bmatrix} = \begin{bmatrix} \frac{2}{\sqrt{6}} & 0 & -\frac{1}{\sqrt{3}} \\ \frac{1}{\sqrt{6}} & -\frac{1}{\sqrt{2}} & \frac{1}{\sqrt{3}} \\ \frac{1}{\sqrt{6}} & \frac{1}{\sqrt{2}} & \frac{1}{\sqrt{3}} \end{bmatrix} \begin{bmatrix} 3 & 0 & 0 \\ 0 & 1 & 0 \\ 0 & 0 & 0 \end{bmatrix} \begin{bmatrix} \frac{2}{\sqrt{6}} & \frac{1}{\sqrt{6}} & \frac{1}{\sqrt{6}} \\ 0 & -\frac{1}{\sqrt{2}} & \frac{1}{\sqrt{2}} \\ -\frac{1}{\sqrt{3}} & \frac{1}{\sqrt{3}} & \frac{1}{\sqrt{3}} \end{bmatrix}.$$

The Moore–Penrose inverse is

$$\begin{bmatrix} \frac{2}{\sqrt{6}} & 0 & -\frac{1}{\sqrt{3}} \\ \frac{1}{\sqrt{6}} & -\frac{1}{\sqrt{2}} & \frac{1}{\sqrt{3}} \\ \frac{1}{\sqrt{6}} & \frac{1}{\sqrt{2}} & \frac{1}{\sqrt{3}} \end{bmatrix} \begin{bmatrix} \frac{1}{3} & 0 & 0 \\ 0 & 1 & 0 \\ 0 & 0 & 0 \end{bmatrix} \begin{bmatrix} \frac{2}{\sqrt{6}} & \frac{1}{\sqrt{6}} & \frac{1}{\sqrt{6}} \\ 0 & -\frac{1}{\sqrt{2}} & \frac{1}{\sqrt{2}} \\ -\frac{1}{\sqrt{3}} & \frac{1}{\sqrt{3}} & \frac{1}{\sqrt{3}} \end{bmatrix} = \begin{bmatrix} \frac{2}{9} & \frac{1}{9} & \frac{1}{9} \\ \frac{1}{9} & \frac{4}{9} & -\frac{4}{9} \\ \frac{1}{9} & -\frac{4}{9} & \frac{5}{9} \end{bmatrix}.$$

From Corollary 13.1, a reflexive generalized inverse that does not satisfy the other two Moore–Penrose axioms is

$$\begin{bmatrix} \frac{2}{\sqrt{6}} & 0 & -\frac{1}{\sqrt{3}} \\ \frac{1}{\sqrt{6}} & -\frac{1}{\sqrt{2}} & \frac{1}{\sqrt{2}} \\ \frac{1}{\sqrt{6}} & \frac{1}{\sqrt{2}} & \frac{1}{\sqrt{3}} \end{bmatrix} \begin{bmatrix} \frac{1}{3} & 0 & 1 \\ 0 & 1 & 1 \\ 1 & 1 & 4 \end{bmatrix} \begin{bmatrix} \frac{2}{\sqrt{6}} & \frac{1}{\sqrt{6}} & \frac{1}{\sqrt{6}} \\ 0 & -\frac{1}{\sqrt{2}} & \frac{1}{\sqrt{2}} \\ -\frac{1}{\sqrt{3}} & \frac{1}{\sqrt{3}} & \frac{1}{\sqrt{3}} \end{bmatrix}$$

$$= \begin{bmatrix} 0.612747 & -0.578272 & -1.39477 \\ -0.578272 & 1.5438 & 1.36029 \\ -1.39477 & 1.36029 & 3.17679 \end{bmatrix}.$$

We have expressed the answer in numerical form to avoid having sums of dissimilar radicals for elements of the matrix. A generalized inverse that satisfies Penrose axioms 3 and 4 but is not reflexive is

$$
\begin{bmatrix} \frac{2}{\sqrt{6}} & 0 & -\frac{1}{\sqrt{3}} \\ \frac{1}{\sqrt{6}} & -\frac{1}{\sqrt{2}} & \frac{1}{\sqrt{3}} \\ \frac{1}{\sqrt{6}} & \frac{1}{\sqrt{2}} & \frac{1}{\sqrt{3}} \end{bmatrix}
\begin{bmatrix} \frac{1}{3} & 0 & 0 \\ 0 & 1 & 0 \\ 0 & 0 & 4 \end{bmatrix}
\begin{bmatrix} \frac{2}{\sqrt{6}} & \frac{1}{\sqrt{6}} & \frac{1}{\sqrt{6}} \\ 0 & -\frac{1}{\sqrt{2}} & \frac{1}{\sqrt{2}} \\ -\frac{1}{\sqrt{3}} & \frac{1}{\sqrt{3}} & \frac{1}{\sqrt{3}} \end{bmatrix}
= \begin{bmatrix} \frac{14}{9} & -\frac{11}{9} & -\frac{11}{9} \\ -\frac{11}{9} & \frac{17}{9} & \frac{8}{9} \\ -\frac{11}{9} & \frac{8}{9} & \frac{17}{9} \end{bmatrix}.
$$

$\square$

Example 13.4 Relationship between Moore–Penrose Inverse and Other Generalized Inverses

Let

$$
X = \begin{bmatrix} 1_3 & 1_3 & 0 & 0 \\ 1_3 & 0 & 1_3 & 0 \\ 1_3 & 0 & 0 & 1_3 \end{bmatrix}.
$$

Then

$$
X'X = \begin{bmatrix} 9 & 3 \cdot 1_3' \\ 3 \cdot 1_3 & 3I_3 \end{bmatrix}.
$$

A generalized inverse of $X'X$ is

$$
G = \begin{bmatrix} 0 & 0 \\ 0 & \frac{1}{3} \cdot I_3 \end{bmatrix}.
$$

For $X'X$, an eigenvector of zero is $V' = \frac{1}{2}[1 \quad -1 \quad -1 \quad -1]$. Thus,

$$
VV' = \frac{1}{4} \begin{bmatrix} 1 & -1_3' \\ -1_3 & J_3 \end{bmatrix}
$$

and

$$
UU' = I_4 - VV' = \begin{bmatrix} \frac{3}{4} & \frac{1}{4} \cdot 1_3' \\ \frac{1}{4} \cdot 1_3 & I_3 - \frac{1}{4} \cdot J_3 \end{bmatrix}.
$$

The Moore–Penrose inverse of $X'X$ is

$$(X'X)^+ = UU'GUU' = \begin{bmatrix} \dfrac{1}{16} & \dfrac{1}{48} \cdot 1_3' \\[2ex] \dfrac{1}{48} \cdot 1_3 & \dfrac{1}{3}\left(I_3 - \dfrac{5}{16}J_3\right) \end{bmatrix}.$$

For a matrix A, it is possible to express the Moore–Penrose inverse $(A'A)^+$ in terms of powers of $A'A$ in a manner similar to Equation (8.5) using the Cayley–Hamilton theorem. First note that

$$(A'A)^k = U\Lambda U'U\Lambda U' \ldots U\Lambda U' = U\Lambda^k U' \tag{13.16}$$

and

$$(A'A)^+(A'A)^k(A'A)^+ = U\Lambda^{-1}U'U\Lambda^k U'U\Lambda^{-1}U' = U\Lambda^{k-2}U'. \tag{13.17}$$

From the Cayley–Hamilton theorem, $A'A$ satisfies its characteristic equation, so

$$a_0 I + a_1 A'A + \cdots + a_n (A'A)^n = 0. \tag{13.18}$$

Assume that a_r is the first nonzero coefficient. Then we would have

$$a_r (A'A)^r + \cdots + a_n (A'A)^n = 0. \tag{13.19}$$

Multiply Equation (13.19) by $((A'A)^+)^{r-1}$ to obtain

$$a_r A'A + a_{r+1}(A'A)^2 + \cdots + a_n (A'A)^{n-r+1} = 0. \tag{13.20}$$

Pre- and post-multiply (13.20) by $(A'A)^+$ and obtain

$$a_r (A'A)^+ + a_{r+1}UU' + a_{r+2}A'A + \cdots + a_n (A'A)^{n-r-1} = 0. \tag{13.21}$$

Now since $a_r \neq 0$,

$$(A'A)^+ = -\dfrac{1}{a_r}\left(a_{r+1}UU' + a_{r+2}A'A + \cdots + a_n (A'A)^{n-r-1}\right). \tag{13.22}$$

Equation (13.21) is the analogue of Equation (8.5) for the full-rank case.   □

Example 13.5 Using Equation (13.22) to Calculate the Moore–Penrose Inverse of a Matrix

Let $A = \begin{bmatrix} 1 & 1 & 0 \\ 1 & 0 & 1 \end{bmatrix}$.

Then

$$A'A = \begin{bmatrix} 2 & 1 & 1 \\ 1 & 1 & 0 \\ 1 & 0 & 1 \end{bmatrix}$$

$$= \begin{bmatrix} \frac{2}{\sqrt{6}} & 0 & -\frac{1}{\sqrt{3}} \\ \frac{1}{\sqrt{6}} & -\frac{1}{\sqrt{2}} & \frac{1}{\sqrt{2}} \\ \frac{1}{\sqrt{6}} & \frac{1}{\sqrt{2}} & \frac{1}{\sqrt{3}} \end{bmatrix} \begin{bmatrix} 3 & 0 & 0 \\ 0 & 1 & 0 \\ 0 & 0 & 0 \end{bmatrix} \begin{bmatrix} \frac{2}{\sqrt{6}} & \frac{1}{\sqrt{6}} & \frac{1}{\sqrt{6}} \\ 0 & -\frac{1}{\sqrt{2}} & \frac{1}{\sqrt{2}} \\ -\frac{1}{\sqrt{3}} & \frac{1}{\sqrt{3}} & \frac{1}{\sqrt{3}} \end{bmatrix}.$$

We will find the Moore–Penrose inverse of $A'A$ using (13.21). The characteristic equation of $A'A$ is

$$-3\lambda + 4\lambda^2 - \lambda^3 = 0.$$

From Equation (13.21)

$$(A'A)^+ = \frac{1}{3}(4UU' - A'A).$$

Then

$$(A'A)^+ = \frac{1}{3}\left(4 \begin{bmatrix} \frac{2}{\sqrt{6}} & 0 \\ \frac{1}{\sqrt{6}} & -\frac{1}{\sqrt{2}} \\ \frac{1}{\sqrt{6}} & \frac{1}{\sqrt{2}} \end{bmatrix} \begin{bmatrix} \frac{2}{\sqrt{6}} & \frac{1}{\sqrt{6}} & \frac{1}{\sqrt{6}} \\ 0 & -\frac{1}{\sqrt{2}} & \frac{1}{\sqrt{2}} \end{bmatrix} - \begin{bmatrix} 2 & 1 & 1 \\ 1 & 1 & 0 \\ 1 & 0 & 1 \end{bmatrix}\right)$$

$$= \frac{1}{3}\left(4 \begin{bmatrix} \frac{2}{3} & \frac{1}{3} & \frac{1}{3} \\ \frac{1}{3} & \frac{2}{3} & -\frac{1}{3} \\ \frac{1}{3} & -\frac{1}{3} & \frac{2}{3} \end{bmatrix} - \begin{bmatrix} 2 & 1 & 1 \\ 1 & 1 & 0 \\ 1 & 0 & 1 \end{bmatrix}\right)$$

$$= \begin{bmatrix} \frac{2}{9} & \frac{1}{9} & \frac{1}{9} \\ \frac{1}{9} & \frac{5}{9} & -\frac{4}{9} \\ \frac{1}{9} & -\frac{4}{9} & \frac{5}{9} \end{bmatrix}.$$ □

13.4 SUMMARY

We established the existence of the Moore–Penrose inverse by giving an explicit formula using the singular value decomposition. We also established its uniqueness. This was followed by a number of illustrative examples showing how to calculate the Moore–Penrose inverse using our formula. We also showed how to obtain reflexive, least square, and minimum norm generalized inverses and presented some illustrative examples.

EXERCISES

13.1 Show that the matrix A^+ in Example 13.1 satisfies the four Penrose equations.

13.2 **A.** Let A be a matrix with generalized inverse G and B be a matrix with generalized inverse H. Show that $G \otimes H$ is a generalized inverse of $A \otimes B$.

    B. Show that

$$(A \otimes B)^+ = A^+ \otimes B^+.$$

    C. Suppose the singular value decomposition of A is $S'_A \Lambda_A^{1/2} U'_A$ and of B is $S'_B \Lambda_B^{1/2} U'_B$. Show that the singular value decomposition of

$$A \otimes B = \left(S'_A \otimes S'_B\right)\left(\Lambda_A^{1/2} \otimes \Lambda_B^{1/2}\right)\left(U'_A \otimes U'_B\right)$$

and that

$$(A \otimes B)^+ = \left(U_A \otimes U_B\right)\left(\Lambda_A^{-1/2} \otimes \Lambda_B^{-1/2}\right)\left(S_A \otimes S_B\right).$$

13.3 Establish (13.13)–(13.15).

13.4 Let

$$X = \begin{bmatrix} 1 & 1 & 0 & 1 & 0 \\ 1 & 1 & 0 & 0 & 1 \\ 1 & 1 & 0 & 1 & 0 \\ 1 & 0 & 1 & 0 & 1 \\ 1 & 0 & 1 & 1 & 0 \\ 1 & 0 & 1 & 0 & 1 \end{bmatrix}.$$

    A. Find the Moore–Penrose inverse of (1) X (2) X'X (3) XX'.
    B. Find another generalized inverse of X'X.

13.5 Show that
    A. $XX' = S'\Lambda S$.
    B. $(XX')^+ = S'\Lambda^{-1}S$.

13.6 Let A be a nonsingular n-dimensional matrix. Show that $(S'AS)^+ = S'A^{-1}S$.

13.7 **A.** Show that $(X'X + VV')^{-1}$ is a generalized inverse of X'X.
    C. Show that $(X'X + VV')^{-1}$ satisfies Penrose axioms 3 and 4 but is not reflexive.

13.8 Show that if $B_1 = 0$ in Corollary 13.1 defining properties 3 and 4 hold true for a Moore–Penrose inverse. Generalized inverses satisfying these properties whether or not they are reflexive are called least square minimum norm generalized inverses.

13.9   Show that $X^+(X')^+ = (X'X)^+$.

13.10   **A.** Check that the generalized inverses in Example 13.2 satisfy the properties mentioned.

   B. Construct other examples of generalized inverses of the matrix in Example 13.2 that satisfy axioms 1 and 3 only; 1 and 4 only; 1 and 2 only; 1, 2, and 3 but not 4; and 1, 2, and 4 but not 3.

13.11   **A.** Check that the generalized inverses in Example 13.3 satisfy the properties mentioned.

   B. Construct other examples of generalized inverses of the matrix in Example 13.3 that satisfy axioms 1 and 2 only and 1, 3, and 4 but not axiom 2.

13.12   Let

$$A = \begin{bmatrix} 1_2 & 1_2 & 0 & I_2 \\ 1_2 & 0 & 1_2 & I_2 \end{bmatrix}.$$

Using the method of Example 13.5 and (13.21), find $(A'A)^+$.

SECTION 14

LEAST SQUARE AND MINIMUM NORM GENERALIZED INVERSES

14.1 INTRODUCTION

In Section 13, we defined a minimum norm generalized matrix as one that satisfied the third property in Definition 12.4, which stated that if M was a generalized inverse of A, then MA was a symmetric matrix. Likewise, a least square (LS) generalized inverse was one that satisfied the fourth property in Definition 12.4 where AM was a symmetric matrix. In this section, we show that solutions to systems of equations derived using a minimum norm generalized inverse indeed have minimum norm. Likewise, solutions using LS generalized inverse minimize the norm of Ax − y. Thus, we establish two equivalent ways to define minimum norm and LS generalized inverses.

We establish formulae to obtain minimum norm and LS generalized inverses in terms of any generalized inverses. We also establish a formula to obtain the Moore–Penrose inverse in terms of the LS and minimum norm generalized inverse. We give some numerical examples to illustrate the use of these formulae.

Finally, we extend Theorem 7.3 to the case of positive-semi-definite matrices and obtain conditions for $dA - aa'$ to be positive semi-definite where A is a positive-semi-definite matrix, d is a scalar, and a is a column vector. In Theorem 7.3 A was a positive-definite matrix.

Matrix Algebra for Linear Models, First Edition. Marvin H. J. Gruber.
© 2014 John Wiley & Sons, Inc. Published 2014 by John Wiley & Sons, Inc.

14.2 MINIMUM NORM GENERALIZED INVERSES

A minimum norm generalized inverse is one where the solution $x = Gy$ of the system $Ax = y$ has minimum norm as compared to all solutions of the system of equations. Thus, it is the generalized inverse where

$$\min_{Ax=y} \|x\| = \|Gy\|. \tag{14.1}$$

In Theorem 13.4, a minimum norm generalized inverse was one where GA was a symmetric matrix. These definitions are equivalent. This will be shown in Theorem 14.1. The proofs are modeled after those in Rao and Mitra (1971).

Theorem 14.1 A generalized inverse satisfies (14.1) if and only if

$$AGA = A \quad \text{and} \quad (GA)' = (GA).$$

To establish this theorem we prove a simple lemma.

Lemma 14.1 Let A be an $n \times m$ matrix. Let x be an m-dimensional column vector and let y be an n-dimensional column vector. Then

$$(Ax, y) = (x, A'y). \tag{14.2}$$

Proof. The n-dimensional vector Ax is given by

$$Ax = \sum_{j=1}^{m} a_{ij} x_j, \quad 1 \le i \le n. \tag{14.3}$$

The m-dimensional vector $A'y$ is given by

$$A'y = \sum_{i=1}^{n} a_{ij} y_i. \tag{14.4}$$

Then

$$\begin{aligned}
(Ax, y) &= \sum_{i=1}^{n} \left(\sum_{j=1}^{m} a_{ij} x_j \right) y_i = \sum_{i=1}^{n} \sum_{j=1}^{m} a_{ij} x_j y_i \\
&= \sum_{j=1}^{m} \left(\sum_{i=1}^{n} a_{ij} y_i \right) x_j = (A'y, x) \\
&= (x, A'y)
\end{aligned} \tag{14.5}$$

Proof of the Theorem. Recall that the totality of solutions to $Ax = y$ is given by $x = Gy + (I - GA)z$ for arbitrary z. If Gy has minimum norm, then for all y in the column space of A and z in n-dimensional Euclidean space,

$$\|Gy\| \le \|Gy + (I - GA)z\|, \qquad (14.6)$$

or for all b and z,

$$\|GAb\| \le \|GAb + (I - GA)z\|. \qquad (14.7)$$

Now

$$\begin{aligned}\|GAb\|^2 &\le \|GAb + (I - GA)z\|^2 \\ &= \|GAb\|^2 + 2(GAb, (I - GA)Z) + \|(I - GA)z\|^2 \end{aligned} \qquad (14.8)$$

However,

$$(GAb, (I - GA)z) = 0 \qquad (14.9)$$

if and only if GA is symmetric. This holds true because

$$(GAb, (I - GA)z) = (b, (GA)'(I - GA)z). \qquad (14.10)$$

But

$$(GA)'(I - GA) = 0 \qquad (14.11)$$

if and only if

$$(GA)' = (GA)'GA. \qquad (14.12)$$

However,

$$GA = GAGA \qquad (14.13)$$

if and only if AGA=A.
    Furthermore,

$$(GA)' = (GAGA)' = (GA)'(GA) \qquad (14.14)$$

if and only if

$$GA = GAGA = (GA)'(GA), \qquad (14.15)$$

and thus, $(GA) = (GA)'$.
    Observe that

$$\|(I - GA)z\|^2 \ge 0. \qquad (14.16)$$

This completes the proof.   ∎

Theorem 13.4 gave one way of finding a minimum norm generalized inverse. We will give another way in Theorem 14.2.

Theorem 14.2 A minimum norm reflexive generalized inverse of $n \times m$ matrix X is given by $X_m^- = X'(XX')^-$ where $(XX')^-$ is any generalized inverse of XX'.

Proof. We have that

$$XX' = S'\Lambda^{1/2}U'U\Lambda^{1/2}S = S'\Lambda S, \qquad (14.17)$$

and since by the definition of a generalized inverse

$$XX'(XX')^- XX' = XX', \qquad (14.18a)$$

$$S'\Lambda S(XX')^- S'\Lambda S = S'\Lambda S. \qquad (14.18b)$$

Premultiply (14.18b) by $\Lambda^{-1}S$ and postmultiply by $S'\Lambda^{-1}$ to obtain

$$S(XX')^- S' = \Lambda^{-1}. \qquad (14.19)$$

Then using (14.19)

$$\begin{aligned} XX_m^- X &= XX'(XX')^- X = S'\Lambda S(XX')^- S'\Lambda^{1/2}U' \\ &= S'\Lambda\Lambda^{-1}\Lambda^{1/2}U' = S'\Lambda^{1/2}U' = X \end{aligned}.$$

Thus, X_m^- is a generalized inverse.
 Observe that

$$\begin{aligned} X_m^- X &= X'(XX')^- X = U\Lambda^{1/2}SS'\Lambda^{-1}SS'\Lambda^{1/2}U' \\ &= UU' \end{aligned}, \qquad (14.20)$$

a symmetric matrix.
 Also,

$$\begin{aligned} X_m^- X X_m^- &= UU'X'(XX')^- = UU'U\Lambda^{1/2}S(XX')^- \\ &= U\Lambda^{1/2}S(XX')^- = X'(XX')^- = X_m^- \end{aligned},$$

so X_m^- is reflexive. ∎

As will be shown in Example 14.2, it is not generally true that $X_m^- X$ is a symmetric matrix. There is one choice of generalized inverse of XX' that leads to the Moore–Penrose inverse of X.

Corollary 14.1 Suppose that $X_m^- = X'(XX')^+$. Then $X_m^- = X^+$.

Proof. Observe that $X_m^- = U\Lambda^{1/2}SS'\Lambda^{-1}S = U\Lambda^{-1/2}S$, the Moore–Penrose inverse of X. ∎

Example 14.1 Example of a Reflexive Minimum Norm Generalized Inverse

Let

$$X = \begin{bmatrix} 1 & 1 & 0 \\ 1 & 1 & 0 \\ 1 & 0 & 1 \\ 1 & 0 & 1 \end{bmatrix}.$$

Then

$$XX' = \begin{bmatrix} 2 & 2 & 1 & 1 \\ 2 & 2 & 1 & 1 \\ 1 & 1 & 2 & 2 \\ 1 & 1 & 2 & 2 \end{bmatrix}.$$

One generalized inverse of XX' is

$$(XX')^- = \begin{bmatrix} 0 & 0 & 0 & 0 \\ 0 & \frac{2}{3} & -\frac{1}{3} & 0 \\ 0 & -\frac{1}{3} & \frac{2}{3} & 0 \\ 0 & 0 & 0 & 0 \end{bmatrix}.$$

One reflexive minimum norm generalized inverse of X is

$$X_m^- = \begin{bmatrix} 1 & 1 & 1 & 1 \\ 1 & 1 & 0 & 0 \\ 0 & 0 & 1 & 1 \end{bmatrix} \begin{bmatrix} 0 & 0 & 0 & 0 \\ 0 & \frac{2}{3} & -\frac{1}{3} & 0 \\ 0 & -\frac{1}{3} & \frac{2}{3} & 0 \\ 0 & 0 & 0 & 0 \end{bmatrix} = \begin{bmatrix} 0 & \frac{1}{3} & \frac{1}{3} & 0 \\ 0 & \frac{2}{3} & -\frac{1}{3} & 0 \\ 0 & -\frac{1}{3} & \frac{2}{3} & 0 \end{bmatrix}.$$

Now

$$X_m^- X = \begin{bmatrix} 0 & \frac{1}{3} & \frac{1}{3} & 0 \\ 0 & \frac{2}{3} & -\frac{1}{3} & 0 \\ 0 & -\frac{1}{3} & \frac{2}{3} & 0 \end{bmatrix} \begin{bmatrix} 1 & 1 & 0 \\ 1 & 1 & 0 \\ 1 & 0 & 1 \\ 1 & 0 & 1 \end{bmatrix} = \begin{bmatrix} \frac{2}{3} & \frac{1}{3} & \frac{1}{3} \\ \frac{1}{3} & \frac{2}{3} & -\frac{1}{3} \\ \frac{1}{3} & -\frac{1}{3} & \frac{2}{3} \end{bmatrix},$$

a symmetric matrix. Moreover,

$$XX_m^- X = \begin{bmatrix} 1 & 1 & 0 \\ 1 & 1 & 0 \\ 1 & 0 & 1 \\ 1 & 0 & 1 \end{bmatrix} \begin{bmatrix} \frac{2}{3} & \frac{1}{3} & \frac{1}{3} \\ \frac{1}{3} & \frac{2}{3} & -\frac{1}{3} \\ \frac{1}{3} & -\frac{1}{3} & \frac{2}{3} \end{bmatrix} = \begin{bmatrix} 1 & 1 & 0 \\ 1 & 1 & 0 \\ 1 & 0 & 1 \\ 1 & 0 & 1 \end{bmatrix}.$$

Notice that

$$XX_m^- = \begin{bmatrix} 1 & 1 & 0 \\ 1 & 1 & 0 \\ 1 & 0 & 1 \\ 1 & 0 & 1 \end{bmatrix} \begin{bmatrix} 0 & \frac{1}{3} & \frac{1}{3} & 0 \\ 0 & \frac{2}{3} & -\frac{1}{3} & 0 \\ 0 & -\frac{1}{3} & \frac{2}{3} & 0 \end{bmatrix} = \begin{bmatrix} 0 & 1 & 0 & 0 \\ 0 & 1 & 0 & 0 \\ 0 & 0 & 1 & 0 \\ 0 & 0 & 1 & 0 \end{bmatrix}$$

is not a symmetric matrix. Also

$$X_m^- X X_m^- = \begin{bmatrix} 0 & \frac{1}{3} & \frac{1}{3} & 0 \\ 0 & \frac{2}{3} & -\frac{1}{3} & 0 \\ 0 & -\frac{1}{3} & \frac{2}{3} & 0 \end{bmatrix} \begin{bmatrix} 0 & 1 & 0 & 0 \\ 0 & 1 & 0 & 0 \\ 0 & 0 & 1 & 0 \\ 0 & 0 & 1 & 0 \end{bmatrix} = \begin{bmatrix} 0 & \frac{1}{3} & \frac{1}{3} & 0 \\ 0 & \frac{2}{3} & -\frac{1}{3} & 0 \\ 0 & -\frac{1}{3} & \frac{2}{3} & 0 \end{bmatrix} = X_m^-,$$

so that X_m^- is a reflexive generalized inverse. $\qquad\qquad\qquad\qquad$ □

The result of Theorem 14.1 tells us that $X_m^- = X'(XX')^-$ is the generalized inverse where a solution to an equation $Xa = b$ has the minimum norm. However, this result can also be obtained by solving the constrained optimization problem of minimizing $\|a\|^2$ subject to the constraint $Xa = b$. This will be done in Section 22 because it requires the use of Lagrange multipliers, which will be taken up there.

14.3   LEAST SQUARE GENERALIZED INVERSES

Consider an equation $Ax = y$. A solution $\hat{x}$ is an LS solution if

$$\|A\hat{x} - y\| = \inf_x \|Ax - y\|. \tag{14.21}$$

The necessary and sufficient condition for $\hat{x}$ to be an LS solution is given by Theorem 14.3.

Theorem 14.3 Let G be a matrix (not necessarily a generalized inverse) where $\hat{x} = Gy$ is an LS solution of $Ax = y$ for all y in m-dimensional Euclidean space. The necessary and sufficient conditions for $\hat{x} = Gy$ to be an LS solution to $Ax = y$ are $AGA = A$ and that AG is a symmetric matrix.

Outline of the Proof. Since $\hat{x} = Gy$ is an LS solution, it follows that for all x and y,

$$\|AGy - y\| \le \|Ax - y\| \le \|AGy - y + Aw\| \tag{14.22}$$

for all y, where $w = x - Gy$. Inequality (14.22) holds true if and only if for all y and w,

$$(Aw, (AG - I)y) = 0. \tag{14.23}$$

Equation (14.23) holds true if and only if

$$A'AG = A'. \tag{14.24}$$

The reader may now complete the proof by doing Exercise 14.2. ∎

We now show a way to obtain an LS inverse of a matrix X using a generalized inverse of X'X.

Theorem 14.4 An LS reflexive generalized inverse of $n \times m$ matrix X is given by $X_{LS}^- = (X'X)^- X'$ where $(XX')^-$ is any generalized inverse of XX'.

Proof. First, we establish that $X_{LS}^- = (X'X)^- X'$ is a generalized inverse of X. Observe that

$$XX_{LS}^- X = X(X'X)^- X'X = S'\Lambda^{1/2}U'(X'X)^- U\Lambda U'$$
$$= S'\Lambda^{1/2}\Lambda^{-1}\Lambda U' = S'\Lambda^{1/2}U' = X$$

To establish the reflexive property, we have

$$X_{LS}^- XX_{LS}^- = (X'X)^- X'X(X'X)^- X' = (X'X)^- U\Lambda U'(X'X)^- U\Lambda^{1/2}S$$
$$= (X'X)^- U\Lambda\Lambda^{-1}\Lambda^{1/2}S = (X'X)^- U\Lambda^{1/2}S = (X'X)^- X' = X_{LS}^-$$

We must establish that XX_{LS}^- is symmetric. Notice that

$$XX_{LS}^- = X(X'X)^- X' = S'\Lambda^{1/2}U'(X'X)^- U\Lambda^{1/2}S,$$
$$= S'\Lambda^{1/2}\Lambda^{-1}\Lambda^{1/2}S = S'S,$$

a symmetric matrix. ∎

The definition of LS generalized inverse in Theorem 14.4 seems natural in light of the fact that when we minimize

$$m = (Y - X\beta)'(Y - X\beta),$$

we get

$$\hat{\beta} = (X'X)^- X'Y.$$

Again if we use the Moore–Penrose inverse of X'X in $X_{LS}^- = (X'X)^- X'$, we obtain the Moore–Penrose inverse of X. This will be shown in Corollary 14.2.

Corollary 14.2 If $X_{LS}^- = (X'X)^+ X'$, then $X_{LS}^- = X^+$.

Proof. Observe that

$$X_{LS}^- = (X'X)^+ X' = U\Lambda^{-1}U'U\Lambda^{1/2}S = U\Lambda^{-1/2}S = X^+. \qquad ∎$$

The Moore–Penrose inverse may be computed from the minimum norm generalized inverse and the LS generalized inverse as a result of Theorem 14.5.

Theorem 14.5 The Moore–Penrose inverse of any matrix X is given by

$$X^+ = X_m^- X X_{LS}^-. \tag{14.25}$$

Proof. Substituting the formulae for X_m^- and X_{LS}^- into (14.25) and using the singular value decomposition of the matrices, we have

$$X_m^- X X_{LS}^- = X'(XX')^- X(X'X)^- X' = U\Lambda^{1/2} S(XX')^- S'\Lambda^{1/2} U'(X'X)^- U\Lambda^{1/2} S$$
$$= U\Lambda^{1/2}\Lambda^{-1}\Lambda^{1/2}\Lambda^{-1}\Lambda^{1/2} S = U\Lambda^{-1/2} S = X^+ \qquad ∎$$

Example 14.2 Least Square and Moore–Penrose Inverse

We will obtain the LS generalized inverse of

$$X = \begin{bmatrix} 1 & 1 & 0 \\ 1 & 1 & 0 \\ 1 & 0 & 1 \\ 1 & 0 & 1 \end{bmatrix}$$

and then obtain its Moore–Penrose inverse using (14.25) and the result of Example 14.1. Now

$$X'X = \begin{bmatrix} 4 & 2 & 2 \\ 2 & 2 & 0 \\ 2 & 0 & 2 \end{bmatrix}.$$

One choice of generalized inverse of $X'X$ is

$$(X'X)^- = \begin{bmatrix} 0 & 0 & 0 \\ 0 & \frac{1}{2} & 0 \\ 0 & 0 & \frac{1}{2} \end{bmatrix}$$

so that

$$X_L^- = \begin{bmatrix} 0 & 0 & 0 \\ 0 & \frac{1}{2} & 0 \\ 0 & 0 & \frac{1}{2} \end{bmatrix}\begin{bmatrix} 1 & 1 & 1 & 1 \\ 1 & 1 & 0 & 0 \\ 0 & 0 & 1 & 1 \end{bmatrix} = \begin{bmatrix} 0 & 0 & 0 & 0 \\ \frac{1}{2} & \frac{1}{2} & 0 & 0 \\ 0 & 0 & \frac{1}{2} & \frac{1}{2} \end{bmatrix}.$$

The reader should check that X_L^- does indeed satisfy the axioms for an LS generalized inverse.

Then the Moore–Penrose inverse is

$$X^+ = X_m^- X X_L^- = \begin{bmatrix} 0 & \frac{1}{3} & \frac{1}{3} & 0 \\ 0 & \frac{2}{3} & -\frac{1}{3} & 0 \\ 0 & -\frac{1}{3} & \frac{2}{3} & 0 \end{bmatrix} \begin{bmatrix} 1 & 1 & 0 \\ 1 & 1 & 0 \\ 1 & 0 & 1 \\ 1 & 0 & 1 \end{bmatrix} \begin{bmatrix} 0 & 0 & 0 & 0 \\ \frac{1}{2} & \frac{1}{2} & 0 & 0 \\ 0 & 0 & \frac{1}{2} & \frac{1}{2} \end{bmatrix} .$$

$$= \begin{bmatrix} \frac{1}{6} & \frac{1}{6} & \frac{1}{6} & \frac{1}{6} \\ \frac{1}{3} & \frac{1}{3} & -\frac{1}{6} & -\frac{1}{6} \\ -\frac{1}{6} & -\frac{1}{6} & \frac{1}{3} & \frac{1}{3} \end{bmatrix}$$

□

In addition to being reflexive, a Moore–Penrose inverse is both a minimum norm and an LS generalized inverse.

14.4 AN EXTENSION OF THEOREM 7.3 TO POSITIVE- SEMI-DEFINITE MATRICES

The existence of the Moore–Penrose inverse enables the following extension of Theorem 7.3 due to Baksalary and Kala (1983).

Theorem 14.6 Let A be a positive-semi-definite matrix. Let a be an n-dimensional column vector and d be a positive scalar. Then $dA - aa'$ is positive semi-definite if and only if:

1. The vector a belongs to the range of A.
2. The inequality $a'A^+a \le d$ holds true.

Proof. The singular value decomposition of A is given by

$$A = \begin{bmatrix} P & Q \end{bmatrix} \begin{bmatrix} \Lambda & 0 \\ 0 & 0 \end{bmatrix} \begin{bmatrix} P' \\ Q' \end{bmatrix} = P\Lambda P'. \tag{14.26}$$

Then $dA - aa'$ is positive semi-definite if and only if

$$M = d \begin{bmatrix} \Lambda & 0 \\ 0 & 0 \end{bmatrix} - \begin{bmatrix} P'aa'P & P'aa'Q \\ Q'aa'P & Q'aa'Q \end{bmatrix} \tag{14.27}$$

is positive semi-definite. If the matrix M is positive semi-definite, then $d\Lambda - P'aa'P$ also is. Thus, by Theorem 7.3 $a'A^+a \le d$. Furthermore, $Q'aa'Q = 0$. This implies that $Q'a = 0$ and that a belongs to the range of A. On the other hand, if a belongs to the range of A $Q'a = 0$ and the inequality $a'A^+a \le d$ holds true, then $d\Lambda - P'aa'P$ is positive semi-definite and M is positive semi-definite. ■

14.5 SUMMARY

We have established the equivalence of two definitions on minimum norm and LS generalized inverses. We have given formulae for these kinds of inverses and examples of their use. A formula for the Moore–Penrose inverse was established in terms of minimum norm and LS generalized inverses. Finally, we extended a condition for $dA - aa'$ to be positive semi-definite from the case where matrix A was positive definite to the case where it was positive semi-definite.

EXERCISES

14.1 Show that

$$G_{min} = \begin{bmatrix} 0.455342 & -0.122008 & -0.122008 & 0.455342 \\ 0.333333 & 0.333333 & -0.744017 & 0.410684 \\ -0.744017 & 0.410684 & -0.244017 & 0.910684 \end{bmatrix}$$

is a minimum norm generalized inverse of

$$X = \begin{bmatrix} 1 & 1 & 0 \\ 1 & 1 & 0 \\ 1 & 0 & 1 \\ 1 & 0 & 1 \end{bmatrix}.$$

Do the other two axioms for the Moore–Penrose inverse hold true?
A computer algebra system like Mathematica or Maple would be helpful for this problem. There may be some slight differences from what you should get due to rounding off.

14.2 Complete the proof of Theorem 14.2 by filling in the details and showing that AG is symmetric if and only if (14.24) holds true.

14.3 Show that

$$G_{ls} = \begin{bmatrix} 0.402369 & 0.0690356 & -0.0690356 & -0.402369 \\ 0.0976311 & 0.430964 & 0.0690356 & 0.402369 \\ -0.402369 & -0.0690356 & 0.569036 & 0.902369 \end{bmatrix}$$

is an LS generalized inverse of

$$X = \begin{bmatrix} 1 & 1 & 0 \\ 1 & 1 & 0 \\ 1 & 0 & 1 \\ 1 & 0 & 1 \end{bmatrix}$$

Do the other two axioms for the Moore–Penrose inverse hold true?

A computer algebra system like Mathematica or Maple would be helpful for this problem.

14.4 Show that the generalized inverse of the form

$$G = (X'X)^+ + VCV'$$

is both an LS and a minimum norm generalized inverse of $X'X$ that is not reflexive when $C \neq 0$.

14.5 **A.** Show that the generalized inverse

$$(X'X)^- = \frac{1}{18} \begin{bmatrix} 14 & -11 & -11 \\ -11 & 17 & 8 \\ -11 & 8 & 17 \end{bmatrix}$$

of $X'X$ where X is given in Exercise 14.1 is a minimum norm G inverse and an LS G inverse but not reflexive.

B. Show that $H = (X'X)^- X'X(X'X)^-$ is reflexive and is in fact the Moore–Penrose inverse of $X'X$. (Note: In general it is true that H is reflexive (see Exercise 14.9) but it may not be an LS or minimum norm generalized inverse.)

14.6 Using Theorem 13.4, construct another generalized inverse for $X'X$ where X is given in Exercises 14.1 and 14.2 that is a minimum norm G inverse and an LS G inverse but not reflexive.

14.7 **A.** Use the result of Theorem 14.2 to obtain a reflexive minimum norm generalized inverse and a reflexive LS generalized inverse of

1. $X = \begin{bmatrix} 1 & 1 & -1 \\ 1 & 1 & -1 \\ 1 & -1 & 1 \\ 1 & -1 & 1 \end{bmatrix}$

2. $A = \begin{bmatrix} -1 & 1 \\ 1 & -1 \end{bmatrix}.$

B. Use Theorem 14.5 to obtain the Moore–Penrose inverse of X and A.

14.8 Let

$$X = \begin{bmatrix} 1 & 1 & 0 \\ 1 & 0 & 1 \end{bmatrix}.$$

Show that the minimum norm generalized inverse obtained by the method of Theorem 14.2 turns out to be the Moore–Penrose inverse of X.

14.9   **A.** Show that for any matrix A, $A^r = A^- A A^-$ is reflexive generalized inverse of A for any matrix A.

    B. Show that if A is a minimum norm generalized inverse, A^r is also.

    C. Show that if A is an LS generalized inverse, A^r is also.

    D. Show that if A is an LS generalized inverse and a minimum norm generalized inverse not necessarily reflexive, then A^r is the Moore–Penrose inverse of A.

14.10 Show that

    A. $(X')^-_m = (X^-_{LS})'$.

    B. $(X')^-_{LS} = (X^-_m)'$.

SECTION 15

MORE REPRESENTATIONS
OF GENERALIZED INVERSES

15.1 INTRODUCTION

Two characterizations of the Moore–Penrose inverse are given in this section. The first one makes use of the generalized singular value decomposition. The second one makes use of the reduction of a matrix to a diagonal matrix by pre- and post-multiplication by elementary matrices. We give several illustrative examples.

Finally, we give, without proof (see Rhode (1965) for a proof), a formula for finding the generalized inverse of a partitioned matrix. This formula is the same as that of an ordinary inverse with generalized inverses replacing ordinary inverses. Two illustrative examples are given.

15.2 ANOTHER CHARACTERIZATION OF THE MOORE–PENROSE INVERSE

Let A and B be two matrices as in Theorem 11.4. Assume $m_b \geq n$. Recall that the generalized singular value decompositions were

$$A = UD_A X^{-1} \tag{15.1}$$

and

$$B = VD_B X^{-1}. \tag{15.2}$$

Matrix Algebra for Linear Models, First Edition. Marvin H. J. Gruber.
© 2014 John Wiley & Sons, Inc. Published 2014 by John Wiley & Sons, Inc.

It is not hard to show that the Moore–Penrose inverses of D_A and D_B may be obtained by finding their transposes and then replacing the nonzero elements of those matrices by their reciprocals. By verification of the axioms for the Moore–Penrose inverse, it can then be shown that

$$A^+ = XD_A^+ U' \tag{15.3}$$

and

$$B^+ = XD_B^+ V'. \tag{15.4}$$

The verification that A is a reflexive generalized inverse and that AA^+ is symmetric is straightforward. Observe that

$$AA^+A = UD_A X^{-1} XD_A^+ U'UD_A X^{-1} = UD_A X^{-1} = A \tag{15.5}$$

so that A is indeed a generalized inverse.

The matrix A^+ is a reflexive generalized inverse since

$$A^+AA^+ = XD_A^+ U'UD_A X^{-1} XD_A^+ U' = XD_A^+ U' = A^+. \tag{15.6}$$

Furthermore,

$$AA^+ = UD_A X^{-1} XD_A^+ U' = UD_A D_A^+ U' \tag{15.7}$$

and since $D_A D_A^+$ is symmetric it follows that

$$(AA^+)' = U(D_A D_A^+)'U' = UD_A D_A^+ U' = AA^+. \tag{15.8}$$

Showing that A^+A is symmetric is more difficult. Observe that

$$A^+A = XD_A^+ U'UD_A X^{-1} = XD_A^+ D_A X^{-1}. \tag{15.9}$$

Let Δ_A be the $k \times k$ diagonal matrix of positive elements of D_A. Then

$$D_A = \begin{bmatrix} \Delta_A & 0_{k \times n-k} \\ 0_{(m_a-k) \times k} & 0_{(m_a-k) \times n-k} \end{bmatrix} \tag{15.10}$$

and

$$D_A^+ = \begin{bmatrix} \Delta_A^{-1} & 0_{k \times (m_a-k)} \\ 0_{(n-k) \times k} & 0_{(n-k) \times (m_a-k)} \end{bmatrix} \tag{15.11}$$

so that

$$D_A^+ D_A = \begin{bmatrix} I_{k\times k} & 0_{k\times(m_a-k)} \\ 0_{(m_a-k)\times k} & 0_{(m_a-k)\times(m_a-k)} \end{bmatrix}. \qquad (15.12)$$

From the proof of Theorem 11.4,

$$X = Z_1 \begin{bmatrix} D^{-1}Z_2 & 0_{k\times(n-k)} \\ 0_{(n-k)\times k} & I_{(n-k)\times(n-k)} \end{bmatrix} = [Z_{11}D^{-1}Z_2 \quad Z_{12}],$$

$$X^{-1} = \begin{bmatrix} Z_2'D & 0_{k\times(n-k)} \\ 0_{(n-k)\times k} & I_{(n-k)\times(n-k)} \end{bmatrix} Z_1' = \begin{bmatrix} Z_2'DZ_{11}' \\ Z_{12}' \end{bmatrix}. \qquad (15.13)$$

Then

$$A^+ A = XD_A^+ D_A X^{-1} = \begin{bmatrix} Z_{11}D^{-1}Z_2\Delta_A^{-1} & 0_{n\times(n-k)} \end{bmatrix} \begin{bmatrix} \Delta_A Z_2'DZ_{11}' \\ 0_{(n-k)\times n} \end{bmatrix}$$

$$= Z_{11}Z_{11}',$$

a symmetric matrix.

When $m_b \geq n$, the argument is similar for B.

Example 15.1 Illustration of Finding the Moore–Penrose Inverse Using the Generalized Singular Value Decomposition

For

$$A = \begin{bmatrix} 1 & 1 & 0 \\ 1 & 0 & 1 \\ 1 & 1 & 0 \\ 1 & 0 & 1 \end{bmatrix} \quad \text{and} \quad B = \frac{1}{\sqrt{2}} \begin{bmatrix} 2 & 3 & -1 \\ 2 & -1 & 3 \\ 2 & 3 & -1 \\ 2 & -1 & 3 \end{bmatrix}$$

the generalized singular value decomposition of A and B gives

$$A = \begin{bmatrix} -\frac{1}{2} & \frac{1}{2} & -\frac{1}{\sqrt{2}} & 0 \\ -\frac{1}{2} & \frac{1}{2} & \frac{1}{\sqrt{2}} & 0 \\ \frac{1}{2} & \frac{1}{2} & 0 & \frac{1}{\sqrt{2}} \\ \frac{1}{2} & \frac{1}{2} & 0 & -\frac{1}{\sqrt{2}} \end{bmatrix} \begin{bmatrix} \frac{1}{\sqrt{2}} & 0 & 0 \\ 0 & \frac{1}{\sqrt{2}} & 0 \\ 0 & 0 & 0 \\ 0 & 0 & 0 \end{bmatrix} \begin{bmatrix} 0 & -\sqrt{2} & \sqrt{2} \\ 2\sqrt{2} & \sqrt{2} & \sqrt{2} \\ -\frac{1}{\sqrt{3}} & \frac{1}{\sqrt{3}} & \frac{1}{\sqrt{3}} \end{bmatrix}$$

and

$$B = \begin{bmatrix} -\frac{1}{2} & \frac{1}{2} & 0 & -\frac{1}{\sqrt{2}} \\ \frac{1}{2} & \frac{1}{2} & -\frac{1}{\sqrt{2}} & 0 \\ -\frac{1}{2} & \frac{1}{2} & 0 & \frac{1}{\sqrt{2}} \\ \frac{1}{2} & \frac{1}{2} & \frac{1}{\sqrt{2}} & 0 \end{bmatrix} \begin{bmatrix} 2 & 0 & 0 \\ 0 & 1 & 0 \\ 0 & 0 & 0 \\ 0 & 0 & 0 \end{bmatrix} \begin{bmatrix} 0 & -\sqrt{2} & \sqrt{2} \\ 2\sqrt{2} & \sqrt{2} & \sqrt{2} \\ -\frac{1}{\sqrt{3}} & \frac{1}{\sqrt{3}} & \frac{1}{\sqrt{3}} \end{bmatrix}.$$

The Moore–Penrose inverses are

$$A^+ = XD_A^+ U'$$

$$= \begin{bmatrix} 0 & \frac{1}{3\sqrt{2}} & -\frac{1}{\sqrt{3}} \\ -\frac{1}{2\sqrt{2}} & \frac{1}{6\sqrt{2}} & \frac{1}{\sqrt{3}} \\ \frac{1}{2\sqrt{2}} & \frac{1}{6\sqrt{2}} & \frac{1}{\sqrt{3}} \end{bmatrix} \begin{bmatrix} \sqrt{2} & 0 & 0 & 0 \\ 0 & \sqrt{2} & 0 & 0 \\ 0 & 0 & 0 & 0 \end{bmatrix} \begin{bmatrix} -\frac{1}{2} & \frac{1}{2} & -\frac{1}{2} & \frac{1}{2} \\ \frac{1}{2} & \frac{1}{2} & \frac{1}{2} & \frac{1}{2} \\ -\frac{1}{\sqrt{2}} & -\frac{1}{\sqrt{2}} & 0 & 0 \\ 0 & 0 & \frac{1}{\sqrt{2}} & -\frac{1}{\sqrt{2}} \end{bmatrix}$$

$$= \begin{bmatrix} \frac{1}{6} & \frac{1}{6} & \frac{1}{6} & \frac{1}{6} \\ \frac{1}{3} & \frac{1}{3} & -\frac{1}{6} & -\frac{1}{6} \\ -\frac{1}{6} & -\frac{1}{6} & \frac{1}{3} & \frac{1}{3} \end{bmatrix}.$$

and

$$B^+ = XD_B^+ V'$$

$$= \begin{bmatrix} 0 & \frac{1}{3\sqrt{2}} & -\frac{1}{\sqrt{3}} \\ -\frac{1}{2\sqrt{2}} & \frac{1}{6\sqrt{2}} & \frac{1}{\sqrt{3}} \\ \frac{1}{2\sqrt{2}} & \frac{1}{6\sqrt{2}} & \frac{1}{\sqrt{3}} \end{bmatrix} \begin{bmatrix} \frac{1}{2} & 0 & 0 & 0 \\ 0 & 1 & 0 & 0 \\ 0 & 0 & 0 & 0 \end{bmatrix} \begin{bmatrix} -\frac{1}{2} & \frac{1}{2} & -\frac{1}{2} & \frac{1}{2} \\ \frac{1}{2} & \frac{1}{2} & \frac{1}{2} & \frac{1}{2} \\ 0 & -\frac{1}{\sqrt{2}} & 0 & \frac{1}{\sqrt{2}} \\ -\frac{1}{\sqrt{2}} & 0 & \frac{1}{\sqrt{2}} & 0 \end{bmatrix}$$

$$= \frac{1}{\sqrt{2}} \begin{bmatrix} \frac{1}{6} & \frac{1}{6} & \frac{1}{6} & \frac{1}{6} \\ \frac{5}{24} & -\frac{1}{24} & \frac{5}{25} & -\frac{1}{24} \\ -\frac{1}{24} & \frac{5}{24} & -\frac{1}{24} & \frac{5}{24} \end{bmatrix}.$$

$\square$

Since $(A'A)^+ = A^+(A')^+$ and $(B'B)^+ = B^+(B')^+$, it follows that

$$(A'A)^+ = XD_A^+D_A^{+'}X' \qquad (15.14a)$$

and

$$(B'B)^+ = XD_B^+D_B^{+'}X'. \qquad (15.14b)$$

15.3   STILL ANOTHER REPRESENTATION OF THE GENERALIZED INVERSE

In Section 3, we showed how to find the inverse of a matrix by reducing it to the identity matrix by elementary row operations. These elementary row operations were equivalent to pre-multiplication of the matrix by what we called elementary matrices. By the same token we could have done analogous operations on the columns. These would be equivalent to post-multiplication by elementary matrices. Suppose we have an $n \times m$ matrix X of rank r. By pre- and post-multiplication by elementary matrices, we have nonsingular matrices P and Q where

$$PXQ = \Delta$$

where

$$\Delta = \begin{bmatrix} D_r & 0 \\ 0 & 0 \end{bmatrix}$$

where D_r is an $r \times r$ diagonal matrix. Then

$$X = P^{-1}\Delta Q^{-1}.$$

Now a generalized inverse of X would be

$$X^- = Q\Delta^- P \qquad (15.15a)$$

where

$$\Delta^- = \begin{bmatrix} D_r^{-1} & 0 \\ 0 & 0 \end{bmatrix}'. \qquad (15.15b)$$

Please observe that the generalized inverse is the transpose of the matrix within the brackets. It is easy to verify that Δ^- is a generalized inverse of Δ; in fact it is the Moore–Penrose inverse. Notice that

$$XX^-X = P^{-1}\Delta Q^{-1}Q\Delta^- PP^{-1}\Delta Q^{-1} = X.$$

Furthermore, X^- is reflexive:

$$X^-XX^- = Q\Delta^-PP^{-1}\Delta Q^{-1}Q\Delta^-P = Q\Delta^-P = X^-.$$

However, the other two properties of a Moore–Penrose inverse may not hold true.

Example 15.2 Representation of a Reflexive Generalized Inverse

Consider the rank 2 3×3 matrix

$$M = \begin{bmatrix} 1 & 2 & 4 \\ 2 & 1 & 8 \\ 4 & 2 & 16 \end{bmatrix}.$$

In this case,

$$P_3P_2P_1MQ_1Q_2 = \begin{bmatrix} 1 & 0 & 0 \\ 0 & 3 & 0 \\ 0 & 0 & 0 \end{bmatrix}$$

where

$$P_1 = \begin{bmatrix} 1 & 0 & 0 \\ -2 & 1 & 0 \\ 0 & 0 & 1 \end{bmatrix}, \quad P_2 = \begin{bmatrix} 1 & 0 & 0 \\ 0 & 1 & 0 \\ -4 & 0 & 1 \end{bmatrix}, \quad P_3 = \begin{bmatrix} 1 & 0 & 0 \\ 0 & 1 & 0 \\ 0 & -2 & 1 \end{bmatrix}$$

and

$$Q_1 = \begin{bmatrix} 1 & 0 & -4 \\ 0 & 1 & 0 \\ 0 & 0 & 1 \end{bmatrix}, \quad Q_2 = \begin{bmatrix} 1 & 2 & 0 \\ 0 & -1 & 0 \\ 0 & 0 & 1 \end{bmatrix}.$$

Then after diagonalizing M we get after performing the matrix multiplications where $P = P_3P_2P_1$ and $Q = Q_1Q_2$

$$M = \begin{bmatrix} 1 & 0 & 0 \\ -2 & 1 & 0 \\ 0 & -2 & 1 \end{bmatrix}^{-1} \begin{bmatrix} 1 & 0 & 0 \\ 0 & 3 & 0 \\ 0 & 0 & 0 \end{bmatrix} \begin{bmatrix} 1 & 2 & -4 \\ 0 & -1 & 0 \\ 0 & 0 & 1 \end{bmatrix}^{-1}$$

$$= \begin{bmatrix} 1 & 0 & 0 \\ 2 & 1 & 0 \\ 4 & 2 & 1 \end{bmatrix} \begin{bmatrix} 1 & 0 & 0 \\ 0 & 3 & 0 \\ 0 & 0 & 0 \end{bmatrix} \begin{bmatrix} 1 & 2 & 4 \\ 0 & -1 & 0 \\ 0 & 0 & 1 \end{bmatrix}$$

and

$$
M^- = \begin{bmatrix} 1 & 2 & -4 \\ 0 & -1 & 0 \\ 0 & 0 & 1 \end{bmatrix} \begin{bmatrix} 1 & 0 & 0 \\ 0 & \frac{1}{3} & 0 \\ 0 & 0 & 0 \end{bmatrix} \begin{bmatrix} 1 & 0 & 0 \\ -2 & 1 & 0 \\ 0 & -2 & 1 \end{bmatrix}
$$

$$
= \begin{bmatrix} -\frac{1}{3} & \frac{2}{3} & 0 \\ \frac{2}{3} & -\frac{1}{3} & 0 \\ 0 & 0 & 0 \end{bmatrix}.
$$

By straightforward matrix multiplication, it can be verified that the M^- is a reflexive generalized inverse that does not satisfy the other two Moore–Penrose axioms.   □

Example 15.3 Representation of Another Reflexive Generalized Inverse

Let

$$
X = \begin{bmatrix} 1_3 & 1_3 & 0 & 0 \\ 1_3 & 0 & 1_3 & 0 \\ 1_3 & 0 & 0 & 1_3 \end{bmatrix}.
$$

Then

$$
X'X = \begin{bmatrix} 9 & 3 & 3 & 3 \\ 3 & 3 & 0 & 0 \\ 3 & 0 & 3 & 0 \\ 3 & 0 & 0 & 3 \end{bmatrix}.
$$

Premultiply $X'X$ by

$$
P = \begin{bmatrix} 0 & 1 & 0 & 0 \\ 0 & 0 & 1 & 0 \\ 1 & 0 & 0 & 0 \\ 0 & 0 & 0 & 1 \end{bmatrix} \begin{bmatrix} 0 & 0 & 0 & 1 \\ 0 & 1 & 0 & 0 \\ 0 & 0 & 1 & 0 \\ 1 & 0 & 0 & 0 \end{bmatrix} \begin{bmatrix} 1 & -1 & -1 & -1 \\ 0 & 1 & 0 & 0 \\ 0 & 0 & 1 & 0 \\ 0 & 0 & 0 & 1 \end{bmatrix}
$$

$$
= \begin{bmatrix} 0 & 1 & 0 & 0 \\ 0 & 0 & 1 & 0 \\ 0 & 0 & 0 & 1 \\ 1 & -1 & -1 & -1 \end{bmatrix}
$$

and postmultiply X'X by

$$Q = \begin{bmatrix} 1 & 0 & 0 & 0 \\ -1 & 1 & 0 & 0 \\ -1 & 0 & 1 & 0 \\ -1 & 0 & 0 & 1 \end{bmatrix} \begin{bmatrix} 0 & 0 & 0 & 1 \\ 1 & 0 & 0 & 0 \\ 0 & 1 & 0 & 0 \\ 0 & 0 & 1 & 0 \end{bmatrix} = \begin{bmatrix} 0 & 0 & 0 & 1 \\ 1 & 0 & 0 & -1 \\ 0 & 1 & 0 & -1 \\ 0 & 0 & 1 & -1 \end{bmatrix}$$

to obtain the diagonal matrix

$$\Delta = \begin{bmatrix} 3 & 0 & 0 & 0 \\ 0 & 3 & 0 & 0 \\ 0 & 0 & 3 & 0 \\ 0 & 0 & 0 & 0 \end{bmatrix}.$$

Then

$$\Delta^- = \begin{bmatrix} \frac{1}{3} & 0 & 0 & 0 \\ 0 & \frac{1}{3} & 0 & 0 \\ 0 & 0 & \frac{1}{3} & 0 \\ 0 & 0 & 0 & 0 \end{bmatrix},$$

and the resulting generalized inverse is

$$(X'X)^- = \begin{bmatrix} 0 & 0 & 0 & 0 \\ 0 & \frac{1}{3} & 0 & 0 \\ 0 & 0 & \frac{1}{3} & 0 \\ 0 & 0 & 0 & \frac{1}{3} \end{bmatrix}.$$

The reader may verify that this is a reflexive generalized inverse that does not satisfy the other two Penrose axioms.   □

15.4   THE GENERALIZED INVERSE OF A PARTITIONED MATRIX

The inverse of a partitioned matrix was derived in Subsection 3.4. This formula may be extended to a generalized inverse. Let M be a symmetric but singular matrix written in the form

$$M = \begin{bmatrix} X' \\ Z' \end{bmatrix} \begin{bmatrix} X & Z \end{bmatrix} = \begin{bmatrix} A & B \\ B' & D \end{bmatrix}. \tag{15.16}$$

Then a generalized inverse of M would be written as

$$M^- = \begin{bmatrix} A^- + A^- BQ^- B'A^- & -A^- BQ^- \\ -Q^- B'A^- & Q^- \end{bmatrix}$$

$$= \begin{bmatrix} A^- & 0 \\ 0 & 0 \end{bmatrix} + \begin{bmatrix} -A^- B \\ I \end{bmatrix} Q^- \begin{bmatrix} -B'A^- & I \end{bmatrix}$$

(15.17)

where $Q = D - B'A^- B$.

Example 15.4 Illustration of Generalized Inverse of a Partitioned Matrix

$$X = \begin{bmatrix} 1_3 & 1_3 & 0 & 0 \\ 1_3 & 0 & 1_3 & 0 \\ 1_3 & 0 & 0 & 1_3 \end{bmatrix}.$$

Then

$$M = X'X = \begin{bmatrix} 9 & 3 & 3 & 3 \\ 3 & 3 & 0 & 0 \\ 3 & 0 & 3 & 0 \\ 3 & 0 & 0 & 3 \end{bmatrix}.$$

Let

$$A = [9], \quad B = \begin{bmatrix} 3 & 3 & 3 \end{bmatrix}, \quad D = \begin{bmatrix} 3 & 0 & 0 \\ 0 & 3 & 0 \\ 0 & 0 & 3 \end{bmatrix}.$$

Then

$$Q = \begin{bmatrix} 3 & 0 & 0 \\ 0 & 3 & 0 \\ 0 & 0 & 3 \end{bmatrix} - \begin{bmatrix} 3 \\ 3 \\ 3 \end{bmatrix} \begin{bmatrix} \dfrac{1}{9} \end{bmatrix} \begin{bmatrix} 3 & 3 & 3 \end{bmatrix}$$

$$= \begin{bmatrix} 2 & -1 & -1 \\ -1 & 2 & -1 \\ -1 & -1 & 2 \end{bmatrix},$$

and

$$
M^- = \begin{bmatrix} \frac{1}{9} & 0 & 0 & 0 \\ 0 & 0 & 0 & 0 \\ 0 & 0 & 0 & 0 \\ 0 & 0 & 0 & 0 \end{bmatrix} + \begin{bmatrix} -\frac{1}{3} & -\frac{1}{3} & -\frac{1}{3} \\ 1 & 0 & 0 \\ 0 & 1 & 0 \\ 0 & 0 & 1 \end{bmatrix} \begin{bmatrix} 0 & 0 & 0 \\ 0 & \frac{2}{3} & \frac{1}{3} \\ 0 & \frac{1}{3} & \frac{2}{3} \end{bmatrix} \begin{bmatrix} -\frac{1}{3} & 1 & 0 & 0 \\ -\frac{1}{3} & 0 & 1 & 0 \\ -\frac{1}{3} & 0 & 0 & 1 \end{bmatrix}
$$

$$
= \begin{bmatrix} \frac{1}{3} & 0 & -\frac{1}{3} & -\frac{1}{3} \\ 0 & 0 & 0 & 0 \\ -\frac{1}{3} & 0 & \frac{2}{3} & \frac{1}{3} \\ -\frac{1}{3} & 0 & \frac{1}{3} & \frac{2}{3} \end{bmatrix}.
$$

Both M^- and Q^- are reflexive generalized inverses. Rhode (1965) points out that if Q^- is a reflexive generalized inverse, M^- is also. □

Example 15.5 Another Generalized Inverse of a Partitioned Matrix

Let $X = [X_1 \ Z_1]$ where

$$
X_1 = \begin{bmatrix} 1_3 & 1_3 & 0 & 0 \\ 1_3 & 0 & 1_3 & 0 \\ 1_3 & 0 & 0 & 1_3 \end{bmatrix}, \quad Z_1 = \begin{bmatrix} I_3 \\ I_3 \\ I_3 \end{bmatrix}.
$$

Then

$$
M = X'X
$$

$$
= \begin{bmatrix} 9 & 3 & 3 & 3 & 3 & 3 & 3 \\ 3 & 3 & 0 & 0 & 1 & 1 & 1 \\ 3 & 0 & 3 & 0 & 1 & 1 & 1 \\ 3 & 0 & 0 & 3 & 1 & 1 & 1 \\ 3 & 1 & 1 & 1 & 3 & 0 & 0 \\ 3 & 1 & 1 & 1 & 0 & 3 & 0 \\ 3 & 1 & 1 & 1 & 0 & 0 & 3 \end{bmatrix}.
$$

We use the generalized inverse of $X_1' X_1$ from Example 15.4. The matrix

$$Q = \begin{bmatrix} 3 & 0 & 0 \\ 0 & 3 & 0 \\ 0 & 0 & 3 \end{bmatrix} - \begin{bmatrix} 3 & 1 & 1 & 1 \\ 3 & 1 & 1 & 1 \\ 3 & 1 & 1 & 1 \end{bmatrix} \begin{bmatrix} \frac{1}{3} & 0 & -\frac{1}{3} & \frac{1}{3} \\ 0 & 0 & 0 & 0 \\ -\frac{1}{3} & 0 & \frac{2}{3} & \frac{1}{3} \\ -\frac{1}{3} & 0 & \frac{1}{3} & \frac{2}{3} \end{bmatrix} \begin{bmatrix} 3 & 3 & 3 \\ 1 & 1 & 1 \\ 1 & 1 & 1 \\ 1 & 1 & 1 \end{bmatrix}$$

$$= \begin{bmatrix} 2 & -1 & -1 \\ -1 & 2 & -1 \\ -1 & -1 & 2 \end{bmatrix}.$$

Then

$$M^- = \begin{bmatrix} \frac{1}{3} & 0 & -\frac{1}{3} & -\frac{1}{3} & 0 & 0 & 0 \\ 0 & 0 & 0 & 0 & 0 & 0 & 0 \\ -\frac{1}{3} & 0 & \frac{2}{3} & \frac{1}{3} & 0 & 0 & 0 \\ -\frac{1}{3} & 0 & \frac{1}{3} & \frac{2}{3} & 0 & 0 & 0 \\ 0 & 0 & 0 & 0 & 0 & 0 & 0 \\ 0 & 0 & 0 & 0 & 0 & 0 & 0 \\ 0 & 0 & 0 & 0 & 0 & 0 & 0 \end{bmatrix}$$

$$+ \begin{bmatrix} -\frac{1}{3} & -\frac{1}{3} & -\frac{1}{3} \\ 0 & 0 & 0 \\ 0 & 0 & 0 \\ 0 & 0 & 0 \\ 1 & 0 & 0 \\ 0 & 1 & 0 \\ 0 & 0 & 1 \end{bmatrix} \begin{bmatrix} 0 & 0 & 0 \\ 0 & \frac{2}{3} & \frac{1}{3} \\ 0 & \frac{1}{3} & \frac{2}{3} \end{bmatrix} \begin{bmatrix} -\frac{1}{3} & 0 & 0 & 0 & 1 & 0 & 0 \\ -\frac{1}{3} & 0 & 0 & 0 & 0 & 1 & 0 \\ -\frac{1}{3} & 0 & 0 & 0 & 0 & 0 & 1 \end{bmatrix}$$

$$= \begin{bmatrix} \frac{5}{9} & 0 & -\frac{1}{3} & -\frac{1}{3} & 0 & -\frac{1}{3} & -\frac{1}{3} \\ 0 & 0 & 0 & 0 & 0 & 0 & 0 \\ -\frac{1}{3} & 0 & \frac{2}{3} & \frac{1}{3} & 0 & 0 & 0 \\ -\frac{1}{3} & 0 & \frac{1}{3} & \frac{2}{3} & 0 & 0 & 0 \\ 0 & 0 & 0 & 0 & 0 & 0 & 0 \\ -\frac{1}{3} & 0 & 0 & 0 & 0 & \frac{2}{3} & \frac{1}{3} \\ -\frac{1}{3} & 0 & 0 & 0 & 0 & \frac{1}{3} & \frac{2}{3} \end{bmatrix}$$

The generalized inverse obtained is reflexive. However, it does not satisfy the other two conditions to be a Moore–Penrose inverse. □

15.5 SUMMARY

We gave two characterizations of the Moore–Penrose inverse using the generalized singular value decomposition and the diagonalization of a matrix using nonsingular elementary matrices. We presented a formula for the generalized inverse of a portioned matrix. Several illustrative examples were given.

EXERCISES

15.1 Show that Δ^- as defined in Equation 15.15 is the Moore–Penrose inverse of Δ.

15.2 Let X be a matrix with decomposition $X = P^{-1}\Delta Q^{-1}$ as above. Then for any matrices A, B, and C of the appropriate size,

$$G = Q \begin{bmatrix} D_r^{-1} & A \\ B & C \end{bmatrix} P$$

is a generalized inverse of X.

15.3 Using the method of Subsection 15.2, find the generalized inverses for the following matrices:

A. $X_1 = \begin{bmatrix} 1 & 1 & 0 \\ 1 & 1 & 0 \\ 1 & 0 & 1 \\ 1 & 0 & 1 \end{bmatrix}$.

B. $M = \begin{bmatrix} 2 & 1 & 1 \\ 1 & 1 & 0 \\ 1 & 0 & 1 \end{bmatrix}$.

C. $H = \begin{bmatrix} 4 & 4 & 2 & 2 \\ 4 & 4 & 2 & 2 \\ 2 & 2 & 4 & 4 \\ 2 & 2 & 4 & 4 \end{bmatrix}$.

D. $Z = \begin{bmatrix} 1 & 1 & -1 & -1 \\ 1 & 1 & -1 & -1 \\ -1 & -1 & 1 & 1 \\ -1 & -1 & 1 & 1 \end{bmatrix}$.

15.4 Consider the sequence of matrices

$$X_1 = \begin{bmatrix} 1 & 1 & 0 \\ 1 & 1 & 0 \\ 1 & 0 & 1 \\ 1 & 0 & 1 \end{bmatrix}, \quad X_2 = \begin{bmatrix} 1 & 1 & 0 & 1 & 0 \\ 1 & 1 & 0 & 0 & 1 \\ 1 & 0 & 1 & 1 & 0 \\ 1 & 0 & 1 & 0 & 1 \end{bmatrix},$$

$$X_3 = \begin{bmatrix} 1 & 1 & 0 & 1 & 0 & 1 & 0 & 0 & 0 \\ 1 & 1 & 0 & 0 & 1 & 0 & 1 & 0 & 0 \\ 1 & 0 & 1 & 1 & 0 & 0 & 0 & 1 & 0 \\ 1 & 1 & 1 & 0 & 1 & 0 & 0 & 0 & 1 \end{bmatrix}.$$

Find the generalized inverse of $M_i = X_i'X_i$, $i = 1, 2, 3$. Follow the pattern in Examples 15.4 and 15.5. Use [4] for the matrix A in M_1. For M_2 use the generalized inverse you found for M_1. For M_3 use the generalized inverse you found for M_2.

15.5 Find a generalized inverse for H in 15.3 C and Z in 15.3 D using the formula for the generalized inverse of a partitioned matrix.

15.6 Use Equation (15.14) in Example 15.1 to compute the Moore–Penrose inverses of $A'A$ and $B'B$.

SECTION 16

LEAST SQUARE ESTIMATORS FOR LESS THAN FULL-RANK MODELS

16.1 INTRODUCTION

This section will show how to obtain least square (LS) estimators for a non-full-rank model. First, in Subsection 16.2 a few preliminary results will be derived from the expressions for generalized inverses in terms of the Moore–Penrose inverse in Section 13. These results will be used in Subsection 16.2 to extend the derivation of the LS estimator in Subsection 3.6 to the non-full-rank model. The LS estimators that are obtained in this section will not be unique. Subsection 16.2 will give conditions for parametric functions of the LS estimator to be unique.

16.2 SOME PRELIMINARIES

Theorem 16.1 will be needed to derive the LS estimator for the non-full-rank case.

Theorem 16.1 The following relationships hold true:

1. $XUU' = X$.   (16.1)
2. $X'X(X'X)^+ X' = X'$.   (16.2)
3. For any generalized inverse G,

$$X'XGX' = X'. \qquad (16.3)$$

Matrix Algebra for Linear Models, First Edition. Marvin H. J. Gruber.
© 2014 John Wiley & Sons, Inc. Published 2014 by John Wiley & Sons, Inc.

Proof. 1. Using the singular value decomposition,

$$XUU' = S'\Lambda^{1/2}U'UU' = S'\Lambda^{1/2}U' = X$$

since $U'U = I$.

2. Using (13.7) and (13.8) since $U'U = I$,

$$X'X(X'X)^+ X' = U\Lambda U'U\Lambda^{-1}U'U\Lambda^{1/2}S = U\Lambda^{1/2}S = X. \qquad (16.4)$$

3. Using (13.2), (13.10), and (13.3),

$$X'XGX' = X'XUU'GUU'X' = X'X(X'X)^+ X' = X'. \qquad \blacksquare$$

16.3 OBTAINING THE LS ESTIMATOR

The derivation of the LS estimator will be similar to that of Subsection 3.5. The goal is to minimize

$$F(\beta) = (Y - X\beta)'(Y - X\beta). \qquad (16.5)$$

Observe that by the argument in Subsection 3.5 making use of (16.3) for any generalized inverse of $X'X$,

$$
\begin{aligned}
(Y - X\beta)'(Y - X\beta) &= Y'Y - \beta'X'Y - Y'X\beta + \beta'X'X\beta \\
&= Y'Y - \beta'(X'X)GX'Y - Y'XG(X'X)\beta \\
&\quad + Y'XG(X'X)GX'Y - Y'XG(X'X)GX'Y \qquad (16.6) \\
&\quad + \beta'X'X\beta \\
&= Y'(I - XGX')Y + (\beta - GX'Y)'(X'X)(\beta - GX'Y).
\end{aligned}
$$

The expression in (16.6) is minimized for

$$\widehat{\beta} = GX'Y. \qquad (16.7)$$

The LS estimator may also be derived by matrix differentiation, as was done in Subsection 4.7. From Equation (4.22) we get the normal equation

$$X'X\beta = X'Y, \qquad (16.8)$$

and solving for β using a generalized inverse in place of an ordinary inverse, we obtain Equation (16.7).

For the non-full-rank case, the LS estimator is not unique because it depends on the choice of the generalized inverse. Let p be an m-dimensional column vector. However, certain parametric functions $p'\widehat{\beta}$ of the LS estimators are independent of the choice of the generalized inverses. These are the estimable parametric functions.

A parametric function $p'\beta$ is estimable if p is in the column space of the matrix of exploratory variables X. Three equivalent ways to state this are:

1. The relationship $p'V = 0$ holds true.
2. There exists an s-dimensional vector d such that $p = Ud$.
3. There exists an n-dimensional vector t such that $p' = t'X$.

The first statement ensures the uniqueness of the LS estimator. From (13.11) for any generalized inverse G,

$$p'b = p'(X'X)^+X'Y + p'VV'GUU'X'Y \qquad (16.9)$$

because $XV = 0$. Since a condition for estimability is that $p'V = 0$, estimable parametric functions of the LS estimator have the unique representation

$$p'b = p'(X'X)^+ X'Y. \qquad (16.10)$$

The representation (16.10) also holds true for non-estimable parametric functions if $V'GU = 0$, that is, when G is an LS minimum norm generalized inverse of $X'X$.

The second statement simply says that p is in the vector space formed by the linear combinations of the columns of X. The columns of U are a basis for that vector space.

The third statement would hold true that there exists an unbiased estimator that is a linear combination of the observations. Most linear model textbooks would define $p'\beta$ as being estimable if there exists an n-dimensional vector t such that

$$p'\beta = E[t'Y].$$

Since $E[t'Y] = t'X\beta = p'\beta$, condition 3 holds true.

To show the equivalence of the three statements, observe that

$$p' = p'(UU' + VV') = p'UU', \text{ since } p'V = 0.$$

This implies statement 2. Statement 2 implies statement 3 because

$$p = Ud = U\Lambda^{1/2}SS'\Lambda^{-1/2}d,$$

so $t = S'\Lambda^{-1/2}d$. Statement 3 implies statement 1 because $XV = 0$.

Example 16.1 A Characterization of Estimability in a Particular Model

Consider the linear model

$$
\begin{bmatrix} y_{11} \\ y_{12} \\ y_{13} \\ y_{21} \\ y_{22} \\ y_{23} \\ y_{31} \\ y_{32} \\ y_{33} \end{bmatrix} = \begin{bmatrix} 1_3 & 1_3 & 0 & 0 \\ 1_3 & 0 & 1_3 & 0 \\ 1_3 & 0 & 0 & 1_3 \end{bmatrix} \begin{bmatrix} \mu \\ \tau_1 \\ \tau_2 \\ \tau_3 \end{bmatrix} + \varepsilon.
$$

The normal equation would be

$$
\begin{bmatrix} 9 & 3 & 3 & 3 \\ 3 & 3 & 0 & 0 \\ 3 & 0 & 3 & 0 \\ 3 & 0 & 0 & 3 \end{bmatrix} \begin{bmatrix} \mu \\ \tau_1 \\ \tau_2 \\ \tau_3 \end{bmatrix} = \begin{bmatrix} y_{..} \\ y_{1.} \\ y_{2.} \\ y_{3.} \end{bmatrix}.
$$

Using the generalized inverse

$$
G_1 = \begin{bmatrix} 0 & 0 \\ 0 & \dfrac{1}{3} \cdot I_3 \end{bmatrix},
$$

the LS estimators are

$$
\hat{\mu} = 0
$$
$$
\hat{\tau}_i = \bar{y}_{i.}, \quad 1 \le i \le 3
$$

where

$$
\bar{y}_{i.} = \frac{1}{3} \sum_{j=1}^{3} y_{ij}.
$$

For the generalized inverse

$$
G_2 = \begin{bmatrix} \frac{1}{9} & 0 \\ -\frac{1}{9} 1_3 & \frac{1}{3} I_3 \end{bmatrix},
$$

the LS estimator is

$$\hat{\mu} = \overline{y}..$$
$$\hat{\tau}_i = \overline{y}_{i.} - \overline{y}.., \quad i = 1, 2, 3.$$

Notice the difference in the LS estimators for the individual parameters. Consider the parametric function τ_2. An attempt to find t where $p = Xt$ would mean solving the system of equations

$$0 = \sum_{i=1}^{9} t_i$$
$$0 = \sum_{i=1}^{3} t_i$$
$$1 = \sum_{i=4}^{6} t_i$$
$$0 = \sum_{i=7}^{9} t_i$$

an inconsistent system of equations. Thus, τ_2 would not be an estimable parametric function. On the other hand, if we consider the parametric function $\mu + \tau_2$, we have to solve the system

$$1 = \sum_{i=1}^{9} t_i$$
$$0 = \sum_{i=1}^{3} t_i$$
$$1 = \sum_{i=4}^{6} t_i$$
$$0 = \sum_{i=7}^{9} t_i$$

There are many solutions to this system. One possibility is

$$t' = \left(0, 0, 0, \tfrac{1}{3}, \tfrac{1}{3}, \tfrac{1}{3}, 0, 0, 0\right).$$

The equations are consistent. The condition for estimability says that a t exists. It need not be unique.

Using the Moore–Penrose inverse and the fact that

$$X'Y = \begin{bmatrix} y.. \\ y_{1.} \\ y_{2.} \\ y_{3.} \end{bmatrix},$$

the LS estimator is

$$
\begin{bmatrix} \hat{\mu} \\ \hat{\tau}_1 \\ \hat{\tau}_2 \\ \hat{\tau}_3 \end{bmatrix} = \begin{bmatrix} \frac{1}{16} & 0 & 0 & 0 \\ \frac{1}{48} & \frac{11}{48} & -\frac{5}{48} & -\frac{5}{48} \\ \frac{1}{48} & -\frac{5}{48} & \frac{11}{48} & -\frac{5}{48} \\ \frac{1}{48} & -\frac{5}{48} & -\frac{5}{48} & \frac{11}{48} \end{bmatrix} \begin{bmatrix} y_{..} \\ y_{1.} \\ y_{2.} \\ y_{3.} \end{bmatrix}.
$$

Notice that the LS estimators for the two generalized inverses are quite different.

The parametric functions may be written in the form

$$
\theta = g\mu + h_1\tau_1 + h_2\tau_2 + h_3\tau_3.
$$

The condition $p'VV' = 0$ would be

$$
g - h_1 - h_2 - h_3 = 0.
$$

Examples of parametric functions that satisfy these conditions are
$\mu + \tau_i$, $1 \le i \le 3$, and $\tau_i - \tau_j$. It is not hard to show that using both LS estimators above that the estimable parametric functions are the same. For example, for the first generalized inverse

$$
\hat{\tau}_1 - \hat{\tau}_3 = \bar{y}_{1.} - \bar{y}_{3.}.
$$

For the second generalized inverse

$$
\hat{\tau}_1 - \hat{\tau}_3 = (\bar{y}_{1.} - y_{..}) - (\bar{y}_{3.} - y_{..}) = \bar{y}_{1.} - \bar{y}_{3.}.
$$

For the Moore–Penrose inverse,

$$
\begin{aligned}
\hat{\tau}_1 - \hat{\tau}_3 &= \left(\frac{1}{48}y_{..} + \frac{11}{48}y_{1.} - \frac{5}{48}y_{2.} - \frac{5}{48}y_{3.} \right) \\
&\quad - \left(\frac{1}{48}y_{..} - \frac{5}{48}y_{1.} - \frac{5}{48}y_{2.} + \frac{11}{48}y_{3.} \right) \\
&= \frac{1}{3}(y_{1.} - y_{3.}) \\
&= \bar{y}_{1.} - \bar{y}_{3.}.
\end{aligned}
$$

□

Example 16.2 A Characterization of Estimability in Another Particular Model

Consider the linear model

$$
\begin{bmatrix} Y_{111} \\ Y_{121} \\ Y_{112} \\ Y_{122} \\ Y_{211} \\ Y_{212} \\ Y_{221} \\ Y_{222} \end{bmatrix} = \begin{bmatrix} 1_2 & 1_2 & 0 & 1_2 & 0 & I_2 \\ 1_2 & 1_2 & 0 & 0 & 1_2 & I_2 \\ 1_2 & 0 & 1_2 & 1_2 & 0 & I_2 \\ 1_2 & 0 & 1_2 & 0 & 1_2 & I_2 \end{bmatrix} \begin{bmatrix} \mu \\ \tau_1 \\ \tau_2 \\ \gamma_1 \\ \gamma_2 \\ \theta_1 \\ \theta_2 \end{bmatrix} + \varepsilon.
$$

The normal equation is

$$
\begin{bmatrix} 8 & 4 & 4 & 4 & 4 & 4 & 4 \\ 4 & 4 & 0 & 2 & 2 & 2 & 2 \\ 4 & 0 & 4 & 2 & 2 & 2 & 2 \\ 4 & 2 & 2 & 4 & 0 & 2 & 2 \\ 4 & 2 & 2 & 0 & 4 & 2 & 2 \\ 4 & 2 & 2 & 2 & 2 & 4 & 0 \\ 4 & 2 & 2 & 2 & 2 & 0 & 4 \end{bmatrix} \begin{bmatrix} \mu \\ \tau_1 \\ \tau_2 \\ \gamma_1 \\ \gamma_2 \\ \theta_1 \\ \theta_2 \end{bmatrix} = \begin{bmatrix} Y_{...} \\ Y_{1..} \\ Y_{2..} \\ Y_{.1.} \\ Y_{.2.} \\ Y_{..1} \\ Y_{..2} \end{bmatrix}.
$$

Using the formula for the generalized inverse of a partitioned matrix with

$$
A = \begin{bmatrix} 8 & 4 & 4 \\ 4 & 4 & 0 \\ 4 & 0 & 4 \end{bmatrix},\ B = \begin{bmatrix} 4 & 4 & 4 & 4 \\ 2 & 2 & 2 & 2 \\ 2 & 2 & 2 & 2 \end{bmatrix},\ D = \begin{bmatrix} 4 & 0 & 2 & 2 \\ 0 & 4 & 2 & 2 \\ 2 & 2 & 4 & 0 \\ 2 & 2 & 0 & 4 \end{bmatrix}
$$

and

$$
A^- = \begin{bmatrix} 0 & 0 & 0 \\ 0 & \frac{1}{4} & 0 \\ 0 & 0 & \frac{1}{4} \end{bmatrix},
$$

we obtain the generalized inverse

$$
G = \begin{bmatrix}
0 & 0 & 0 & 0 & 0 & 0 & 0 \\
0 & \frac{1}{4} & 0 & 0 & 0 & 0 & 0 \\
0 & 0 & \frac{1}{4} & 0 & 0 & 0 & 0 \\
0 & 0 & 0 & \frac{1}{8} & -\frac{1}{8} & 0 & 0 \\
0 & 0 & 0 & -\frac{1}{8} & \frac{1}{8} & 0 & 0 \\
0 & 0 & 0 & 0 & 0 & \frac{1}{8} & -\frac{1}{8} \\
0 & 0 & 0 & 0 & 0 & -\frac{1}{8} & \frac{1}{8}
\end{bmatrix},
$$

and the LS estimators are

$$
\hat{\mu} = 0, \hat{\tau}_1 = \bar{y}_{1..}, \hat{\tau}_2 = \bar{y}_{2..}, \hat{\gamma}_1 = \tfrac{1}{2}(\bar{y}_{.1.} - \bar{y}_{.2.}), \hat{\gamma}_2 = \tfrac{1}{2}(\bar{y}_{.2.} - \bar{y}_{.1.}),
$$
$$
\hat{\theta}_1 = \tfrac{1}{2}(\bar{y}_{..1} - \bar{y}_{..2}), \quad \text{and} \quad \hat{\theta}_2 = \tfrac{1}{2}(\bar{y}_{..2} - \bar{y}_{..1}).
$$

In order for a parametric function to be estimable, there must exist a t where $p = Xt$. Consider the parametric function $\mu + \tau_1 + \beta_2 + \theta_1$. A solution to the system of equations

$$
\begin{aligned}
t_1 + t_2 + t_3 + t_4 + t_5 + t_6 + t_7 + t_8 &= 1 \\
t_1 + t_2 + t_3 + t_4 &= 1 \\
t_5 + t_6 + t_7 + t_8 &= 0 \\
t_1 + t_2 + t_5 + t_6 &= 0 \\
t_3 + t_4 + t_7 + t_8 &= 1 \\
t_1 + t_3 + t_5 + t_7 &= 0 \\
t_2 + t_4 + t_6 + t_8 &= 1
\end{aligned}
$$

is $t_1 = 0$, $t_2 = \frac{1}{4}$, $t_3 = \frac{1}{4}$, $t_4 = \frac{1}{2}$, $t_5 = -\frac{1}{4}$, $t_6 = 0$, $t_7 = 0$, and $t_8 = \frac{1}{4}$. Thus, $\mu + \tau_1 + \beta_2 + \theta_1$ is an estimable parametric function. On the other hand, for example, the system of equations

$$
\begin{aligned}
t_1 + t_2 + t_3 + t_4 + t_5 + t_6 + t_7 + t_8 &= 0 \\
t_1 + t_2 + t_3 + t_4 &= 1 \\
t_5 + t_6 + t_7 + t_8 &= 0 \\
t_1 + t_2 + t_5 + t_6 &= 0 \\
t_3 + t_4 + t_7 + t_8 &= 0 \\
t_1 + t_3 + t_5 + t_7 &= 0 \\
t_2 + t_4 + t_6 + t_8 &= 0
\end{aligned}
$$

is inconsistent. Adding the second and third equation gives $1=0$. Thus, the parametric function τ_1 is not estimable.   □

16.4 SUMMARY

We defined and gave examples of estimable parametric functions. We showed how to derive the LS estimator for less than full-rank models and gave some examples.

EXERCISES

16.1 Show that a linear combination of two estimable parametric functions is estimable.

16.2 In Example 16.1, which of the following parametric functions are estimable? Justify your answer:

  A. $\sum_{i=1}^{3} \tau_i$

  B. $2\tau_1 - \tau_2 - \tau_3$

  C. τ_3

  D. $\tau_1 - \tau_3$

16.3 A. For Example 16.1, obtain LS and minimum norm generalized inverses of $X'X$.

  B. Find the LS estimator in Example 16.1 for both the LS and the minimum norm generalized inverse. Show that the basic estimable functions of the LS are the same as that obtained in Example 16.1 for the other two generalized inverses.

16.4 Consider the model

$$Y = \begin{bmatrix} 1_3 & 1_3 & 0 & 0 & I_3 \\ 1_3 & 0 & 1_3 & 0 & I_3 \\ 1_3 & 0 & 0 & 1_3 & I_3 \end{bmatrix} \begin{bmatrix} \mu \\ \tau_1 \\ \tau_2 \\ \tau_3 \\ \beta_1 \\ \beta_2 \\ \beta_3 \end{bmatrix} + \varepsilon.$$

  A. Find two different LS estimators.

  B. What are the estimators of the basic estimable functions $\mu + \tau_i + \beta_j$, $i = 1, 2, 3$, $j = 1, 2, 3$ for each one of them?

  C. Show that the basic estimable functions are the same for your two different LS estimators.

16.5   Given the linear model

$$
\begin{bmatrix}
Y_{111} \\
Y_{112} \\
Y_{211} \\
Y_{212} \\
Y_{121} \\
Y_{122} \\
Y_{221} \\
Y_{222}
\end{bmatrix}
=
\begin{bmatrix}
1_2 & 1_2 & 0 & 1_2 & 0 \\
1_2 & 1_2 & 0 & 0 & 1_2 \\
1_2 & 0 & 1_2 & 1_2 & 0 \\
1_2 & 0 & 1_2 & 0 & 1_2
\end{bmatrix}
\begin{bmatrix}
\mu \\
\tau_1 \\
\tau_2 \\
\beta_1 \\
\beta_2
\end{bmatrix}
+ \varepsilon :
$$

A. Find the LS estimators using two different generalized inverses.

B. Show that the parametric functions $\theta_{ij} = \mu + \tau_i + \beta_j$, $i = 1, 2$, $j = 1, 2$ are estimable but that the individual τ_i and β_j are not.

16.6   Consider the linear model

$$
\begin{bmatrix}
y_{11} \\
y_{12} \\
y_{21} \\
y_{22} \\
y_{23} \\
y_{31} \\
y_{32} \\
y_{33} \\
y_{34}
\end{bmatrix}
=
\begin{bmatrix}
1_2 & 1_2 & 0 & 0 \\
1_3 & 0 & 1_3 & 0 \\
1_4 & 0 & 0 & 1_4
\end{bmatrix}
\begin{bmatrix}
\mu \\
\tau_1 \\
\tau_2 \\
\tau_3
\end{bmatrix}
+ \varepsilon.
$$

Find a set of LS estimators of the individual parameters and the basic estimable parametric functions.

PART IV

QUADRATIC FORMS AND THE ANALYSIS OF VARIANCE

This part of the book considers the quadratic forms important for testing hypothesis about the regression parameters in a general linear model and different special cases of this model. It will be assumed that the error terms of the regression models are normally distributed with mean zero and common variance. Given this assumption, the quadratic forms will follow a chi-square distribution under certain conditions. We will tell:

1. what these conditions are;
2. when quadratic forms of normal random variables are independent;
3. how many degrees of freedom they have, or in other words, how many independent pieces of information they represent.

This will be done for the general linear model and the different analysis of variance (ANOVA) models.

In Section 17, we define what quadratic forms are, give conditions for them to be chi-square distributions, and give conditions for statistical independence. A quadratic form may always be written as a sum of squares. We show how to find a transformation to do this. The number of summands is the rank of its matrix and when the quadratic form is chi-square the number of degrees of freedom.

Section 18 presents the ANOVA for a general linear model and for experimental design models with one or two factors. To develop tests of hypothesis to determine

Matrix Algebra for Linear Models, First Edition. Marvin H. J. Gruber.
© 2014 John Wiley & Sons, Inc. Published 2014 by John Wiley & Sons, Inc.

the statistical significance of the regression models and the factors they represent we apply the results obtained in Section 17.

Section 19 presents models with two factors where either there is an interaction between them or one factor is nested within another factor.

Testing whether linear combinations of regression parameters is different from zero is taken up in Section 20. This is the general linear hypothesis. We give examples of how the hypothesis tested for the different ANOVA models are special cases of the general linear hypothesis.

SECTION 17

QUADRATIC FORMS AND THEIR PROBABILITY DISTRIBUTIONS

17.1 INTRODUCTION

Quadratic forms are very useful in the analysis of data using analysis of variance. The test statistics for tests of hypothesis to determine whether different factors are statistically significant are based on ratios of independent quadratic forms. Thus, we need some conditions to determine whether quadratic forms are independent. The main goal of this section is the development of these conditions.

In Subsection 17.2, we tell what a quadratic form is and show how quadratic forms may be reduced to sums of squares. The number of summands is equal to the rank of the matrix of the quadratic form. Subsection 17.3 establishes that the sums of squares of independent standard normal variables are a chi-square distribution. Subsection 17.4 gives conditions for a quadratic form to be a chi-square distribution. Subsection 17.5 tells when two quadratic forms of normal random variables are independent.

17.2 EXAMPLES OF QUADRATIC FORMS

Let x be an n-dimensional vector and A a symmetric $n \times n$ matrix. Then $x'Ax$ is a quadratic form. Observe that

$$x'Ax = \sum_{i=1}^{n}\sum_{j=1}^{n}a_{ij}x_ix_j = \sum_{i=1}^{n}a_{ii}x_i^2 + 2\sum_{i<j}a_{ij}x_ix_j.$$

Matrix Algebra for Linear Models, First Edition. Marvin H. J. Gruber.
© 2014 John Wiley & Sons, Inc. Published 2014 by John Wiley & Sons, Inc.

Quadratic forms used for the analysis of variance are always positive semi-definite, meaning that their matrix is positive semi-definite. For this reason, we will restrict the discussion to positive-semi-definite quadratic forms.

Example 17.1 Reducing a Quadratic Form to a Sum of Squares

Observe that

$$
\begin{aligned}
\begin{bmatrix} x_1 & x_2 \end{bmatrix} \begin{bmatrix} 2 & -1 \\ -1 & 2 \end{bmatrix} \begin{bmatrix} x_1 \\ x_2 \end{bmatrix} &= \begin{bmatrix} x_1 & x_2 \end{bmatrix} \begin{bmatrix} 2x_1 - x_2 \\ -x_1 + 2x_2 \end{bmatrix} \\
&= x_1(2x_1 - x_2) + x_2(-x_1 + 2x_2) = 2x_1^2 - 2x_1 x_2 + 2x_2^2 \\
&= 2\left(x_1 - \tfrac{1}{2}x_2\right)^2 + \tfrac{3}{2}x_2^2 = y_1^2 + y_2^2
\end{aligned}
$$

where $y_1 = \sqrt{2}\left(x_1 - \tfrac{1}{2}x_2\right), y_2 = \sqrt{\tfrac{3}{2}}x_2$.

Now the singular value decomposition of

$$
\begin{bmatrix} 2 & -1 \\ -1 & 2 \end{bmatrix} = \begin{bmatrix} -\tfrac{1}{\sqrt{2}} & \tfrac{1}{\sqrt{2}} \\ \tfrac{1}{\sqrt{2}} & \tfrac{1}{\sqrt{2}} \end{bmatrix} \begin{bmatrix} 3 & 0 \\ 0 & 1 \end{bmatrix} \begin{bmatrix} -\tfrac{1}{\sqrt{2}} & \tfrac{1}{\sqrt{2}} \\ \tfrac{1}{\sqrt{2}} & \tfrac{1}{\sqrt{2}} \end{bmatrix}.
$$

Let

$$
z_1 = \frac{1}{\sqrt{2}}(-x_1 + x_2), \; z_2 = \frac{1}{\sqrt{2}}(x_1 + x_2).
$$

Then

$$
2x_1^2 - 2x_1 x_2 + 2x_2^2 = \left(\sqrt{3}z_1\right)^2 + z_2^2. \qquad \square
$$

Example 17.2 Another Reduction of a Quadratic Form to a Sum of Squares

Consider the quadratic form

$$
\begin{aligned}
\begin{bmatrix} x_1 & x_2 & x_3 \end{bmatrix} &\begin{bmatrix} \dfrac{2}{3} & -\dfrac{1}{3} & -\dfrac{1}{3} \\ -\dfrac{1}{3} & \dfrac{2}{3} & -\dfrac{1}{3} \\ -\dfrac{1}{3} & -\dfrac{1}{3} & \dfrac{2}{3} \end{bmatrix} \begin{bmatrix} x_1 \\ x_2 \\ x_3 \end{bmatrix} \\
&= \frac{2}{3}\left(x_1^2 + x_2^2 + x_3^2\right) - \frac{2}{3}\left(x_1 x_2 + x_1 x_3 + x_2 x_3\right) \\
&= x_1^2 + x_2^2 + x_3^2 - \frac{1}{3}\left(x_1 + x_2 + x_3\right)^2.
\end{aligned}
$$

The singular value decomposition of the matrix of the above quadratic form is

$$\begin{bmatrix} \frac{2}{3} & -\frac{1}{3} & -\frac{1}{3} \\ -\frac{1}{3} & \frac{2}{3} & -\frac{1}{3} \\ -\frac{1}{3} & -\frac{1}{3} & \frac{2}{3} \end{bmatrix} = \begin{bmatrix} -\frac{1}{\sqrt{2}} & -\frac{1}{\sqrt{6}} & \frac{1}{\sqrt{3}} \\ 0 & \frac{2}{\sqrt{6}} & \frac{1}{\sqrt{3}} \\ \frac{1}{\sqrt{2}} & -\frac{1}{\sqrt{6}} & \frac{1}{\sqrt{3}} \end{bmatrix} \begin{bmatrix} 1 & 0 & 0 \\ 0 & 1 & 0 \\ 0 & 0 & 0 \end{bmatrix} \begin{bmatrix} -\frac{1}{\sqrt{2}} & 0 & \frac{1}{\sqrt{2}} \\ -\frac{1}{\sqrt{6}} & \frac{2}{\sqrt{6}} & -\frac{1}{\sqrt{6}} \\ \frac{1}{\sqrt{3}} & \frac{1}{\sqrt{3}} & \frac{1}{\sqrt{3}} \end{bmatrix}.$$

Observe that

$$\begin{bmatrix} -\frac{1}{\sqrt{2}} & -\frac{1}{\sqrt{6}} & \frac{1}{\sqrt{3}} \\ 0 & \frac{2}{\sqrt{6}} & \frac{1}{\sqrt{3}} \\ \frac{1}{\sqrt{2}} & -\frac{1}{\sqrt{6}} & \frac{1}{\sqrt{3}} \end{bmatrix} \begin{bmatrix} 1 & 0 & 0 \\ 0 & 1 & 0 \\ 0 & 0 & 0 \end{bmatrix} = \begin{bmatrix} -\frac{1}{\sqrt{2}} & -\frac{1}{\sqrt{6}} & 0 \\ 0 & \frac{2}{\sqrt{6}} & 0 \\ \frac{1}{\sqrt{2}} & -\frac{1}{\sqrt{6}} & 0 \end{bmatrix}.$$

Let

$$y_1 = \frac{1}{\sqrt{2}}(-x_1 + x_3), \quad y_2 = \frac{1}{\sqrt{6}}(-x_1 + 2x_2 - x_3).$$

Then

$$y_1^2 + y_2^2 = \frac{2}{3}\left(x_1^2 + x_2^2 + x_3^2\right) - \frac{2}{3}(x_1 x_2 + x_1 x_3 + x_2 x_3).$$

$\square$

Example 17.3 A Helmert Matrix

The singular value decomposition of

$$I - \frac{1}{n} = \begin{bmatrix} \tilde{H}'_n & \frac{1}{\sqrt{n}} 1_n \end{bmatrix} \begin{bmatrix} I_{n-1 \times n-1} & 0_{n-1 \times 1} \\ 0_{1 \times n-1} & 0_{1 \times 1} \end{bmatrix} \begin{bmatrix} \tilde{H}_n \\ \frac{1}{\sqrt{n}} 1'_n \end{bmatrix}.$$

$$= \tilde{H}'_n \tilde{H}_n$$

Let x be an n-dimensional column vector. Let $y = \tilde{H}_n x$. The dimension of the vector y is $n-1$. Then the quadratic form

$$\sum_{i=1}^{n}(x_i - \bar{x})^2 = x'\left(I - \frac{1}{n}J\right)x = x'\tilde{H}'_n \tilde{H}_n x = y'y = \sum_{i=1}^{n-1} y_i^2.$$

The matrix H_n is the Helmert matrix that was discussed in Subsection 4.3. $\square$

Quadratic forms with a positive-semi-definite matrix can always be expressed as a sum of squares. The number of terms in the sum of squares is equal to the rank of the matrix. This is expressed in Theorem 17.1.

Theorem 17.1 Let $x'Mx$ be a quadratic form where M is an $n \times n$ positive-semi-definite matrix of rank r. Then $x'Mx = \sum_{i=1}^{r} y_i^2$.

Proof. The singular value decomposition of

$$M = \begin{bmatrix} U & V \end{bmatrix} \begin{bmatrix} \Lambda_{r \times r} & 0_{r \times n-r} \\ 0_{n-r \times r} & 0_{r \times r} \end{bmatrix} \begin{bmatrix} U' \\ V' \end{bmatrix} = U\Lambda_{r \times r}U'.$$

Let $y = \Lambda^{1/2}U'x$ where $\Lambda^{1/2}$ is the diagonal matrix with elements $\lambda_i^{1/2}$ the square roots of the eigenvalues of M. Then

$$x'Mx = x'U\Lambda^{1/2}\Lambda^{1/2}U'x = y'y = \sum_{i=1}^{r} y_i^2. \qquad \blacksquare$$

17.3   THE CHI-SQUARE DISTRIBUTION

The sum of squares of n-independent standard normal random variables is a chi-square distribution with n degrees of freedom. The eventual goal is to obtain conditions for a quadratic form of normal random variables to be a chi-square distribution. The chi-square distribution for n degrees of freedom denoted by $\chi^2(n)$ takes the form

$$f(x) = \frac{1}{\Gamma\left(\frac{n}{2}\right)2^{\frac{n}{2}}} x^{n/2-1} e^{-x/2}, \quad x > 0. \tag{17.1}$$

The gamma function is defined by

$$\Gamma(\alpha) = \int_{0}^{\infty} x^{\alpha-1} e^{-x} dx. \tag{17.2}$$

A few useful facts about the gamma function that the reader may verify are

1. $\Gamma(\alpha+1) = \alpha\Gamma(\alpha)$
2. $\Gamma(n+1) = n!$
3. $\Gamma\left(\frac{1}{2}\right) = \sqrt{\pi}$

The standard normal distribution $N(0,1)$ takes the form

$$g(x) = \frac{1}{\sqrt{2\pi}} e^{-x^2/2}, \quad -\infty < x < \infty. \tag{17.3}$$

We shall also have use for the beta function that has two arguments. It takes the form

$$B(\alpha,\beta) = \frac{\Gamma(\alpha+\beta)}{\Gamma(\alpha)\Gamma(\beta)} \int_0^1 x^{\alpha-1}(1-x)^{\beta-1} dx. \tag{17.4}$$

We first show in Theorem 17.2 that the square of a standard normal random variable is in fact a chi-square distribution.

Theorem 17.2 Let X be a random variable that follows a standard normal distribution (17.3). Then $Y = X^2 \sim \chi^2(1)$.

Proof. The cumulative distribution of Y is obtained by calculating

$$H(y) = P(Y \le y) = P\left(X^2 \le y\right) = P\left(-\sqrt{y} \le X \le \sqrt{y}\right)$$

$$= \frac{1}{\sqrt{2\pi}} \int_{-\sqrt{y}}^{\sqrt{y}} e^{-x^2/2} dx.$$

The derivative

$$h(y) = H'(y) = \frac{1}{\sqrt{2\pi y}} e^{-y/2} = \frac{1}{2^{1/2}\Gamma\left(\frac{1}{2}\right)} e^{-y/2}, \quad y > 0.$$

Clearly, h(y) is a $\chi^2(1)$ distribution. ∎

The sum of n-independent chi-square distributions with one degree of freedom is a chi-square distribution with n degrees of freedom. More formally, we have the following.

Theorem 17.3 Let $Y = \sum_{i=1}^n Y_i$ where $Y_i = X_i^2$ and $X_i \sim N(0,1)$. The X_i and hence the Y_i are independent random variables. Then $Y \sim \chi^2(n)$.
    This result will follow easily after we prove the following lemma.

Lemma 17.1 Assume that U and V are independent random variables where $U \sim \chi^2(r)$ and $V \sim \chi^2(s)$. Then $Z = U + V \sim \chi^2(r+s)$.

Proof. The joint distribution of U and V is

$$h(u)k(v) = \frac{1}{2^{\frac{r+s}{2}}\Gamma(r/2)\Gamma(s/2)} u^{(r/2)-1} v^{(s/2)-1} e^{-\frac{u+v}{2}}, \quad u > 0, \ v > 0.$$

Let $w = v/(u+v)$, $z = u+v$. Then $v = wz$ and $u = z(1-w)$. The Jacobian of this transformation is

$$J = \begin{vmatrix} \dfrac{\partial u}{\partial w} & \dfrac{\partial v}{\partial w} \\[2mm] \dfrac{\partial u}{\partial z} & \dfrac{\partial v}{\partial z} \end{vmatrix} = \begin{vmatrix} -z & z \\ 1-w & w \end{vmatrix} = -z$$

$$|J| = z.$$

The joint distribution of w and z is

$$m(w, z) = \frac{1}{2^{(r+s)/2}\,\Gamma(r/2)\Gamma(s/2)}\, z^{((r+s)/2)-1} e^{-z/2} w^{(s/2)-1}(1-w)^{(r/2)-1}, \quad 0 < z < \infty,\ 0 < w < 1.$$

The marginal distribution of z is obtained by integrating $m(w,z)$ with respect to w. From (17.4)

$$\int_0^1 w^{s/2-1}(1-w)^{r/2-1}\,dw = \frac{\Gamma(r/2)\Gamma(s/2)}{\Gamma((r+s)/2)}.$$

The marginal distribution

$$p(z) = \frac{1}{2^{(r+s)/2}\,\Gamma((r+s)/2)}\, z^{((r+s)/2)-1} e^{-z/2}, \quad z > 0, \tag{17.5}$$

a $\chi^2(r+s)$ distribution. ∎

The proof may now be completed by mathematical induction. This will be left to the reader in Exercise 17.3.

17.4 WHEN DOES THE QUADRATIC FORM OF A RANDOM VARIABLE HAVE A CHI-SQUARE DISTRIBUTION?

Let x be an n-dimensional random variable with a standard normal distribution. Consider the quadratic form $x'Ax$. Suppose that the singular value decomposition of

$$A = [U \quad V] \begin{bmatrix} \Lambda_{r \times r} & 0_{r \times n-r} \\ 0_{n-r \times r} & 0_{r \times r} \end{bmatrix} \begin{bmatrix} U' \\ V' \end{bmatrix} = U\Lambda_{r \times r}U'.$$

Let $y = \Lambda_{r \times r}^{1/2}U'x$. Then $x'Ax = y'y = \sum_{i=1}^{r} y_i^2$. Now y is normal with mean zero and variance $\sigma^2\Lambda$. Then the components y_i are normally distributed with mean zero and variance λ_i^2. Thus, the quadratic form $x'Ax$ is a weighted sum of chi-square random variables with one degree of freedom unless all of the nonzero eigenvalues are one. If all the nonzero eigenvalues are one, then $x'Ax$ is a chi-square distribution with $r = \text{rank}(A)$ degrees of freedom. This may be summarized by Theorem 17.4.

Theorem 17.4 Let $X \sim N(0,1)$. Then $x'Ax$ has a chi-square distribution with $r = \text{rank}(A)$ degrees of freedom if and only if all of the nonzero eigenvalues of A are one.

A square matrix A is idempotent if and only if $A^2 = A$. ∎

Theorem 17.5 The eigenvalues of a symmetric idempotent matrix are zero or one.

Proof. Let A be an idempotent matrix. Using the singular value decomposition of A, $A^2 = U\Lambda U'U\Lambda U' = U\Lambda^2 U'$. But $U\Lambda^2 U' = U\Lambda U'$ if and only if $\Lambda^2 = \Lambda$. This is only possible if all of the diagonal elements of Λ are zero or one. ∎

Thus, we have the following corollary to Theorem 17.5.

Corollary 17.1 Let $x \sim N(0,I)$. Then $x'Ax$ has a chi-square distribution if and only if A is idempotent. ∎

17.5 WHEN ARE TWO QUADRATIC FORMS WITH THE CHI-SQUARE DISTRIBUTION INDEPENDENT?

A test to determine whether two quadratic forms $x'Ax$ and $x'Bx$ are independent where $x \sim N(0,I)$ would be to determine whether $AB = 0$. For our purposes, assume that A and B are idempotent matrices. If $AB = 0$, then after finding the singular value decomposition of A and B, $U_A\Lambda_A U'_A U_B\Lambda_B U'_B = 0$. Multiply the left-hand side by $\Lambda_A^{-1/2}U'_A$ and the right-hand side by $U_B\Lambda_B^{-1/2}$ to get $\Lambda_A^{1/2}U'_A U_B\Lambda_B^{-1/2} = 0$. Let $w = U_B\Lambda_B^{1/2}x$ and $z = U_A\Lambda_A^{1/2}x$ so that

$$\text{cov}(z, w) = \Lambda_A^{1/2}U'_A \text{var}(x)U_B\Lambda_B^{1/2} = \Lambda_A^{1/2}U'_A U_B\Lambda_B^{1/2} = 0. \tag{17.6}$$

Since z and w have a joint normal distribution, they are independent, and thus, $z'z$ and $w'w$ are independent and $x'Ax$ and $x'Bx$ are independent.

The above discussion may be summarized by the following theorem.

Theorem 17.6 Let $x \sim N(0,I)$. Let two quadratic forms $Q_A = x'Ax$ and $Q_B = x'Bx$ be such that $AB = 0$. Then Q_A and Q_B are independent.

The converse of Theorem 17.6 is also true, namely, that if the quadratic forms Q_A and Q_B are independent, $AB = 0$. See, for example, Hogg et al. (2005) for a proof. ∎

The following theorem and its corollary will be helpful for analysis of variance. It is due to Hogg and Craig.

Theorem 17.7 Let x be a vector of random variables where $x \sim N(0,I)$. Let $x'Qx = x'Q_1x + x'Q_2x$ where $x'Qx \sim \chi^2(r)$ and $x'Q_1x \sim \chi^2(r_1)$. Assume that Q_1 and Q_2 are positive semi-definite. Then Q_1 and Q_2 are independent and $x'Q_2x$ is $\chi^2(r-r_1)$.

Proof. From Corollary 17.1, Q and Q_1 are idempotent matrices. Thus, their eigenvalues are zero or one, and there exists an orthogonal matrix P such that

$$P'QP = \begin{bmatrix} I_r & 0 \\ 0 & 0 \end{bmatrix}. \qquad (17.7)$$

Since $Q = Q_1 + Q_2$,

$$\begin{bmatrix} I_r & 0 \\ 0 & 0 \end{bmatrix} = P'Q_1P + P'Q_2P = P'QP. \qquad (17.8)$$

Since Q_1 and Q_2 are positive semi-definite, the same is true for $P'Q_1P$ and $P'Q_2P$. For a positive-semi-definite matrix, each element on the principal diagonal is nonnegative. From Exercise 11.9, if an element on the principal diagonal is zero, all elements in the row and column that the zero element belongs to are zero. Thus, Equation (17.8) can be written as

$$\begin{bmatrix} I_r & 0 \\ 0 & 0 \end{bmatrix} = \begin{bmatrix} G_r & 0 \\ 0 & 0 \end{bmatrix} + \begin{bmatrix} H_r & 0 \\ 0 & 0 \end{bmatrix}. \qquad (17.9)$$

Since Q_1 is idempotent, $P'Q_1P$ is also idempotent so that

$$(P'Q_1P)^2 = P'Q_1P\begin{bmatrix} G_r & 0 \\ 0 & 0 \end{bmatrix}. \qquad (17.10)$$

Multiplication of Equation (17.9) on the left by $P'Q_1P$ yields

$$\begin{bmatrix} G_r & 0 \\ 0 & 0 \end{bmatrix} = \begin{bmatrix} G_r & 0 \\ 0 & 0 \end{bmatrix} + \begin{bmatrix} G_rH_r & 0 \\ 0 & 0 \end{bmatrix} \qquad (17.11)$$

or

$$P'Q_1P = P'Q_1P + P'Q_1PP'Q_2P \qquad (17.12)$$

so that

$$P'Q_1PP'Q_2P = 0 \qquad (17.13)$$

and $Q_1Q_2 = 0$. Thus, $x'Q_1x$ and $x'Q_2x$ are independent. Now

$$
\begin{aligned}
Q_2^2 &= (Q - Q_1)^2 \\
&= Q^2 - QQ_1 - Q_1Q + Q_1^2 \\
&= Q - 2Q_1 + Q_1 \\
&= Q - Q_1 = Q_2
\end{aligned}
\tag{17.14}
$$

so that Q_2 is idempotent and $x'Q_2x$ is chi-square with $r - r_1$ degrees of freedom. ∎

The following theorem is important to the sum of squares decomposition in analysis of variance.

Theorem 17.8 (Cochran's Theorem). Let $X_1, X_2, \ldots X_n$ be a random sample from an $N(0,1)$ distribution. Let x represent the vector of these observations. Suppose that

$$
x'x = \sum_{i=1}^{k} x'Q_i x.
$$

where the matrix Q_i has rank r_i. The k quadratic forms in the sum are independent and $x'Q_ix \sim \chi^2(r_i)$, $i = 1, 2, \ldots, k$ if and only if $\sum_{i=1}^{k} r_i = n$.

Proof. First assume that $x'x = \sum_{i=1}^{k} x'Q_ix$ and $\sum_{i=1}^{k} r_i = n$ is satisfied. Then $I = \sum_{i=1}^{k} Q_i$. Let W_j be the sum of the matrices Q_i that do not include Q_j. That is $W_j = I - Q_j$. Let R_j denote the rank of W_j. The rank of the sum of several matrices is less than or equal to the sum of the ranks. Thus, $R_j \leq \sum_{v}^{k} r_j - r_j = n - r_j$. However, $I = W_j + Q_j$. The eigenvalues of W_j are the roots of the characteristic equation $\det(W_j - \lambda I) = 0$ or equivalently $\det(I - Q_j - \lambda I) = 0$. This is equivalent to saying that each of whose roots of the characteristic equation is one minus an eigenvalue of Q_j. Each W_j has $n - R_j = r_j$ eigenvalues that are zero so that Q_j has r_j nonzero eigenvalues that are one. Thus, Q_i is an idempotent matrix and $x'Q_ix \sim \chi^2(r_i)$, $i = 1, 2, \ldots, k$. From Theorem 17.7, the random variables $x'Q_ix \sim \chi^2(r_i)$, $i = 1, 2, \ldots, k$ are independent.

On the other hand, suppose that

$$
x'x = \sum_{i=1}^{k} x'Q_i x
$$

and $x'Q_ix \sim \chi^2(r_i)$, $i = 1, 2, \ldots, k$ are independent. Then $\sum_{i=1}^{k} x'Q_ix \sim \chi^2\left(\sum_{i=1}^{k} r_i\right)$ and $x'x \sim \chi^2(n)$. Thus, $\sum_{i=1}^{k} r_i = n$. The proof is complete. ∎

Example 17.4 Sum of Squares as Sum of Two Quadratic Forms

Assume $x \sim N(0,I)$

$$
\sum_{i=1}^{n} x_i^2 = x'\left(I - \tfrac{1}{n}J\right)x + x'\tfrac{1}{n}Jx.
$$

The matrix $I-(1/n)J$ is idempotent with rank $n-1$. The matrix $\frac{1}{n}J$ is idempotent with rank 1. The product of the two matrices is zero. Thus, the two quadratic forms are independent with chi-square distributions with $n-1$ and 1 degree of freedom.   □

Example 17.5 Sum of Squares as Sum of Three Independent Quadratic Forms

Assume $x \sim N(0,I_3)$. Then

$$x'x = x'Q_1x + x'Q_2x + x'Q_3x$$

where $Q_1 = \dfrac{1}{3}J_3$

$$Q_2 = \frac{1}{2}\begin{bmatrix} 1 & -1 & 0 \\ -1 & 1 & 0 \\ 0 & 0 & 0 \end{bmatrix}.$$

$$Q_3 = \frac{1}{6}\begin{bmatrix} 1 & 1 & -2 \\ 1 & 1 & -2 \\ -2 & -2 & 4 \end{bmatrix}$$

The reader may check that these three quadratic forms are independent with $\chi^2(1)$ distributions.

A well-known and important result in statistics is the independence of the sample mean and the sample variance $\bar{x}$ and s^2 for normal populations. Suppose that x represents an n-dimensional random sample with observations $x_i \sim N(\mu,\sigma^2)$. Then

$$\sum_{i=1}^{n} \frac{(x_i-\mu)^2}{\sigma^2} = \frac{(n-1)s^2}{\sigma^2} + \frac{n(\bar{x}-\mu)^2}{\sigma^2}$$

$$= \frac{(n-1)}{\sigma^2}x'\left(I-\tfrac{1}{n}J\right)x + \frac{n}{\sigma^2}x'\left(\tfrac{1}{n}J\right)x.$$

Since $(I-(1/n)J)(1/n)J = 0$, the two quadratic forms are independent, and the independence of $\bar{x}$ and s^2 follows.   □

17.6 SUMMARY

We have defined and given examples of quadratic forms. We presented the chi-square distribution. Conditions were given for quadratic forms of normal random variables to be independent chi-square distributions.

EXERCISES

17.1   What transformation will reduce $2x_1^2 - 2x_1x_2 + 2x_2^2$ to the form $w_1^2 + w_2^2$?

17.2   Express the following quadratic forms as a sum of squares with coefficients one by finding the transformations that reduce it to that form. (Hint: Find the singular value decomposition of the matrix of the quadratic form.)

   A. $5x_1^2 - 4x_1x_2 + 2x_2^2$

   B. $4x_1^2 - 4x_1x_2 + x_2^2$

   C. $x_1^2 + x_2^2 + x_3^2 + \frac{1}{2}x_1x_2 + \frac{1}{2}x_1x_3 + \frac{1}{2}x_2x_3$

   D. $4x_1^2 + 2x_2^2 + 2x_3^2 + 4x_1x_2 + 4x_1x_3$

   E. $5x_1^2 + 5x_2^2 + 3x_3^2 + 2x_1x_2 + 6x_1x_3 + 6x_2x_3$

17.3   Complete the proof of Theorem 17.3 using the lemma and mathematical induction.

17.4   The chi-square distribution is a special case of the gamma distribution whose probability density function is

$$1(x) = \frac{1}{\Gamma(\alpha)\beta^\alpha} x^{\alpha-1} e^{-x/\beta}, \quad x > 0.$$

Assume $\beta = 1$. If X is gamma with $\alpha = m$, Y is gamma with $\alpha = n$, and X and Y are independent random variables then $X+Y$ is gamma with parameter $\alpha = m+n$. The random variables X and Y are independent.

17.5   Show that if x is a standard normal random variable, the quadratic form

$$\sum_{i=1}^{n}(x_i - \bar{x})^2 = x'\left(I - \tfrac{1}{n}J\right)x = x'\tilde{H}'_n H_n x = y'y = \sum_{i=1}^{n} y_i^2$$

is chi-square with $n-1$ degrees of freedom.

17.6   Show that $I - (1/n)J$ is idempotent.

17.7   Let a be a scalar. For what value of a is $aJ_{n\times n}$ idempotent?

17.8   Let $x \sim N(0,I)$. Suppose that $x'Ax \sim \chi^2(r)$ where A is an $n\times n$ symmetric matrix. Show that $x'(I-A)x \sim \chi^2(n-r)$.

17.9   Let $x \sim N(\mu,V)$. Then $(x-\mu)'A(x-\mu)$ is chi-square if and only if $V^{1/2}AV^{1/2}$ is idempotent. In particular if $V = \sigma^2 I$, then $(x-\mu)'A(x-\mu)/\sigma^2$ is chi-square if and only if A is idempotent.

17.10 **A.** Show that $(I_{a\times a} - (1/a)J_{a\times a}) \otimes (1/b)J_{b\times b}$ is idempotent.

   B. If $x \sim N(0, I_{ab\times ab})$, how many degrees of freedom does the chi-square random variable $x'(I_{a\times a} - (1/a)J_{a\times a}) \otimes (1/b)J_{b\times b}x$ have?

17.11 Let V be nonsingular. Suppose that $x \sim N(0,V)$. Then if $AVB = 0$, Q_A and Q_B are independent. Would the converse be true based on the fact that the converse of Theorem 17.6 is true?

17.12 Let x be a vector of random variables where $x : N(0,I)$. Suppose that $x'Qx = \sum_{i=1}^{r} x'Q_i x$. Assume that $x'Q_i x \sim \chi^2(r_i)$. Show by mathematical induction and Theorem 17.7 that the r quadratic forms are independent and that

$$x'Q_i x \sim \chi^2 \left(r - \sum_{i=1}^{r-1} r_i \right).$$

SECTION 18

ANALYSIS OF VARIANCE: REGRESSION MODELS AND THE ONE- AND TWO-WAY CLASSIFICATION

18.1 INTRODUCTION

Some examples of experimental design models and their analysis of variance (ANOVA) will be considered. This will show us how to determine if there is a statistically significant difference among more than two factors. It will also illustrate some of the properties of quadratic forms.

First, we will consider the general linear regression model. After that we will consider some of the experimental design models that are important in applications. These include the one-way and the two-way ANOVA without interaction.

18.2 THE FULL-RANK GENERAL LINEAR REGRESSION MODEL

The purpose of the ANOVA is to determine whether the linear model accounts for a statistically significant portion of the variation. Thus, the objective is to see whether a large enough portion of the variation is due to the regression model in comparison with the variation due to error. To determine this, the total sum of squares is partitioned into two statistically independent parts, a sum of squares due to regression and a sum of squares due to error.

The notation in Example 3.12 will be used here. Consider the partitioning of the corrected sum of squares

Matrix Algebra for Linear Models, First Edition. Marvin H. J. Gruber.
© 2014 John Wiley & Sons, Inc. Published 2014 by John Wiley & Sons, Inc.

$$\begin{aligned} \mathrm{SST} &= \mathbf{y}'\left(\mathbf{I} - \tfrac{1}{n}\mathbf{J}\right)\mathbf{y} \\ &= \mathrm{SSE} + \mathrm{SSR}_m \\ &= \mathbf{y}'\left(\mathbf{I} - \tilde{\mathbf{X}}(\tilde{\mathbf{X}}'\tilde{\mathbf{X}})^{-1}\tilde{\mathbf{X}}' - \tfrac{1}{n}\mathbf{J}\right)\mathbf{y} + \mathbf{y}'(\tilde{\mathbf{X}}(\tilde{\mathbf{X}}'\tilde{\mathbf{X}})^{-1}\tilde{\mathbf{X}}')\mathbf{y}. \end{aligned} \tag{18.1}$$

Let $Q_R = \tilde{\mathbf{X}}(\tilde{\mathbf{X}}'\tilde{\mathbf{X}})^{-1}\tilde{\mathbf{X}}'$ and $Q_E = \mathbf{I} - \tilde{\mathbf{X}}'(\tilde{\mathbf{X}}'\tilde{\mathbf{X}})^{-1}\tilde{\mathbf{X}}' - (1/n)\mathbf{J}$. The following may be shown by direct computation (see Exercise 18.1):

1. The matrices Q_R and Q_E are idempotent.
2. The matrix $\tilde{\mathbf{X}}(\tilde{\mathbf{X}}'\tilde{\mathbf{X}})^{-1}\tilde{\mathbf{X}}'$ has rank $m-1$ and applying Theorem 17.7, Q_E has rank $n-m$.
3. The matrix product $Q_R Q_E = 0$, so for $\mathbf{y} \sim N(0,\mathbf{I})$, the quadratic forms $\mathbf{y}'Q_R\mathbf{y}$ and $\mathbf{y}'Q_E\mathbf{y}$ are independent.
4. If $\mathbf{y} : N(0, \sigma^2\mathbf{I})$, then $\mathbf{y}'Q_R\mathbf{y}/\sigma^2 \sim x^2(m-1)$ and $\mathbf{y}'Q_E\mathbf{y}/\sigma^2 \sim x^2(n-m)$.

The distribution of the ratio of two chi-square random variables each divided by the number is the Snedecor F distribution. Thus,

$$F = \frac{Q_R/(m-1)}{Q_E/(n-m)}$$

follows an F distribution with $m-1$ degrees of freedom in the numerator and $n-m$ degrees of freedom in the denominator. This is usually worked out in a table as presented below.

Source of variation	df	Sum of squares	Mean square	F ratio
Regression	$m-1$	$\mathbf{y}'Q_R\mathbf{y}$	$\dfrac{\mathbf{y}'Q_R\mathbf{y}}{m-1}$	$F = \dfrac{\mathbf{y}'Q_R\mathbf{y}/(m-1)}{\mathbf{y}'Q_E\mathbf{y}/(n-m)}$
Error	$n-m$	$\mathbf{y}'Q_E\mathbf{y}$	$\dfrac{\mathbf{y}'Q_E\mathbf{y}}{n-m}$	
Total	$n-1$	$\mathbf{y}'\left(\mathbf{I} - \tfrac{1}{n}\mathbf{J}\right)\mathbf{y}$		

For a numerical problem, the value of the F ratio may be compared with a tabulated value at a given level of significance, for example, $\alpha = 0.05$ or $\alpha = 0.01$ to determine whether there is a statistically significant difference among the factors.

Example 18.1 ANOVA Table for a Regression Model

For the data of Example 3.13, the total sum of squares

$$\mathbf{y}'\left(\mathbf{I} - \tfrac{1}{5}\mathbf{J}\right)\mathbf{y} = 49168.3.$$

The regression sum of squares

$$\mathbf{y}'Q_R\mathbf{y} = 49149.1.$$

The error sum of squares is found by subtracting the regression sum of squares from the total sum of squares. The ANOVA table is given below.

Source of variation	df	SS	MS	F
Regression	3	49149.1	16383	853.28*
Error	1	19.2	19.2	
Total	4	49168.3		

The star indicates statistical significance at $\alpha = 0.05$. $\qquad$ □

Example 18.2 Deriving Formulae of Sums of Squares in a Particular Model

Consider the model

$$\begin{bmatrix} y_{11} \\ y_{12} \\ y_{21} \\ y_{22} \end{bmatrix} = \begin{bmatrix} 1 & 1 & 0 \\ 1 & 1 & 0 \\ 1 & 0 & 1 \\ 1 & 0 & 1 \end{bmatrix} \begin{bmatrix} \mu \\ \tau_1 \\ \tau_2 \end{bmatrix} + \varepsilon.$$

Now, partitioning the matrix as was done in Example 3.11,

$$Z = \begin{bmatrix} 1 & 0 \\ 1 & 0 \\ 0 & 1 \\ 0 & 1 \end{bmatrix}$$

and

$$\tilde{X} = Z - \frac{1}{4} JZ$$

$$= \begin{bmatrix} 1 & 0 \\ 1 & 0 \\ 0 & 1 \\ 0 & 1 \end{bmatrix} - \frac{1}{4} \begin{bmatrix} 1 & 1 & 1 & 1 \\ 1 & 1 & 1 & 1 \\ 1 & 1 & 1 & 1 \\ 1 & 1 & 1 & 1 \end{bmatrix} \begin{bmatrix} 1 & 0 \\ 1 & 0 \\ 0 & 1 \\ 0 & 1 \end{bmatrix}$$

$$= \begin{bmatrix} \frac{1}{2} & -\frac{1}{2} \\ \frac{1}{2} & -\frac{1}{2} \\ -\frac{1}{2} & \frac{1}{2} \\ -\frac{1}{2} & \frac{1}{2} \end{bmatrix}.$$

Observe that

$$\tilde{X}'\tilde{X} = \begin{bmatrix} 1 & -1 \\ -1 & 1 \end{bmatrix}$$

is singular, so the Moore–Penrose inverse is used to obtain the regression estimator. Thus,

$$
b = \begin{bmatrix} \hat{\tau}_1 \\ \hat{\tau}_2 \end{bmatrix} = \begin{bmatrix} \frac{1}{4} & -\frac{1}{4} \\ -\frac{1}{4} & \frac{1}{4} \end{bmatrix} \begin{bmatrix} \frac{1}{2} & \frac{1}{2} & -\frac{1}{2} & -\frac{1}{2} \\ -\frac{1}{2} & -\frac{1}{2} & \frac{1}{2} & \frac{1}{2} \end{bmatrix} \begin{bmatrix} y_{11} \\ y_{12} \\ y_{21} \\ y_{22} \end{bmatrix}
$$

and

$$
\hat{\tau}_1 = \frac{1}{4}(y_{11} + y_{12} - y_{21} - y_{22})
$$

$$
\hat{\tau}_2 = \frac{1}{4}(-y_{11} - y_{12} + y_{21} + y_{22}).
$$

Now

$$
\bar{y} = \frac{1}{4}(y_{11} + y_{12} + y_{21} + y_{22})
$$

and

$$
\bar{z} = \begin{bmatrix} 1 \\ 1 \end{bmatrix}.
$$

Then

$$
a = \hat{\mu} = \frac{1}{4}(y_{11} + y_{12} + y_{21} + y_{22})
$$

$$
- \left[\frac{1}{4}(y_{11} + y_{12} - y_{21} - y_{22}) \quad \frac{1}{4}(-y_{11} - y_{12} + y_{21} + y_{22}) \right] \begin{bmatrix} 1 \\ 1 \end{bmatrix}
$$

$$
= \frac{1}{4}(y_{11} + y_{12} + y_{21} + y_{22}).
$$

Now

$$
Q_R = \tilde{X}(\tilde{X}'X)^+ \tilde{X}'
$$

$$
= \frac{1}{4} \begin{bmatrix} 1 & 1 & -1 & -1 \\ 1 & 1 & -1 & -1 \\ -1 & -1 & 1 & 1 \\ -1 & -1 & 1 & 1 \end{bmatrix}
$$

and

$$
SSR = y'Q_R y = \frac{1}{4}(y_{1.} - y_{2.})^2. \qquad \square
$$

18.3   ANALYSIS OF VARIANCE: ONE-WAY CLASSIFICATION

The objective in performing a one-way ANOVA is to determine whether the mean response is different for different treatments. For example, do different fertilizers result in different yields of a particular crop? Is there a difference in the mean time before failure (MTBF) of computer hard drives made by different manufacturers? Is there a difference in average performance of students in the same course with different instructors on a common examination?

The methods used in Example 16.2 may be used to derive the estimates of the parameters and the sum of squares for the one-way ANOVA. With this in mind, consider the linear model

$$Y = \begin{bmatrix} 1_a \otimes 1_n & I_a \otimes 1_n \end{bmatrix} \begin{bmatrix} \mu \\ \tau_1 \\ \vdots \\ \tau_a \end{bmatrix} + \varepsilon. \tag{18.2}$$

The hypothesis to be tested is $H_0 \ \tau_1 = \tau_2 = \ldots = \tau_a$ versus the alternative hypothesis H_1 that at least one equality fails to hold true. It seems reasonable that if one effect is significantly different from another, the model sum of squares will be large compared to the error sum of squares.

We shall view (18.2) as a partitioned model with

$$Z = I_a \otimes 1_n.$$

Then

$$\tilde{X} = I_a \otimes 1_n - \frac{1}{na}(J_a \otimes J_n)(I_a \otimes 1_n)$$

$$= \left(I_a - \frac{1}{a} J_a \right) \otimes 1_n$$

and

$$\tilde{X}'\tilde{X} = n\left(I_a - \frac{1}{a} J_a \right). \tag{18.3}$$

The Moore–Penrose inverse of 18.3 is

$$(\tilde{X}'\tilde{X})^+ = \frac{1}{n}\left(I_a - \frac{1}{a} J_a \right) \tag{18.4}$$

exploiting the fact that $\left(I_a - \frac{1}{a} J_a \right)$ is a symmetric idempotent matrix and hence its own Moore–Penrose inverse.

Then

$$\hat{\tau} = \left(I_a - \frac{1}{a}J_a\right) \otimes \frac{1_n}{n}y.$$

Componentwise,

$$\tilde{\tau}_i = \overline{y}_i - \overline{y}, \quad 1 \le i \le a. \tag{18.5a}$$

Also

$$\hat{\mu} = \overline{y}.. - \hat{\tau}1_n = \overline{y}.. \tag{18.5b}$$

Now

$$Q_R = \tilde{X}(\tilde{X}'\tilde{X})^+\tilde{X}' = \left(I_a - \frac{1}{a}J_a\right) \otimes \frac{1}{n}J_n$$

so

$$SSR = y'Q_R y = y'\left(\frac{1}{n}(I_a \otimes J_n) - \frac{1}{na}(J_a \otimes J_n)\right)y.$$
$$= \frac{1}{n}\sum_{i=1}^{a}y_i^2 - \frac{1}{na}y^2... \tag{18.6}$$

Furthermore,

$$Q_E = I_a \otimes I_n - \frac{1}{n}(I_a \otimes J_n)$$

so that

$$SSE = y'Q_E y = y'\left(I_a \otimes I_n - I_a \otimes \frac{1}{n}J_n\right)y$$
$$= \sum_{i=1}^{a}\sum_{j=1}^{n}y_{ij}^2 - \frac{1}{n}\sum_{i=1}^{a}y_i^2.$$

The ANOVA table would be

Source of variation	df	Sum of squares	Mean square	F ratio
Regression	$a-1$	$y'Q_R y$	$\dfrac{y'Q_R y}{a-1}$	$F = \dfrac{y'Q_R y/(a-1)}{y'Q_E y/(na-a)}$
Error	$na-a$	$y'Q_E y$	$\dfrac{y'Q_E y}{na-a}$	
Total	$na-1$	$y'(Q_R + Q_E)y$		

Example 18.3 ANOVA for Comparison of Student Performance at Three Different Universities

The data below is taken from Exercise 10.1.3 in Gruber (2010). Five sophomore Mathematics students at three universities are given a standard Calculus test. Is there a difference in the average performance of these three groups of students?

University	A	B	C
1	85	81	90
2	82	84	65
3	86	76	82
4	85	79	84
5	78	81	76

The model is

$$
Y = \begin{bmatrix} 1_5 & 1_5 & 0 & 0 \\ 1_5 & 0 & 1_5 & 0 \\ 1_5 & 0 & 0 & 1_5 \end{bmatrix} \begin{bmatrix} \mu \\ \tau_1 \\ \tau_2 \\ \tau_3 \end{bmatrix} + \varepsilon.
$$

Now

$$
y_{1.} = 416, \, y_{2.} = 401, \, y_{3.} = 397, \, y_{..} = 1214.
$$

The least square estimates are

$$
\hat{\mu} = 80.933, \, \hat{\tau}_1 = 2.26667, \, \hat{\tau}_2 = -0.73333, \, \hat{\tau}_3 = -1.5333.
$$

The regression sum of squares also called the between sum of squares is

$$
SS \, (\text{Universities}) = \frac{y_{1.}^2}{5} + \frac{y_{2.}^2}{5} + \frac{y_{3.}^2}{5} - \frac{y_{..}^2}{15}.
$$

We have, after substitution of the values

$$
SS \, (\text{Universities}) = \frac{416^2}{5} + \frac{401^2}{5} + \frac{397^2}{5} - \frac{1214^2}{15} = 40.133.
$$

The total sum of squares is found by adding up the squares of all fifteen observations and subtracting the correction for the mean. Thus,

$$
TSS = 476.933.
$$

Then the ANOVA table is as follows.

Source of variation	df	Sum of squares	Mean square	F ratio
Regression (Universities)	2	40.133	20.061	$F = 0.55 < 89$
Error	12	436.6	36.6	
Total	14	476.933		

We conclude that there is not a statistically significant difference in the performance of the students at the three different universities at $\alpha = 0.10$.   □

18.4   ANALYSIS OF VARIANCE: TWO-WAY CLASSIFICATION

An experiment may have more than one factor to be analyzed. For example, we might want to compare the effect of five different kinds of fertilizer on the yield in five different parts of the United States. We might like to study the yield of a process for five different levels of temperature and five different levels of pressure. Example 18.4 illustrates how we can compare students' GPA based on two factors, their major, and their class status. These kinds of problems can be studied by two-factor ANOVA.

The methodology of 18.2 with a few additions may be used to derive the least square estimators and the associated sums of squares for two-factor ANOVA. The linear model to be considered is

$$Y = X\tau + \varepsilon \tag{18.7}$$
$$= \left[1_a \otimes 1_b \quad I_a \otimes 1_b \quad 1_a \otimes 1_b \right] \tau + \varepsilon$$

where

$$\tau' = [\mu \quad \alpha' \quad \beta']$$
$$= [\mu, \alpha_1, \alpha_2, ..., \alpha_a, \beta_1, \beta_2, ..., \beta_b].$$

The hypothesis to be tested is

$$H_0 \; \alpha_1 = \alpha_2 = ... = \alpha_a$$

versus an alternative hypothesis that at least one of these $a(a-1)/2$ equalities does not hold true. Likewise we can test

$$H_0 \; \beta_1 = \beta_2 = ... = \beta_b$$

versus an alternative hypothesis that at least one of these $b(b-1)/2$ equalities does not hold true.

We shall partition the X matrix in a manner similar to what was done for the one-way ANOVA.

Thus, we have

$$Z = [I_a \otimes 1_b \quad 1_a \otimes I_b]$$

and

$$\tilde{X} = Z - \frac{1}{ab}(J_a \otimes J_b) Z$$

$$= \left[\left(I_a - \frac{1}{a} J_a \right) \otimes 1_b \quad 1_a \otimes \left(I_b - \frac{1}{b} J_b \right) \right].$$

Then

$$\tilde{X}'\tilde{X} = \begin{bmatrix} b\left(I_a - \frac{1}{a} J_a \right) \otimes 1 & 0 \\ 0 & 1 \otimes a\left(I_b - \frac{1}{a} J_b \right) \end{bmatrix}$$

and

$$\tilde{X}'\tilde{X} = \begin{bmatrix} \frac{1}{b}\left(I_a - \frac{1}{a} J_a \right) \otimes 1 & 0 \\ 0 & 1 \otimes \frac{1}{a}\left(I_b - \frac{1}{a} J_b \right) \end{bmatrix}.$$

Then

$$\begin{bmatrix} \hat{\alpha} \\ \hat{\beta} \end{bmatrix} = (\tilde{X}'\tilde{X})^+ \tilde{X}'Y$$

$$= \begin{bmatrix} \frac{1}{b}\left(I_a - \frac{1}{a} J_a \right) \otimes 1 & 0 \\ 0 & 1 \otimes \frac{1}{a}\left(I_b - \frac{1}{a} J_b \right) \end{bmatrix} \begin{bmatrix} \left(I_a - \frac{1}{a} J_a \right) \otimes 1_b' \\ 1_a' \otimes \left(I_b - \frac{1}{a} J_b \right) \end{bmatrix} Y$$

$$= \begin{bmatrix} \left(I_a - \frac{1}{a} J_a \right) \otimes \frac{1_b'}{b} \\ \frac{1_a'}{a} \otimes \left(I_b - \frac{1}{a} J_b \right) \end{bmatrix} Y.$$

The components are

$$\hat{\alpha}_i = \bar{y}_{i.} - \bar{y}.., \quad 1 \le i \le a$$

$$\hat{\beta}_j = \bar{y}_{j.} - \bar{y}.., \quad 1 \le j \le b. \tag{18.8}$$

Again

$$\hat{\mu} = \bar{y}.. - \hat{\tau}1_n = \bar{y}.. \, .$$

Now

$$Q_R = \tilde{X}(\tilde{X}'\tilde{X})^+\tilde{X}'$$

$$= \left[\left(I_a - \frac{1}{a}J_a\right) \otimes 1_b \quad 1_a \otimes \left(I_b - \frac{1}{b}J_b\right)\right]\left[\begin{array}{c}\left(I_a - \frac{1}{a}J_a\right) \otimes \frac{1'_b}{b} \\ \frac{1'_a}{a} \otimes \left(I_b - \frac{1}{a}J_b\right)\end{array}\right]$$

$$= \left(I_a - \frac{1}{a}J_a\right) \otimes \frac{J_b}{b} + \frac{J_a}{a} \otimes \left(I_b - \frac{1}{b}J_b\right).$$

Then the model sum of squares, the sum of squares due to regression, is

$$\text{SSR} = y'\left(I_a - \frac{1}{a}J_a\right) \otimes \frac{J_b}{b}y + y'\frac{J_a}{a}\left(I_b - \frac{1}{b}J_b\right)y. \qquad (18.9)$$

For the submodel

$$Y = \begin{bmatrix} 1_a \otimes 1_b & I_a \otimes 1_b \end{bmatrix}\begin{bmatrix} \mu \\ \alpha_1 \\ \vdots \\ \alpha_a \end{bmatrix} + \varepsilon$$

the first term of (18.9) is the regression sum of squares due to the α parameters. For the model where the α coefficients are omitted, the second term is the sum of squares due to the β parameters. Thus,

$$\text{SSA} = y'Q_A y = y'\left(I_a - \frac{1}{a}J_a\right) \otimes \frac{J_b}{b}y$$

$$= \frac{1}{b}\sum_{i=1}^{a}y_{i.}^2 - \frac{1}{ab}y^2..$$

and

$$\text{SSB} = y'Q_B y = y'\frac{J_a}{a} \otimes \left(I_b - \frac{1}{b}J_b\right)y$$

$$= \frac{1}{a}\sum_{j=1}^{b}y_{.j}^2 - \frac{1}{ab}y^2..$$

Also

$$SSE = SST - SSA - SSB$$
$$= y'Q_E y$$
$$= y'\left(I_a \otimes I_b - I_a \otimes \frac{1}{b} J_b - \frac{1}{a} J_a \otimes I_b + \frac{1}{a} J_a \otimes \frac{1}{b} J_b \right) y.$$
$$= \sum_{i=1}^{a} \sum_{j=1}^{b} y_{ij}^2 - \frac{1}{b} \sum_{i=1}^{a} y_{i.}^2 - \frac{1}{a} \sum_{j=1}^{b} y_{.j}^2 + \frac{1}{ab} y_{..}^2.$$

The ANOVA table follows.

Source of variation	df	Sum of squares	Mean square	F
A	$a-1$	$y'Q_A y$	$\dfrac{y'Q_A y}{a-1}$	$\dfrac{y'Q_A y/(a-1)}{y'Q_E y/(a-1)(b-1)}$
B	$b-1$	$y'Q_B y$	$\dfrac{y'Q_B y}{b-1}$	$\dfrac{y'Q_B y/(b-1)}{y'Q_E y/(a-1)(b-1)}$
Error	$(a-1)(b-1)$	$y'Q_E y$	$\dfrac{y'Q_E y}{(a-1)(b-1)}$	
Total	$ab-1$	$y(Q_A + Q_B + Q_E)y$		

Example 18.4 Two-Way ANOVA Example: Student's GPA Based on Major and Class Status

The data below looks at student's GPA based on their major and class status. It was formulated by Larry Green, a Professor of Mathematics at Lake Tahoe Community College, and is being used with his permission.

Major	Freshman	Class Sophomore	Status Junior	Senior
Science	2.8	3.1	3.2	2.7
Humanities	3.3	3.5	3.6	3.1
Other	3.0	3.2	2.9	3.0

The model would be

$$Y = \begin{bmatrix} 1_4 & 1_4 & 0 & 0 & 1_4 \\ 1_4 & 0 & 1_4 & 0 & 1_4 \\ 1_4 & 0 & 0 & 1_4 & 1_4 \end{bmatrix} \begin{bmatrix} \mu \\ \tau_1 \\ \tau_2 \\ \tau_3 \\ \beta_1 \\ \beta_2 \\ \beta_3 \\ \beta_4 \end{bmatrix} + \varepsilon.$$

We have

$$y_{1.} = 11.8, \ y_{2.} = 13.5, \ y_{3.} = 12.1,$$
$$y_{.1} = 9.1, \ y_{.2} = 9.8, \ y_{.3} = 9.7, \text{ and}$$
$$y_{..} = 37.4.$$

Then

$$SS \ (\text{Major}) = \frac{y_{1.}^2}{4} + \frac{y_{2.}^2}{4} + \frac{y_{3.}^2}{4} - \frac{y_{..}^2}{12}$$
$$= \frac{11.8^2}{4} + \frac{13.5^2}{4} + \frac{12.1^2}{4} - \frac{37.4^2}{12}$$
$$= 0.411667,$$

$$SS \ (\text{Class}) = \frac{y_{.1}^2}{3} + \frac{y_{.2}^2}{3} + \frac{y_{.3}^2}{3} - \frac{y_{..}^2}{12}$$
$$= \frac{9.1^2}{3} + \frac{9.8^2}{3} + \frac{9.7^2}{3} - \frac{8.8^2}{3} \frac{37.4^2}{12}$$
$$= 0.23,$$

$$TSS = \sum_{i=1}^{3} \sum_{j=1}^{4} y_{ij}^2 - \frac{y_{..}^2}{12} = 0.77667,$$

and

$$SSE = TSS - SS(\text{Major}) - SS(\text{Class})$$
$$= 0.77667 - 0.411667 - 0.23$$
$$= 0.135.$$

The ANOVA table is given below.

Source of variation	Degrees of Freedom	Sum of squares	Mean square	F
Major	2	0.411667	0.205833	$9.1485 > 5.14$
Class	3	0.23	0.0766667	$3.40741 < 4.76$
Error	6	0.135	0.0225	
Total	11	0.77667		

At $\alpha = 0.05$, there is a significant difference in GPA for majors but not for the year of study.   □

18.5 SUMMARY

We showed how to determine whether there was significant variation to be attributed to a regression model for three cases, the full-rank linear regression model, the one-way ANOVA model, and the two-way ANOVA model without interaction. Using the formulae for the inverse of a partitioned matrix developed in Section 3, we showed how to obtain least square estimators for the ANOVA models. We gave three numerical examples to illustrate the results.

EXERCISES

18.1 Show that

    A. Q_R and Q_E are idempotent matrices.

    B. The matrix $\tilde{X}(\tilde{X}'\,\tilde{X})^{-1}\tilde{X}'$ has rank $m-1$ and using Theorem 17.7 Q_E has rank $n-m$.

    C. $Q_R Q_E = 0$, so for $y \sim N(0,I)$, the quadratic forms $y'Q_R y$ and $y'Q_E y$ are independent.

    D. If $y \sim N(0, \sigma^2 I)$, then $y'Q_R y/\sigma^2 \sim x^2(m-1)$ and $y'Q_E y/\sigma^2 \sim x^2(n-m)$.

18.2 Show that in Example 18.2

$$\text{SSR} = y'Q_R y = \frac{1}{4}(y_{1.} - y_{2.})^2 = \frac{y_{1.}^2}{2} + \frac{y_{2.}^2}{2} - \frac{y_{..}^2}{4}.$$

18.3 Redo Example 18.2 for the model

$$Y = \begin{bmatrix} 1_3 & 1_3 & 0 & 0 \\ 1_3 & 0 & 1_3 & 0 \\ 1_3 & 0 & 0 & 1_3 \end{bmatrix} \begin{bmatrix} \mu \\ \tau_1 \\ \tau_2 \\ \tau_3 \end{bmatrix} + \varepsilon.$$

18.4 **A.** Redo Example 18.2 for the model

$$Y = \begin{bmatrix} 1 & 1 & 0 & 1 & 0 \\ 1 & 1 & 0 & 0 & 1 \\ 1 & 0 & 1 & 1 & 0 \\ 1 & 0 & 1 & 0 & 1 \end{bmatrix} \begin{bmatrix} \mu \\ \tau_1 \\ \tau_2 \\ \beta_1 \\ \beta_2 \end{bmatrix} + \varepsilon.$$

    B. Obtain an expression for $SS(\beta)$ as the difference between the regression sum of squares and the sum of squares in Example 18.2.

18.5 The data on which this problem is based was compiled by the National Center for Statistics and Analysis. For five states in different parts of the country, the data represents the number of speeding-related fatalities by road type and speed limit in miles per hour during 2003 on non-interstate highways.

State / speed limit	55	50	45	40	35	<35
California	397	58	142	107	173	156
Florida	80	13	150	49	80	75
Illinois	226	3	22	47	69	88
New York	177	10	23	30	23	80
Washington	16	38	15	18	53	43

Perform a one-way ANOVA to determine if there is a significant difference in the average number of fatalities for the different speed limits.

18.6 For the data below, the number of speeding-related fatalities was divided by the number of licensed drivers in millions. These statistics were obtained from the Department of Transportation Highway Statistics. The following was obtained.

State/speed limit	55	50	45	40	35	<35
California	17.52	2.56	6.27	4.72	7.74	6.89
Florida	6.20	1.01	11.62	3.80	6.20	5.81
Illinois	28.1	0.372	2.73	5.84	8.57	10.93
New York	15.59	0.881	2.03	2.64	2.03	7.04
Washington	3.63	8.62	3.40	4.08	12.03	9.76

Do a two-way ANOVA to determine

A. Whether there is significant per capita difference in speeding-related highway fatalities between the different states

B. Whether there is significant per capita difference in speeding-related highway fatalities for different speed limits

18.7 Show that the quadratic forms SSA, SSB, and SSE are independent each having a chi-square distribution with $(a-1)$, $(b-1)$, and $(a-1)(b-1)$ degrees of freedom.

18.8 Show that for a least square estimator b, $b'X'Xb$ is independent of the choice of generalized inverse so that the model sum of squares is the same regardless of which least square estimator is found.

18.9 This exercise is based on the approach taken to the one-way classification on pp. 253–255 of Gruber (2010).

Consider the model

$$Y = X \begin{bmatrix} \mu \\ \tau \end{bmatrix} + \varepsilon$$

$$= \begin{bmatrix} 1_a \otimes 1_n & I_a \otimes 1_n \end{bmatrix} \begin{bmatrix} \mu \\ \tau_1 \\ \vdots \\ \tau_a \end{bmatrix} + \varepsilon.$$

A. Show that

$$X'X = \begin{bmatrix} an & n1'_a \\ n1_a & nI_a \end{bmatrix}.$$

B. Show that a generalized inverse of $X'X$ is

$$G = \begin{bmatrix} \frac{1}{na} & 0 \\ -\frac{1}{na}1_a & \frac{1}{n}I_a \end{bmatrix}.$$

C. Show that the least square estimator is given by (18.5a) and (18.5b).
D. Let

$$b = \begin{bmatrix} \hat{\mu} \\ \hat{\tau}_1 \\ \vdots \\ \hat{\tau}_a \end{bmatrix}.$$

Show that

$$SSR = b'X'Xb - \frac{Y^2_{..}}{na}.$$

18.10 Consider the linear model

$$Y = X\beta + \varepsilon = \begin{bmatrix} 1_{n_1} & 1_{n_1} & 0 & \cdots & 0 \\ 1_{n_2} & 0 & 1_{n_2} & \cdots & 0 \\ \vdots & \vdots & \vdots & \ddots & 0 \\ \vdots & \vdots & \vdots & \vdots & \vdots \\ 1_{n_a} & 0 & 0 & 0 & 1_{n_a} \end{bmatrix} \begin{bmatrix} \mu \\ \tau_1 \\ \tau_2 \\ \vdots \\ \tau_a \end{bmatrix} + \varepsilon$$

where $N = \sum_{i=1}^a n_a$. Show that

$$SSR = \sum_{i=1}^a \frac{y_{i.}^2}{n_i} - \frac{y_{..}^2}{N}.$$

18.11 The life lengths of four different brands of lightbulbs are being compared. Do an ANOVA using the results of Exercise 18.9 to determine if there is a difference in the life of the bulbs of the different brands.

A	B	C	D
915	1011	989	1055
912	1001	979	1048
903	1003	1004	1061
893	992	992	1068
910	981	1008	1053
890	1001	1009	1063
879	989	996	
	1003	998	
		997	

18.12 This exercise is based on the discussion on page 268 of Gruber (2010). Consider the model (18.7).

A. Re-establish that

$$X'X = \begin{bmatrix} ab & bl'_a & al'_b \\ bl_a & bI_a & J_{a\times b} \\ al_b & J_{b\times a} & aI_b \end{bmatrix}.$$

B. Re-establish that

$$G = \begin{bmatrix} \frac{1}{ab} & 0 & 0 \\ -\frac{1}{ab}1_a & \frac{1}{b}I_a & 0 \\ -\frac{1}{ab}1_b & 0 & \frac{1}{b}I_a \end{bmatrix}$$

is a generalized inverse of $X'X$.

C. Using the generalized inverse in B, Establish that the least square estimators are

$$\hat{\mu} = \bar{y}..$$
$$\hat{\alpha}_i = \bar{y}_{i.} - \bar{y}.., \quad 1 \le i \le a$$
$$\hat{\beta}_j = \bar{y}_{.j} - \bar{y}.., \quad 1 \le j \le b.$$

D. Using the results of A, B, and C, Establish the formulae for SSR, SSA, and SSB.

SECTION 19

MORE ANOVA

19.1 INTRODUCTION

This section will extend the work on the two-way analysis of variance (ANOVA) to situations where there is interaction between the factors and to the case where one factor is nested inside another. In running an experiment the response may in addition to depending on two factors also depend on the interaction of the two factors. For example, if two factors for laundry are the brand of detergent and the temperature of the water in the washing machine and the response is the amount of dirt removed, maybe brand A removes more dirt for a low temperature than a high temperature, but brand B removes less dirt at a low temperature than a high temperature. Then there might be a significant interaction between the brand of the detergent and the temperature of the water. We consider two-way ANOVA with interaction in Subsection 19.2.

As an example of a nested design, consider a company that purchases raw material from several different suppliers. The company wishes to determine if the purity of the raw material is the same for each supplier. Suppose there are four batches available from each supplier and several determinations are taken for each batch. The factors are suppliers and batches. However, the batches are unique for each supplier. That is why the factors are nested instead of crossed. The nested design is considered in Subsection 19.3.

Matrix Algebra for Linear Models, First Edition. Marvin H. J. Gruber.
© 2014 John Wiley & Sons, Inc. Published 2014 by John Wiley & Sons, Inc.

19.2   THE TWO-WAY CLASSIFICATION WITH INTERACTION

In order to study interaction, the levels of the experiment have to be run more than once. Thus, we consider a linear model of the form

$$Y = [1_a \otimes 1_b \otimes 1_c \quad I_a \otimes 1_b \otimes 1_c \quad 1_a \otimes I_b \otimes 1_c \quad I_a \otimes I_b \otimes 1_c]\tau + \varepsilon,$$

where

$$\tau = \begin{bmatrix} \mu \\ \alpha \\ \beta \\ \alpha\beta \end{bmatrix}.$$

The a-dimensional vector has elements α_i, $1 \le i \le a$. Likewise, the b-dimensional vector has elements $\beta_j, 1 \le j \le b$. The interaction vector is of dimension ab and has elements $(\alpha\beta)_{ij}, 1 \le i \le a, 1 \le j \le b$. The error vector ε has elements $\varepsilon_{ijk}, 1 \le i \le a, 1 \le j \le b,$ $1 \le k \le c$. Componentwise the model may be written

$$y_{ijk} = \mu + \alpha_i + \beta_j + (\alpha\beta)_{ij} + \varepsilon_{ijk}, \quad 1 \le i \le a, 1 \le j \le b, 1 \le k \le c.$$

The methodology of the last two subsections becomes a bit cumbersome, so a different approach will be used here.

We will estimate $\Theta = X\tau$. The components are

$$\theta_{ij} = \mu + \alpha_i + \beta_j + (\alpha\beta)_{ij}, \quad 1 \le i \le a, 1 \le j \le b.$$

So we reparameterize the model to

$$Y = (I_a \otimes I_b \otimes 1_c)\Theta + \varepsilon. \tag{19.1}$$

The least square estimator for Θ is

$$\hat{\theta}_{ij} = \bar{y}_{ij}, \quad 1 \le i \le a, 1 \le j \le b.$$

Using the estimators for $\hat{\alpha}_i$ and $\hat{\beta}_j$ from the previous section and noting that $\hat{\mu} = \bar{y}_{...}$, we get

$$\widehat{(\alpha\beta)}_{ij} = \bar{y}_{ij.} - \bar{y}_{i..} - \bar{y}_{.j.} + \bar{y}_{...}, \quad 1 \le i \le a, 1 \le j \le b.$$

The model sum of squares is for 19.1 given by

$$y'(I_a \otimes I_b \otimes 1_c)\left(I_a \otimes I_b \otimes \frac{1}{c}1\right)(I_a \otimes I_b \otimes 1'_c)y.$$

$$= y'\left(I_a \otimes I_b \otimes \frac{1}{c}J_c\right)y$$

$$= \frac{1}{c}\sum_{j=1}^{k} y_{ij.}^2.$$

With adjustment for the means, the model sum of squares is

$$SSM = y'\left(I_a \otimes I_b \otimes \frac{1}{c}J_c\right)y - y'\left(\frac{1}{a}J_a \otimes \frac{1}{b}J_b \otimes \frac{1}{c}J_c\right)y.$$

$$= \frac{1}{c}\sum_{i=1}^{a}\sum_{j=1}^{b} y_{ij.}^2 - \frac{y_{...}^2}{abc}$$

For an experiment with replications, it can be shown that

$$SSA = y'Q_A y$$

$$= y'\left[\left(I_a - \frac{1}{a}J_a\right) \otimes \frac{J_b}{b} \otimes \frac{J_c}{c}\right]y$$

$$= \frac{1}{bc}\sum_{i=1}^{a} y_{i..}^2 - \frac{1}{abc}y_{...}^2$$

and

$$SSB = y'Q_B y$$

$$= y'\left[\frac{J_a}{a} \otimes \left(I_b - \frac{1}{b}J_b\right) \otimes \frac{1}{c}J_c\right]y$$

$$= \frac{1}{ac}\sum_{j=1}^{b} y_{.j.}^2 - \frac{1}{abc}y_{...}^2.$$

Then the sum of squares due to interactions is

$$SS(AB) = SSM - SSA - SSB.$$

As a quadratic form

$$SS(AB) = y'Q_I y,$$

where

$$Q_I = \left(I_a - \frac{1}{a}J_a\right) \otimes \left(I_b - \frac{1}{b}J_b\right) \otimes \frac{1}{c}J_c.$$

The sum of squares due to error takes the form

$$SSE = y'Q_E y,$$

where

$$Q_E = I_a \otimes I_b \otimes I_c - I_a \otimes I_b \otimes \frac{1}{c}J_c + \frac{1}{a}J_a \otimes \frac{1}{b}J_b \otimes \frac{1}{c}J_c.$$

The ANOVA table takes the form

Source of variation	DF	Sum of squares	Mean square	F
A	$a-1$	$y'Q_A y$	$\dfrac{y'Q_A y}{a-1}$	$\dfrac{y'Q_A y / (a-1)}{y'Q_E y / ab(c-1)}$
B	$b-1$	$y'Q_B y$	$\dfrac{y'Q_B y}{b-1}$	$\dfrac{y'Q_B y / (b-1)}{y'Q_E y / ab(c-1)}$
Interaction $(A \times B)$	$(a-1)(b-1)$	$y'Q_I y$	$\dfrac{y'Q_I y}{(a-1)(b-1)}$	$\dfrac{y'Q_I y / (a-1)(b-1)}{y'Q_E y / ab(c-1)}$
Error	$ab(c-1)$	$y'Q_E y$	$\dfrac{y'Q_E y}{ab(c-1)}$	
Total	$abc-1$	$y'(Q_A + Q_B + Q_I + Q_E)y$		

The F statistics given in this table are for a fixed effects model.

Example 19.1 A Two-Way ANOVA with Interaction

This data is from Section 14 Problem 14.2 of Montgomery and Runger (2007) reproduced with the permission of John Wiley & Sons. An engineer suspects that the surface finish of metal paint is influenced by the type of paint used and the drying times. He selected three drying times, 20, 25, and 30 min, and used two types of paint. Three parts are tested with each combination of paint and drying time. The data are as follows.

	Drying	Time	min
Paint	20	25	30
1	74	73	78
	64	61	85
	50	44	92
2	92	98	66
	86	73	45
	68	88	85

Source: Montgomery, D.C. and G.C. Runger: Applied Statistics and Probability for Engineers Reproduced by permission of John Wiley and Sons.

Factor A is the paint, so a = 2. Factor B is the drying time, so b = 3. There are three replications so that c = 3. Assign i = 1 to paint 1 and i = 2 to paint 2. Assign j = 1 to 20 min,

$j = 2$ to 25 min, and $j = 3$ to 30 min. Then $x_{1..} = 621$ and $x_{2..} = 701$. Also $x.. = 1322$.
Then

$$SSA = \frac{621^2}{9} + \frac{701^2}{9} - \frac{1322^2}{18} = 355.556.$$

To compute SSB, observe that $x_{.1.} = 434$, $x_{.2.} = 437$, and $x_{.3.} = 451$. Then

$$SSB = \frac{434^2}{6} + \frac{437^2}{6} + \frac{451^2}{6} - \frac{1322^2}{18} = 27.4444.$$

For interaction we need to calculate the cell sums. We have $x_{11.} = 188$, $x_{12.} = 178$, $x_{13.} = 255$, $x_{21.} = 246$, $x_{22.} = 259$, and $x_{23.} = 196$. Now

$$SSM = \frac{188^2}{3} + \frac{178^2}{3} + \frac{255^2}{3} + \frac{246^2}{3} + \frac{259^2}{3} + \frac{196^2}{3} - \frac{1322^2}{18} = 2261.78$$

and

$$SSI = SSM - SSA - SSB = 2261.78 - 355.556 - 27.4444 = 1878.78.$$

The total sum of squares corrected for the mean is

$$TSS = 101598 - \frac{1322^2}{18} = 4504.44$$

and the error sum of squares is

$$SSE = SST - SSA - SSB - SSI = 4504.44 - 355.556 - 27.444 - 1878.78$$
$$= 2242.66.$$

Summarizing this in the ANOVA table

Source	DF	Sum of squares	Mean square	F
Paint	1	355.556	355.556	1.90251
Drying time	2	27.4444	13.7222	0.07342
Interaction	2	1878.78	939.39	5.02649*
Error	12	2242.66	186.888	
Total	17	4504.44		

it appears that the finish is not affected by either the paint or the drying time. However, there is a significant interaction between the choice of paint and the drying time. The single asterisk means that the interaction is significant at $\alpha = 0.05$ but not at $\alpha = 0.01$. $\square$

19.3   THE TWO-WAY CLASSIFICATION WITH ONE FACTOR NESTED

In Subsection 19.2, we considered a two-way ANOVA where the factors were crossed. We showed how to determine whether the factors were significant and whether there was significant interaction. This subsection considers the case where the factors are nested.

Consider an experimental design where there are a levels of factor A and b levels of factor B nested in A. There are c replications of the response variable that is being measured. The linear model takes the form

$$Y = [1_a \otimes 1_b \otimes 1_c \quad I_a \otimes 1_b \otimes 1_c \quad I_a \otimes I_b \otimes 1_c]\tau + \varepsilon,$$

where

$$\tau = \begin{bmatrix} \mu \\ \alpha \\ \beta \end{bmatrix}.$$

Componentwise the model may be written

$$y_{ijk} = \mu + \alpha_i + \beta_{j(i)} + \varepsilon_{(ij)k}, \quad 1 \le i \le a,\ 1 \le j \le b,\ 1 \le k \le c.$$

Again we will estimate the basic estimable functions $\Theta = X\tau$, where

$$\theta_{ij} = \mu + \alpha_i + \beta_{j(i)}, \quad 1 \le i \le a,\ 1 \le j \le b.$$

Thus, the model is reparameterized to

$$Y = \Theta + \varepsilon$$

or

$$y_{ijk} = \theta_{ij} + \varepsilon_{ijk}, \quad 1 \le i \le a,\ 1 \le j \le b,\ 1 \le k \le c.$$

The estimator

$$\hat{\theta}_{ij} = \bar{y}_{ij.}, \quad 1 \le i \le a,\ 1 \le j \le b.$$

Then

$$\hat{\mu} = \bar{y}_{...},$$

$$\hat{\alpha}_i = \bar{y}_{i..} - \bar{y}_{...}, \quad 1 \le i \le a,$$

and

$$\hat{\beta}_{j(i)} = \overline{y}_{ij.} - \overline{y}_{i..}, \quad 1 \le i \le a, \ 1 \le j \le b.$$

Again

$$\widehat{(\alpha\beta)}_{ij} = \overline{y}_{ij.} - \overline{y}_{i..} - \overline{y}_{.j.} + \overline{y}_{...}, \quad 1 \le i \le a, \ 1 \le j \le b.$$

The model sum of squares is given by

$$y'(I_a \otimes I_b \otimes 1_c)\left(I_a \otimes I_b \otimes \frac{1}{c}1\right)(I_a \otimes I_b \otimes 1'_c)y$$

$$= y'\left(I_a \otimes I_b \otimes \frac{1}{c}J_c\right)y$$

$$= \frac{1}{c}\sum_{j=1}^{k} y_{ij.}^2$$

With adjustment for the means, the model sum of squares is

$$= \frac{1}{c}\sum_{i=1}^{a}\sum_{j=1}^{b} y_{ij.}^2 - \frac{y_{...}^2}{abc}$$

$$\text{SS(Model)} = y'\left(I_a \otimes I_b \otimes \frac{1}{c}J_c\right)y - y'\left(\frac{1}{a}J_a \otimes \frac{1}{b}J_b \otimes \frac{1}{c}J_c\right)y$$

$$= \frac{1}{c}\sum_{i=1}^{a}\sum_{j=1}^{b} y_{ij.}^2 - \frac{y_{...}^2}{abc}$$

For an experiment with replications, it can be shown that

$$\text{SSA} = y'Q_A y$$

$$= y'\left[\left(I_a - \frac{1}{a}J_a\right) \otimes \frac{J_b}{b} \otimes \frac{J_c}{c}\right]y$$

$$= \frac{1}{bc}\sum_{i=1}^{a} y_{i..}^2 - \frac{1}{abc}y_{...}^2$$

Then the sum of squares for the nested factor B is given by

$$SSB(A) = y'Q_{B(A)}y$$

$$= y'\left(I_a \otimes \left(I_b - \frac{1}{b}J_b\right) \otimes \frac{1}{c}J_c\right)y$$

$$= \frac{1}{c}\sum_{i=1}^{a}\sum_{j=1}^{b}y_{ij.}^2 - \frac{1}{bc}\sum_{i=1}^{a}y_{i..}^2$$

The error sum of squares is given by

$$SSE = y'Q_E y$$

$$= y'\left(I_a \otimes I_b\left(I_c - \frac{1}{c}\right)J_c\right)y$$

$$= \sum_{i=1}^{a}\sum_{j=1}^{b}\sum_{k=1}^{c}y_{ijk}^2 - \frac{1}{c}\sum_{i=1}^{a}\sum_{j=1}^{b}y_{ij.}^2.$$

The results are summarized in the ANOVA table below.

Source	DF	Sum of squares	Mean square	F
A	$a-1$	$y'Q_A y$	$\dfrac{y'Q_A y}{a-1}$	$\dfrac{y'Q_A y/(a-1)}{y'Q_E y/ab(c-1)}$
B(A)	$a(b-1)$	$y'Q_{B(A)}y$	$\dfrac{y'Q_{B(A)}y}{a(b-1)}$	$\dfrac{y'Q_{B(A)}y/a(b-1)}{y'Q_E y/ab(c-1)}$
Error	$ab(c-1)$	$y'Q_E y$	$\dfrac{y'Q_E y}{a(b-1)}$	
Total	$abc-1$	$y'y$		

The F statistics given in this table are for a fixed effects model. Frequently, the factor that is nested consists of items chosen at random. For the example above, the batches might be chosen at random. In this instance the F statistic in the first line of the ANOVA table would be replaced by

$$F = \frac{y'Q_A y/(a-1)}{y'Q_{B(A)}y/a(b-1)}.$$

See, for example, Montgomery (2005) for a discussion of mixed models where one factor is fixed and the other random.

Example 19.2 Two-Way ANOVA with One Factor Nested

This example is taken from Exercise 7.1 in Hicks and Turner (1999, p. 212). The data is reproduced with the permission of Oxford University Press. Porosity readings on condenser paper were recorded for paper from four rolls taken at random from each of three lots. The results are given in the table below.

Lot	I				II				III			
Roll	1	2	3	4	5	6	7	8	9	10	11	12
	1.5	1.5	2.7	3.0	1.9	2.3	1.8	1.9	2.5	3.2	1.4	7.8
	1.7	1.6	1.9	2.4	1.5	2.4	2.9	3.5	2.9	5.5	1.5	5.2
	1.6	1.7	2.0	2.6	2.1	2.4	4.7	2.8	3.3	7.1	3.4	5.0

Source: Hicks, C.R. and K.V. Turner. Fundamental Concepts in the Design of Experiments. Fifth edition. Reproduced by Permission of Oxford University Press.

For this problem a = 3, b = 4, and c = 3. Then

$$y_{1..} = 24.2, y_{2..} = 30.2, y_{3..} = 48.8, y_{...} = 103.2$$

$$SS(Lots) = \frac{24.2^2}{12} + \frac{30.2^2}{12} + \frac{48.8^2}{12} - \frac{103.2^2}{36} = 27.42.$$

To obtain SS(Rolls(Lots)), we need

$$y_{11.} = 4.8, y_{12.} = 4.6, y_{13.} = 6.6, y_{14.} = 8$$
$$y_{21.} = 5.5, y_{22.} = 7.1, y_{23.} = 9.4, y_{24.} = 8.2$$
$$y_{31.} = 8.7, y_{32.} = 15.8, y_{33.} = 6.3, y_{34.} = 18$$
$$\frac{1}{3}\sum_{i=1}^{3}\sum_{j=1}^{4} y_{ij.}^2 = 359.64$$
$$\frac{1}{12}\sum_{i=1}^{3} y_{i..}^2 = 323.26$$
$$SS(Rolls(Lots)) = 359.64 - 323.26 = 36.38.$$

The total sum of squares corrected for the mean is

$$TSS = 381.44 - \frac{103.2^2}{36} = 85.6.$$

The error sum of squares is

$$SSE = TSS - SS(Lots) - SS(Rolls(Lots))$$
$$= 85.6 - 27.42 - 36.38$$
$$= 21.8$$

The ANOVA table is

Source	DF	SS	MS	F
Lots	2	27.42	13.71	3.45
Rolls (Lots)	9	36.38	4.04	4.45**
Error	24	21.8	0.908	
Total	35	85.6		

At $\alpha = 0.05$, there is not a significant difference between lots. At $\alpha = 0.01$, there is a significant difference of rolls within lots. This is denoted in the table by a double asterisk.   □

The next section will consider the general linear hypotheses and how the problems just considered can be formulated as special cases.

19.4 SUMMARY

Two extensions of the two-way classification model that was considered in Section 18 were discussed in this section. The first was one for a replicated experiment that included possible interaction. The second was the case where one factor was nested within another factor. Both cases were illustrated with an example using data.

EXERCISES

19.1 The data for this exercise is taken from Montgomery (2005) with permission from John Wiley & Sons. A golfer recently purchased new clubs in the hope of improving his game. He plays three rounds of golf at three different golf courses with the old and the new clubs. The scores are given below.

	Course		
Clubs	Ahwatukee	Karsten	Foothills
Old	90	91	88
	87	93	86
	86	90	90
New	88	90	86
	87	91	85
	85	88	88

Source: Montgomery, D.C. Design and Analysis of Experiments. Sixth Edition. Reproduced by permission of John Wiley & Sons.

Perform the ANOVA to determine if:

A. The score is different for the old and the new clubs.

B. There is a significant difference among the scores on the three different golf courses.

C. There is significant interaction between the golf courses and the age of the clubs.

19.2 The data for this exercise is through the courtesy of Ingo Ruczinski, Associate Professor, Department of Biostatistics, Johns Hopkins University School of Public Health.

For this example, we have three specific hospitals, four randomly chosen subjects within each hospital, and two independent measurements of blood coagulation time in seconds. Perform the ANOVA to determine whether there is a significant difference between hospitals and significant variability among the subjects.

Hospital I				Hospital II				Hospital III			
1	2	3	4	1	2	3	4	1	2	3	4
58.5	77.8	84.0	70.1	69.8	56.0	50.7	63.8	56.6	77.8	69.9	62.1
59.5	80.9	83.6	68.3	69.8	54.5	49.3	65.8	57.5	79.2	69.2	64.5

SECTION 20

THE GENERAL LINEAR HYPOTHESIS

20.1   INTRODUCTION

The hypotheses tested in the ANOVA examples in Sections 18 and 19 are special cases of the general linear hypothesis of the form

$$H_0 : K'\beta = \eta.$$

This section will explain how these hypotheses are tested and give examples of how the hypotheses considered in Sections 18 and 19 are special cases.

First, the full-rank case will be considered in Subsection 20.2. Then the non-full-rank case will be considered in Subsection 20.3. We also consider examples of estimable linear hypothesis called contrasts in Subsection 20.4. Illustrations of how some of the examples of Sections 18 and 19 fit into this context will be given.

20.2   THE FULL-RANK CASE

The hypothesis being tested takes the form

$$H_0 : K'\beta = \eta,$$

where K is an $s \times m$ matrix, β are the parameters of the regression model, and η is a vector of order s. For example, if in the model

---

Matrix Algebra for Linear Models, First Edition. Marvin H. J. Gruber.
© 2014 John Wiley & Sons, Inc. Published 2014 by John Wiley & Sons, Inc.

$$Y = \begin{bmatrix} 1 & Z \end{bmatrix} \begin{bmatrix} \beta_0 \\ \beta_1 \\ \beta_2 \\ \beta_3 \end{bmatrix} + \varepsilon$$

we would like to test $H_0 : \beta_1 - \beta_2 = 3$, $2\beta_2 - \beta_3 = 0$, and $H_0 : \beta_1 - \beta_2 = 3, 2\beta_2 - \beta_3 = 0$, we would have

$$\begin{bmatrix} 0 & 1 & -1 & 0 \\ 0 & 0 & 2 & -1 \end{bmatrix} \begin{bmatrix} \beta_0 \\ \beta_1 \\ \beta_2 \\ \beta_3 \end{bmatrix} = \begin{bmatrix} 3 \\ 0 \end{bmatrix}.$$

We will show that the F statistic for testing the general linear hypothesis is given by

$$F = \frac{Q/r(K)}{SSE/(n - r(X))}, \tag{20.1}$$

where $r(K)$ is the rank of K, $r(X)$ is the rank of the X matrix, and

$$Q = (K'b - \eta)'[K'(X'X)^{-1}K]^{-1}(K'b - \eta). \tag{20.2}$$

We shall establish that the statistic in (20.1) follows an F distribution. With this in mind observe that under the usual assumptions for a regression model

$$y \sim N(X\beta, \sigma^2 I),$$

the least square estimator

$$b = (X'X)^{-1}X'y \sim N(\beta, (X'X)^{-1}\sigma^2),$$

and as a result

$$K'b - \eta \sim N(K\beta - \eta, K'(X'X)^{-1}K).$$

Then using the result of Exercise 17.9 when the null hypothesis $K\beta = \eta$ holds true, Q/σ^2 with Q given in (20.2) has a chi-square distribution with s degrees of freedom.

To show that the statistic in (20.1) follows an F distribution, it is necessary to prove that Q and SSE are independent chi-square random variables. The reader may show in Exercise 20.1 that first

$$Q = (y - XK(K'K)^{-1}\eta)'X(X'X)^{-1}K(K'(X'X)^{-1}K)^{-1}$$
$$K'(X'X)^{-1}X'(y - XK(K'K)^{-1}\eta), \tag{20.3}$$

second that

$$SSE = (y - XK(K'K)^{-1}\eta)' [I - X(X'X)^{-1}X'](y - XK(K'K)^{-1}\eta), \qquad (20.4)$$

and third that the product of the matrices of these two quadratic forms is zero, thus establishing independence.

Several tests for specific cases of the general linear hypothesis will be given in Example 20.1.

Example 20.1 Illustrations of the General Linear Hypothesis

The data below is about student performance in a class taught by the author that includes topics in multivariate calculus and differential equations. The predictors are the grades on four-hour examinations during the quarter, and the response variable is the grade on the final examination.

Student	Exam 1	Exam 2	Exam 3	Exam 4	Final
1	95	85	97	83	91
2	69	70	70	72	68
3	89	84	96	85	88
4	66	58	100	70	73
5	77	94	100	86	74
6	49	71	94	80	71
7	81	82	100	78	68
8	89	86	100	73	95
9	90	96	97	88	95
10	87	75	93	83	61
11	74	71	88	73	71
12	64	65	89	67	39
13	89	88	93	64	63
14	56	63	83	34	42
15	93	86	92	85	95

The least square equation is

$$y = -29.6748 + 0.366827x_1 + 0.236916x_2 + 0.167127x_3 + 0.535136x_4,$$

and SSE = 1731.53.

Suppose we wish to test $H_0: \beta_1 = 0$, $\beta_2 = 0$, $\beta_3 = 0$, $\beta_4 = 0$. Then we would write in matrix form

$$\begin{bmatrix} 0 & 1 & 0 & 0 & 0 \\ 0 & 0 & 1 & 0 & 0 \\ 0 & 0 & 0 & 1 & 0 \\ 0 & 0 & 0 & 0 & 1 \end{bmatrix} \begin{bmatrix} \beta_0 \\ \beta_1 \\ \beta_2 \\ \beta_3 \\ \beta_4 \end{bmatrix} = \begin{bmatrix} 0 \\ 0 \\ 0 \\ 0 \end{bmatrix}.$$

Then

$$F = \frac{2689.4}{(4(1731.53)/10} = 3.88,$$

which is significant at $\alpha = 0.05$ but not at $\alpha = 0.01$.

Suppose we wish to test $H_0 : \beta_1 - 2\beta_3 = 2, 2\beta_2 - \beta_4 = 3$. In matrix form, this would be

$$\begin{bmatrix} 0 & 1 & 0 & -2 & 0 \\ 0 & 0 & 2 & 0 & -1 \end{bmatrix} \begin{bmatrix} \beta_0 \\ \beta_1 \\ \beta_2 \\ \beta_3 \\ \beta_4 \end{bmatrix} = \begin{bmatrix} 2 \\ 3 \end{bmatrix}.$$

Then

$$F = \frac{2004.38}{2(1731.53)/10} = 5.78787.$$

Again this is significant at $\alpha = 0.05$ but not at $\alpha = 0.01$. $\qquad \square$

20.3   THE NON-FULL-RANK CASE

The test statistic for the non-full-rank case is the same as that for the full-rank case with any generalized inverse of $X'X$ replacing the ordinary inverse provided that $K'\beta$ is estimable. This means that there exists a matrix C where $K' = C'U'$, where U' comes from the singular value decomposition of $X'X$ discussed in Section 9. From considerations in Section 9, the usual linear model

$$Y = X\beta + \varepsilon$$

may be reparameterized to the full-rank model

$$Y = XU\gamma + \varepsilon, \qquad (20.5)$$

where $\gamma = U'\beta$. Observe that

$$K'\beta - \eta = C'U'\beta - \eta = C'\gamma - \eta$$

so that the hypothesis $K'\beta = m$ reduces to $C'\gamma = \eta$ relative to the model (20.5). Since $U'X'XU = \Lambda$ using the statistic (20.3),

$$
\begin{aligned}
Q &= (C'\gamma - \eta)'(C'\Lambda^{-1}C)^{-1}(C'\gamma - \eta) \\
&= (C'U'\beta - \eta)'(C'\Lambda^{-1}C)^{-1}(C'U'\beta - \eta) \\
&= (K'\beta - \eta)'(C'U'GUC)^{-1}(K'\beta - \eta) \\
&= (K'\beta - \eta)'(K'GK)^{+}(K'\beta - \eta)
\end{aligned}
$$

because as was shown in Equation (13.9), $U'GU = \Lambda^{-1}$ for any generalized inverse G. Likewise, using the reparameterized model

$$
\begin{aligned}
SSE &= y'[I - XU\Lambda^{-1}U'X']y \\
&= y'[I - XUU'GUU'X']y \\
&= y'[I - XGX']y
\end{aligned}
$$

using Equation (13.9) of Section 13 and the fact that $XUU' = X$.

Since XGX' is independent of the choice of generalized inverse, the same is true about the SSE.

Example 20.2 Illustrations of Tests of Linear Hypothesis for the Non-full-Rank Case

Consider the data of Example 18.3. For the linear model given there,

$$
X'X = \begin{bmatrix} 15 & 5 & 5 & 0 \\ 5 & 5 & 0 & 0 \\ 5 & 0 & 5 & 0 \\ 5 & 0 & 0 & 5 \end{bmatrix}.
$$

A generalized inverse is

$$
G = \begin{bmatrix} 0 & 0 & 0 & 0 \\ 0 & \frac{1}{5} & 0 & 0 \\ 0 & 0 & \frac{1}{5} & 0 \\ 0 & 0 & 0 & \frac{1}{5} \end{bmatrix}.
$$

Let us consider testing the joint hypothesis

$$
H_0 : \tau_1 - \tau_2 = 0, \ \tau_2 - \tau_3 = 0.
$$

Using the G inverse, we get

$$
\hat{\tau}_1 = \bar{y}_{1.} = 83.2, \ \hat{\tau}_2 = \bar{y}_{2.} = 80.2, \ \hat{\tau}_3 = \bar{y}_{3.} = 79.4.
$$

For the given hypothesis,

$$K' = \begin{bmatrix} 0 & 1 & -1 & 0 \\ 0 & 0 & 1 & -1 \end{bmatrix}$$

so that

$$K'\hat{\tau} = \begin{bmatrix} 0 & 1 & -1 & 0 \\ 0 & 0 & 1 & -1 \end{bmatrix} \begin{bmatrix} 0 \\ 83.2 \\ 80.2 \\ 79.4 \end{bmatrix} = \begin{bmatrix} 3 \\ 0.8 \end{bmatrix},$$

$$K'GK = \begin{bmatrix} 0 & 1 & -1 & 0 \\ 0 & 0 & 1 & -1 \end{bmatrix} \begin{bmatrix} 0 & 0 & 0 & 0 \\ 0 & \frac{1}{5} & 0 & 0 \\ 0 & 0 & \frac{1}{5} & 0 \\ 0 & 0 & 0 & \frac{1}{5} \end{bmatrix} \begin{bmatrix} 0 & 0 \\ 1 & 0 \\ -1 & 1 \\ 0 & -1 \end{bmatrix}$$

$$= \begin{bmatrix} \frac{2}{5} & -\frac{1}{5} \\ -\frac{1}{5} & \frac{2}{5} \end{bmatrix},$$

and

$$(K'GK)^{-1} = \begin{bmatrix} \frac{10}{3} & \frac{5}{3} \\ \frac{5}{3} & \frac{10}{3} \end{bmatrix}.$$

Now

$$Q = \begin{bmatrix} 3 & 0.8 \end{bmatrix} \begin{bmatrix} \frac{10}{3} & \frac{5}{3} \\ \frac{5}{3} & \frac{10}{3} \end{bmatrix} \begin{bmatrix} 3 \\ 0.8 \end{bmatrix} = 40.1333,$$

and

$$F = \frac{40.133/2}{436.6/12} = 0.55153.$$

The hypothesis would not be rejected at $\alpha = 0.05$ or any reasonable level of significance.

Another possible hypothesis test is $H_0 : \tau_1 + \tau_2 - 2\tau_3 = 0$. Here

$$K' = \begin{bmatrix} 0 & 1 & 1 & -2 \end{bmatrix}$$

so that

$$K'\hat{\tau} = \begin{bmatrix} 0 & 1 & 1 & -2 \end{bmatrix} \begin{bmatrix} 0 \\ 83.2 \\ 80.2 \\ 79.4 \end{bmatrix} = 4.6$$

and

$$K'GK = \begin{bmatrix} 0 & 1 & 1 & -2 \end{bmatrix} \begin{bmatrix} 0 & 0 & 0 & 0 \\ 0 & \frac{1}{5} & 0 & 0 \\ 0 & 0 & \frac{1}{5} & 0 \\ 0 & 0 & 0 & \frac{1}{5} \end{bmatrix} \begin{bmatrix} 0 \\ 1 \\ 1 \\ -2 \end{bmatrix} = \frac{6}{5}.$$

Then

$$Q = (4.6)\left(\frac{5}{6}\right)(4.6) = 17.6333, \text{ and}$$

$$F = \frac{17.6333}{36.6} = 0.48175.$$

Again we do not have statistical significance.   □

20.4 CONTRASTS

For balanced data, that is when there is the same number of observations for each treatment linear combinations of the form $\sum_{i=1}^{r} c_i \tau_i$, where $\sum_{i=1}^{r} c_i = 0$ are called contrasts. For the one- and two-way ANOVA, these are estimable parametric functions. Two contrasts $\sum_{i=1}^{r} c_i \tau_i$ and $\sum_{i=1}^{r} d_i \tau_i$ are orthogonal in the case of balanced data if $\sum_{i=1}^{r} c_i d_i = 0$. When the contrasts are orthogonal, the sum of squares will add up to the model sum of squares. For the analogous material on contrasts and orthogonality for unbalanced data, see, for example, Searle (1971).

Example 20.3 Testing Some Contrasts

A factory has three machines that do the same thing but are of different brands. The data below give the number of hours in operation until after the machines were installed needed repair.

A	B	C
5047	5048	5565
5017	5121	5473
5125	4979	5520
4950	5078	5499
5006	4961	5536
5038	4986	5570
5018	5098	5503
5017	5110	5525
4929	4935	5483
4984	5029	5551

We consider the linear model

$$Y = \begin{bmatrix} 1_{10} & 1_{10} & 0 & 0 \\ 1_{10} & 0 & 1_{10} & 0 \\ 1_{10} & 0 & 0 & 1_{10} \end{bmatrix} \begin{bmatrix} \mu \\ \tau_1 \\ \tau_2 \\ \tau_3 \end{bmatrix} + \varepsilon.$$

The least square estimator is the solution to the normal equation in matrix form

$$\begin{bmatrix} 30 & 10 & 10 & 10 \\ 10 & 10 & 0 & 0 \\ 10 & 0 & 10 & 0 \\ 10 & 0 & 0 & 10 \end{bmatrix} \begin{bmatrix} \mu \\ \tau_1 \\ \tau_2 \\ \tau_3 \end{bmatrix} = \begin{bmatrix} 155701 \\ 50131 \\ 50345 \\ 55225 \end{bmatrix}.$$

Then

$$\begin{bmatrix} \hat{\mu} \\ \hat{\tau}_1 \\ \hat{\tau}_2 \\ \hat{\tau}_3 \end{bmatrix} = \begin{bmatrix} \frac{1}{30} & 0 & 0 & 0 \\ -\frac{1}{30} & \frac{1}{10} & 0 & 0 \\ -\frac{1}{30} & 0 & \frac{1}{10} & 0 \\ -\frac{1}{30} & 0 & 0 & \frac{1}{10} \end{bmatrix} \begin{bmatrix} 155701 \\ 50131 \\ 50345 \\ 55225 \end{bmatrix} = \begin{bmatrix} 5190.03 \\ -176.933 \\ -155.533 \\ 332.467 \end{bmatrix}.$$

We are going to test each of the hypotheses

$$H_{01} : \tau_1 - \tau_2 = 0$$
$$H_{02} : \tau_1 + \tau_2 - 2\tau_3 = 0$$

versus the alternative hypothesis of non-equality. These hypotheses are orthogonal contrasts.

First for H_{01}

$$\begin{bmatrix} 0 & 1 & -1 & 0 \end{bmatrix} \begin{bmatrix} 5190.03 \\ -176.933 \\ -155.533 \\ 332.467 \end{bmatrix} = -21.4,$$

$$k_1' G k_1 = \begin{bmatrix} 0 & 1 & -1 & 0 \end{bmatrix} \begin{bmatrix} \frac{1}{30} & 0 & 0 & 0 \\ -\frac{1}{30} & \frac{1}{10} & 0 & 0 \\ -\frac{1}{30} & 0 & \frac{1}{10} & 0 \\ -\frac{1}{30} & 0 & 0 & \frac{1}{10} \end{bmatrix} \begin{bmatrix} 0 \\ 1 \\ -1 \\ 0 \end{bmatrix} = \frac{1}{5},$$

and

$$(k_1' G k_1)^+ = 5.$$

Then

$$Q_1 = (-21.4)(5)(-21.4) = 2289.8.$$

For H_{02} we have

$$\begin{bmatrix} 0 & 1 & 1 & -2 \end{bmatrix} \begin{bmatrix} 5190.03 \\ -176.933 \\ -155.533 \\ 332.467 \end{bmatrix} = -997.4,$$

$$k_2'Gk_2 = \begin{bmatrix} 0 & 1 & 1 & -2 \end{bmatrix} \begin{bmatrix} \frac{1}{30} & 0 & 0 & 0 \\ -\frac{1}{30} & \frac{1}{10} & 0 & 0 \\ -\frac{1}{30} & 0 & \frac{1}{10} & 0 \\ -\frac{1}{30} & 0 & 0 & \frac{1}{10} \end{bmatrix} \begin{bmatrix} 0 \\ 1 \\ 1 \\ -2 \end{bmatrix} = \frac{3}{5}$$

$$(k_2'Gk_2)^+ = \frac{5}{3}.$$

Now

$$Q_2 = (-997.4)\left(\tfrac{5}{3}\right)(-997.4) = 1658010.$$

After performing the ANOVA (see Exercise 20.6) and including Q_1 and Q_2 in the table, we get the following.

Source		SS	MS	F
Machine	2	1660301	830151	293.52 **
Q_1	1	2289.8	2289.8	0.809
Q_2	1	1658010	1658010	586.28 **
Error	27	76364	2828	
Total	29	1736665		

Notice that there is a highly significant difference in time to repair among the three machines as denoted by the ** in the ANOVA table. However, machines 1 and 2 are not significantly different. There is a significant difference in time to repair between machine 3 and the average of the times to repair for machines 1 and 2 denoted by the ** in the table.

Also notice that $SSR = SSQ_1 + SSQ_2$. The very slight difference in the numbers is due to roundoff error.   □

20.5 SUMMARY

We considered the general linear hypothesis for the full-rank and the non-full-rank case. We illustrated the testing of general linear hypothesis for a regression model and a one-way ANOVA. Finally, we defined contrasts for a balanced linear model and illustrated testing their statistical significance.

EXERCISES

20.1 **A.** Establish the equivalence of (20.2) and (20.3).

B. Establish (20.4).

C. Show that the product of the matrices in the quadratic forms (20.3) and (20.4) has product zero, thus showing that Q and SSE are independent and that when the null hypothesis holds true, (20.1) follows an F distribution with s degrees of freedom for the numerator and $N - r$ degrees of freedom for the denominator.

20.2 Test the following hypotheses using the F statistic for the data in Example 20.1:

A. $H_0 : \beta_3 = 0.15$.

B. $H_0 : \beta_1 - 2\beta_3 = 0,\ 2\beta_2 - \beta_4 = 0$.

20.3 For the data of Example 18.3, separately test the hypotheses $H_{01} : \tau_1 - \tau_2 = 0$ and $H_{02} : \tau_1 + \tau_2 - 2\tau_3 = 0$. Show that the Q_1 and Q_2 from these tests add up to the regression sum of squares.

20.4 For the abridged data below in Exercise 18.5

State/speed limit	55	50	40	<35
California	397	58	107	156
Illinois	226	3	47	88
New York	177	20	18	80

A. Do a one-way ANOVA to determine whether there is a significant difference among speed limits.

B. Why would it not be appropriate to do a two-way ANOVA to determine if there is a difference between states?

C. Which of the following contrasts are statistically significant at $\alpha = 0.05$? What do their sums of squares add up to and why?

$$c_1 = \tau_1 - \tau_3$$
$$c_2 = \tau_1 - 2\tau_2 + \tau_3$$
$$c_3 = \tau_1 + \tau_2 + \tau_3 - 3\tau_4$$

20.5 For the data of Example 18.3, attempt to test the hypothesis $H_0 : \tau_1 = 0$. Where do you run into trouble? Notice that this is not an estimable function. Non-estimable functions cannot be tested. (Hint: Using two different generalized inverses, find the numerator of the test statistic.)

20.6   **A.**   For Example 20.3, perform the ANOVA and verify the numbers that were not calculated in the table.

B.   Formulate hypothesis to test for a significant difference between machine B and machine C and to determine whether the time to repair of machine A is different from the average of machines B and C. Test these hypotheses and verify that the sum of squares adds up to the model sum of squares.

C.   Does the sum of squares from testing the hypothesis of equality of machines A and B and that of machines B and C add up to the regression or model sum of squares? Give a reason why your answer is plausible.

20.7   **A.**   For Example 18.4, test the hypotheses

$$H_{01} : \tau_1 + \tau_2 - 2\tau_3 = 0$$
$$H_{02} : \tau_1 - \tau_2 = 0.$$

B.   Eyeballing the data, pick a contrast that you think might be significantly different from zero and perform the test of hypothesis to see if this is actually the case.

PART V

MATRIX OPTIMIZATION PROBLEMS

The goal of this final part of the book is to formulate different kinds of regression estimators as solutions to matrix optimization problems. We will show how the minimization of different quadratic forms subject to no constraints, linear constraints, and quadratic constraints lead to different kinds of optimal estimators. These include the least square estimator, the linear Bayes estimator, the mixed estimator, ridge-type estimators, and the best linear unbiased predictor.

We first consider the unconstrained optimization problems in Section 21. The solution to these problems includes the least square estimator for the univariate and multivariate linear models. In order to consider optimization problems with constraints, a discussion of the method of Lagrange multipliers is given in Section 22. This is a topic often not discussed adequately in multivariate calculus courses. Generally, there are other topics that need to be covered and time constraints. We then consider optimization problems with linear constraints. Using Lagrange multipliers, we show how to obtain a constrained least square estimator and its relationship to the general linear hypothesis.

In Section 23, we prove the well-known Gauss Markov theorem that states that the least square estimator has minimum variance among all unbiased estimators. Given prior knowledge about the mean and the variance, we show how to obtain the linear Bayes estimator. We also obtain the least square estimator as a minimum variance unbiased estimator and show how the linear Bayes estimator is minimum variance unbiased in an extended sense. We also show how canonical correlation is an example of an optimization problem with a linear constraint.

Matrix Algebra for Linear Models, First Edition. Marvin H. J. Gruber.
© 2014 John Wiley & Sons, Inc. Published 2014 by John Wiley & Sons, Inc.

The ridge regression estimator of Hoerl and Kennard (1970) involves minimization of a quadratic form with respect to a quadratic constraint. In Section 24, we show how ridge-type estimators are derived by this method. We then compare the efficiencies of these estimators to that of the least square estimator.

SECTION 21

UNCONSTRAINED OPTIMIZATION PROBLEMS

21.1 INTRODUCTION

First, in Subsection 21.2 we consider the problem of minimizing the sum of a quadratic form, a linear form, and a constant. Then in Subsection 21.3 we show how the optimization problem that leads to the least square estimator as a special case.

21.2 UNCONSTRAINED OPTIMIZATION PROBLEMS

We consider the minimization of forms

$$f(c) = c'Wc + c'd + e \tag{21.1}$$

where W is an $n \times n$ positive-definite matrix, c and d are n-dimensional vectors, and e is a scalar constant.

For f to be minimized, it is necessary that

$$\frac{\partial f}{\partial c} = 0 \tag{21.2}$$

and sufficient that the Hessian matrix whose elements are

Matrix Algebra for Linear Models, First Edition. Marvin H. J. Gruber.
© 2014 John Wiley & Sons, Inc. Published 2014 by John Wiley & Sons, Inc.

$$\frac{\partial^2 f}{\partial c_i \partial c_j} = w_{ij} \qquad (21.3)$$

be positive semi-definite.

We have that

$$\frac{\partial f}{\partial c} = 2Wc + d = 0$$

so that

$$c_{opt} = -\frac{1}{2} W^{-1} d. \qquad (21.4)$$

Since W is positive definite,

$$\frac{\partial^2 f}{\partial c_{11}^2} = w_{11} > 0$$

so that the value of c that minimizes f is given by (21.4). Then the minimum value of f is

$$f(c_{opt}) = e - \tfrac{1}{4} d' W^{-1} d. \qquad (21.5)$$

Example 21.1 A Standard Calculus Problem with a Second-Degree Polynomial Written in Terms of Vectors

Let $v = \begin{bmatrix} x \\ y \end{bmatrix}$.

$$f(v) = [x \quad y] \begin{bmatrix} 5 & -1 \\ -1 & 5 \end{bmatrix} \begin{bmatrix} x \\ y \end{bmatrix} - [x \quad y] \begin{bmatrix} 18 \\ 6 \end{bmatrix}. \qquad (*)$$

$$\frac{\partial f}{\partial v} = \begin{bmatrix} 10 & -2 \\ -2 & 10 \end{bmatrix} \begin{bmatrix} x \\ y \end{bmatrix} - \begin{bmatrix} 18 \\ 6 \end{bmatrix} = 0.$$

Then in matrix form the value where f assumes a minimum is

$$v = \begin{bmatrix} x \\ y \end{bmatrix} = \begin{bmatrix} 10 & -2 \\ -2 & 10 \end{bmatrix}^{-1} \begin{bmatrix} 18 \\ 6 \end{bmatrix}$$

$$= \frac{1}{96} \begin{bmatrix} 10 & 2 \\ 2 & 10 \end{bmatrix} \begin{bmatrix} 18 \\ 6 \end{bmatrix} = \begin{bmatrix} 2 \\ 1 \end{bmatrix}.$$

Now

$$f\left(\begin{bmatrix} 2 \\ 1 \end{bmatrix}\right) = [2 \quad 1]\begin{bmatrix} 5 & -1 \\ -1 & 5 \end{bmatrix}\begin{bmatrix} 2 \\ 1 \end{bmatrix} - [2 \quad 1]\begin{bmatrix} 18 \\ 6 \end{bmatrix} = -21.$$

$\square$

A solution to the above problem may be obtained without using matrix differentiation. To do this, let $W = AA'$. Such an A exists for positive-semi-definite matrices. Also let $h = A^{-1}d$. Then

$$f(c) = c'AA'c + c'd + e$$

$$= c'AA'c + \frac{1}{2}c'd + \frac{1}{2}d'c + e$$

$$= c'AA'c + \frac{1}{2}c'Ah + \frac{1}{2}h'A'c + \frac{1}{4}h'h + e - \frac{1}{4}h'h \tag{21.6}$$

$$= \left(A'c + \frac{1}{2}h'\right)'\left(A'c + \frac{1}{2}h\right) + e - \frac{1}{4}h'h.$$

Optimum

$$c = -\frac{1}{2}A'^{-1}h = -\frac{1}{2}W^{-1}d \tag{21.7a}$$

because the first term of the third expression in (21.6) is zero and the minimum value of f is

$$e - \frac{1}{4}h'h = e - \frac{1}{4}d'A'^{-1}A^{-1}d = e - \frac{1}{4}d'W^{-1}d.$$

Also

$$f\left(-\frac{1}{2}W^{-1}d\right) = \frac{1}{4}d'W^{-1}WW^{-1}d - \frac{1}{2}d'W^{-1}d + e$$

$$= -\frac{1}{4}d'W^{-1}d + e. \tag{21.7b}$$

Suppose that W is positive semi-definite but not necessarily positive definite. The singular value decomposition is given by

$$W = U\Lambda U' = U\Lambda^{1/2}\Lambda^{1/2}U'.$$

Then $A = U\Lambda^{1/2}$. Let $c'U = l'$ and $h = \Lambda^{-1/2}U'd$. Then

$$f(l) = l'\Lambda l + \frac{1}{2}l'\Lambda^{1/2}h + \frac{1}{2}h'\Lambda^{1/2}l + \frac{1}{4}h'h - \frac{1}{4}h'h + e$$

$$= (\Lambda^{1/2}l + \frac{1}{2}h)'(\Lambda^{1/2}l + \frac{1}{2}h) - \frac{1}{4}h'h + e$$

and the value of l that minimizes f(l) is

$$l_{opt} = -\tfrac{1}{2}\Lambda^{-1/2}h = -\tfrac{1}{2}\Lambda^{-1}U'd$$

and

$$f(l_{opt}) = e - \tfrac{1}{4}d'U\Lambda^{-1}U'd = e - \tfrac{1}{4}d'W^{+}d = e - \tfrac{1}{4}d'UU'W^{-}UU'd$$

where W^- is any generalized inverse of w.

Example 21.2 Optimization Problem with Infinitely Many Solutions
Let $v' = \begin{bmatrix} x & y & z \end{bmatrix}$. Then

$$f(v) = \begin{bmatrix} x & y & z \end{bmatrix} \begin{bmatrix} 10 & 5 & 5 \\ 5 & 5 & 0 \\ 5 & 0 & 5 \end{bmatrix} \begin{bmatrix} x \\ y \\ z \end{bmatrix} - \begin{bmatrix} x & y & z \end{bmatrix} \begin{bmatrix} 10 \\ 5 \\ 5 \end{bmatrix}.$$

Then

$$\frac{\partial f}{\partial v} = \begin{bmatrix} 20 & 10 & 10 \\ 10 & 10 & 0 \\ 10 & 0 & 10 \end{bmatrix} \begin{bmatrix} x \\ y \\ x \end{bmatrix} - \begin{bmatrix} 10 \\ 5 \\ 5 \end{bmatrix} = 0.$$

Possible solutions include

$$\begin{bmatrix} x \\ y \\ z \end{bmatrix} = \begin{bmatrix} 0 & 0 & 0 \\ 0 & \tfrac{1}{10} & 0 \\ 0 & 0 & \tfrac{1}{10} \end{bmatrix} \begin{bmatrix} 10 \\ 5 \\ 5 \end{bmatrix} = \begin{bmatrix} 0 \\ \tfrac{1}{2} \\ \tfrac{1}{2} \end{bmatrix}$$

and

$$\begin{bmatrix} x \\ y \\ z \end{bmatrix} = \begin{bmatrix} \tfrac{1}{20} & 0 & -\tfrac{1}{20} \\ -\tfrac{1}{20} & \tfrac{1}{10} & \tfrac{1}{20} \\ \tfrac{1}{20} & -\tfrac{1}{10} & \tfrac{1}{20} \end{bmatrix} \begin{bmatrix} 10 \\ 5 \\ 5 \end{bmatrix} = \begin{bmatrix} \tfrac{1}{4} \\ \tfrac{1}{4} \\ \tfrac{1}{4} \end{bmatrix}.$$

For both cases and for that matter using any generalized inverse, you would get after substitution into the original equation that the minimum value is −5/2. How can you account for this?

To find the minimum, we are solving the system of equations

$$20x + 10y + 10z = 10$$
$$10x + 10y = 5$$
$$10x + 10z = 5$$

and the solutions lie on the line of intersection of the two planes

$$y = \frac{1}{2} - x, \ z = \frac{1}{2} - x.$$ □

Example 21.3 Another Optimization Problem with Infintely Many Solutions

Consider

$$f(x, y) = [x \ \ y] \begin{bmatrix} \frac{3}{2} & \frac{3}{2} \\ \frac{3}{2} & \frac{3}{2} \end{bmatrix} \begin{bmatrix} x \\ y \end{bmatrix} - [x \ \ y] \begin{bmatrix} 6 \\ 6 \end{bmatrix}.$$

Then

$$\frac{\partial f}{\partial v} = \begin{bmatrix} 3 & 3 \\ 3 & 3 \end{bmatrix} \begin{bmatrix} x \\ y \end{bmatrix} - \begin{bmatrix} 6 \\ 6 \end{bmatrix} = 0.$$

The minimum values are solutions to the equation $x + y = 2$. The minimum value turns out to be zero. A picture of the surface is shown in Figure 21.1. The surface intersects the plane at $z = 0$ along the intersection of the plane $x + y = 2$ with the xy plane.   □

21.3   THE LEAST SQUARE ESTIMATOR AGAIN

Consider the linear model

$$Y = X\beta + \varepsilon$$

with the assumptions that $E(\varepsilon) = 0$, $D(\varepsilon) = \sigma^2 I$ and the form (21.1) with $W = X'X$, $c = \beta$, $d = -2X'Y$, and $e = 0$. The form (21.1) may then be written as

$$f(\beta) = \beta'X'X\beta - 2\beta'X'Y$$
$$= \beta'X'X\beta - 2\beta'X'Y + Y'Y - Y'Y \qquad (21.8)$$
$$= (Y - X\beta)'(Y - X\beta) - Y'Y.$$

Differentiating the middle expression in (21.8) with respect to β and setting the result equal to zero,

$$2X'X\beta - 2X'Y = 0,$$

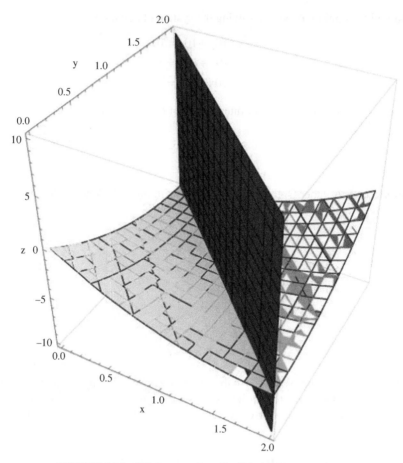

FIGURE 21.1 Minimum values for f(x, y) in Example 21.3.

so when X is of full rank, the minimizing estimator

$$\hat{\beta} = (X'X)^{-1}X'Y \qquad\qquad (21.9a)$$

and for the non-full-rank case,

$$\hat{\beta} = (X'X)^{-}X'Y. \qquad\qquad (21.9b)$$

The problem may also be solved without using matrix differentiation in a manner similar to Subsection 4.6. We shall show how to do this here for the non-full-rank case. Recall that for any generalized inverse of X'X,

$$X = X(X'X)^{-}X'X.$$

Then (21.5) becomes

$$f(\beta) = (Y - X\beta)'(Y - X\beta) - Y'Y, \tag{21.10}$$

and

$$
\begin{aligned}
f(\beta) &= (Y - X\beta)'(Y - X\beta) - Y'Y \\
&= Y'Y - \beta'X'Y - Y'X\beta + \beta'X'X\beta - Y'Y \\
&= Y'Y - \beta'(X'X)(X'X)^{-}X'Y - Y'X(X'X)^{-}(X'X)\beta + Y'X(X'X)^{-}(X'X)(X'X)^{-}X'Y \\
&\quad - Y'X(X'X)^{-}(X'X)(X'X)^{-}X'Y + \beta'XX\beta - Y'Y \\
&= Y'(I - X(X'X)^{-}X')Y + (\beta - (X'X)^{-}X'Y)'(X'X)(\beta - (X'X)^{-}X'Y) - Y'Y \\
&= (\beta - (X'X)^{-}X'Y)'(X'X)(\beta - (X'X)^{-}X'Y) - Y'X(X'X)^{-}X'Y \tag{21.11}
\end{aligned}
$$

Thus, noting that the minimum occurs when the first term of (21.11) is zero,

$$X'X\widehat{\beta} = X'Y$$

and

$$\widehat{\beta} = (X'X)^{-}X'Y.$$

The minimum value of $f(\beta)$ is given by $-Y'X(X'X)^{-}X'Y$.

21.4 SUMMARY

We have shown how to obtain the minimum value of the sum of a linear and a quadratic form. We have given examples to illustrate the properties of these minima.

Finally, we showed how the least square estimator could be obtained by minimizing a second-degree form.

EXERCISES

For the following exercises, assume X has full rank.

21.1 By multiplying out the matrices in (*) in Example 21.1, obtain the second-degree polynomial. Now find the minimum by calculating partial derivatives as would be done in a multivariable calculus course and verify that you get the same results.

21.2 Find the singular value decomposition of W in Example 21.1. Let $A' = \Lambda^{1/2}U'$. Find h and do the appropriate calculations to establish that the minimum value of f is in fact -21.

21.3 Let V be a positive-definite matrix. Suppose that

$$f(\beta) = (Y - X\beta)'V^{-1}(Y - X\beta) - Y'V^{-1}Y.$$

Show that this is minimized when

$$\hat{\beta} = (X'V^{-1}X)^{-1}X'V^{-1}Y.$$

This is the weighted least square estimator where for the linear model

$$Y = X\beta + \varepsilon, \tag{21.12}$$

it is assumed that $E(\varepsilon) = 0$ and $D(\varepsilon) = \sigma^2 V$.

21.4 Minimize

$$f(\beta) = (W - Z\beta)'V^{-1}(W - Z\beta) - W'V^{-1}W,$$

where

$$W = \begin{bmatrix} Y \\ r \end{bmatrix}, \quad V = \begin{bmatrix} \sigma^2 I & 0 \\ 0 & \tau^2 I \end{bmatrix}, \quad \text{and} \quad Z = \begin{bmatrix} X \\ R \end{bmatrix},$$

and show that the minimizing estimator is the mixed estimator

$$\hat{\beta} = (\tau^2 X'X + \sigma^2 R'R)^{-1}(\tau^2 X'Y + \sigma^2 R'r)$$

of Theil and Goldberger (1961). There the linear model is considered together with the stochastic prior assumptions

$$r = R\beta + \eta,$$

assuming that $E(\tau) = 0$ and $D\,E(\eta) = 0$ and $D(\eta) = \tau^2 I$.

21.5 Consider a model with p correlated variables with the same number of observations each one. Let X be an $n \times m$ matrix. Consider the model

$$Y = (I \otimes X)\beta + \varepsilon,$$

where $Y' = [Y_1'\ Y_2'\ \cdots\ Y_p'], \beta' = [\beta_1'\ \beta_2'\ \cdots\ \beta_2']$, and $\varepsilon' = [\varepsilon_1'\ \varepsilon_2'\ \cdots\ \varepsilon_p']$.

Assume that

$$E(\varepsilon) = 0 \quad \text{and} \quad D(\varepsilon) = \Sigma \otimes I.$$

Consider the problem of minimizing

$$f(\beta) = (Y - (I \otimes X)\beta)'(\Sigma^{-1} \otimes I)(Y - (I \otimes X)\beta) - Y'(\Sigma^{-1} \otimes I)Y.$$

Show that

$$\hat{\beta} = (\Sigma \otimes (X'X)^{-1}X')Y.$$

Thus, obtain the least square estimators for the regression parameters of a multivariate model.

21.6 Consider the model

$$
\begin{bmatrix} Y_1 \\ Y_2 \\ \vdots \\ Y_p \end{bmatrix} =
\begin{bmatrix} X_1 & 0 & \cdots & 0 \\ 0 & X_2 & \cdots & 0 \\ \vdots & \vdots & \ddots & \vdots \\ 0 & 0 & \cdots & X_p \end{bmatrix}
\begin{bmatrix} \beta_1 \\ \beta_2 \\ \vdots \\ \beta_p \end{bmatrix} +
\begin{bmatrix} \varepsilon_1 \\ \varepsilon_2 \\ \vdots \\ \varepsilon_p \end{bmatrix},
$$

where Y_i is an n-dimensional vector, X_i is an $n \times m_i$-dimensional matrix, β_i is an m_i-dimensional vector of parameters, and ε_i is an n-dimensional error vector. More compactly,

$$Y = X\beta + \varepsilon,$$

where Y is an n-dimensional vector of the Y_i s, the dimension of matrix X is $np \times \sum_{i=1}^{p} m_i$, the vector β has dimension $\sum_{i=1}^{p} m_i$, and the error vector ε has dimension np. Consider the following optimization problems:

1. $f(\beta) = (Y - X\beta)'(\Sigma \otimes I)^{-1}(Y - X\beta) - Y'(\Sigma \otimes I)^{-1}Y.$
2. $f(\beta) = (Y - X\beta)'(Y - X\beta) - Y'Y.$

Show that the first optimization model has the solution

$$\hat{\beta} = (X'(\Sigma^{-1} \otimes I)X)X'(\Sigma^{-1} \otimes I)Y,$$

while the second optimization problem has the solution

$$\hat{\beta} = (X'X)^{-1}X'Y.$$

What are the components of $\hat{\beta}$ in each case?

21.7 Find the minimum values of each of the following that have minimum values:

A. $f(x, y) = \begin{bmatrix} x & y \end{bmatrix} \begin{bmatrix} 2 & 1 \\ 1 & 4 \end{bmatrix} \begin{bmatrix} x \\ y \end{bmatrix} + \begin{bmatrix} x & y \end{bmatrix} \begin{bmatrix} -2 \\ 4 \end{bmatrix} + 3$

B. $g(x, y) = \begin{bmatrix} x & y \end{bmatrix} \begin{bmatrix} 8 & 4 \\ 4 & 2 \end{bmatrix} \begin{bmatrix} x \\ y \end{bmatrix} - \begin{bmatrix} x & y \end{bmatrix} \begin{bmatrix} 8 \\ 4 \end{bmatrix} + 5$

C. $h(x, y) = \begin{bmatrix} x & y \end{bmatrix} \begin{bmatrix} 2 & 4 \\ 2 & 4 \end{bmatrix} \begin{bmatrix} x \\ y \end{bmatrix} + \begin{bmatrix} x & y \end{bmatrix} \begin{bmatrix} -2 \\ 4 \end{bmatrix} + 3$

D. $k(x, y, z) = \begin{bmatrix} x & y & z \end{bmatrix} \begin{bmatrix} 4 & -2 & -2 \\ -2 & 4 & -2 \\ -2 & -2 & 4 \end{bmatrix} \begin{bmatrix} x \\ y \\ z \end{bmatrix} + \begin{bmatrix} x & y & z \end{bmatrix} \begin{bmatrix} 8 \\ 4 \\ 2 \end{bmatrix} - 6$

E. $m(v) = v'Av + v'b + c$ where $v = \begin{bmatrix} x_1 \\ x_2 \\ x_3 \\ x_3 \end{bmatrix}$, $A = 4J_4$, $b = 4 \cdot 1_4$, $c = 7$

21.8 Consider the form $f(c) = c'Wc + c'd + e$ where W is a positive-semi-definite matrix. Show that:

A. If W is positive definite, there is a unique value of c where f is minimized.

B. If W is positive semi-definite but not positive definite and the matrix $2W$ has the same rank as the matrix $\begin{bmatrix} 2W & -d \end{bmatrix}$ there are infinitely many values of c where f is a minimum.

C. If W is positive semi-definite but not positive definite and the matrix $2W$ has lower rank than the matrix $\begin{bmatrix} 2W & -d \end{bmatrix}$, then f does not have a minimum.

SECTION 22

CONSTRAINED MINIMIZATION PROBLEMS WITH LINEAR CONSTRAINTS

22.1 INTRODUCTION

This section will give some examples of statistical problems that are constrained minimization problems using linear constraints. To accomplish this, an introduction to Lagrange multipliers is given in Subsection 22.2. We use Lagrange multipliers:

1. To show that for a random sample, the linear unbiased estimator of the population mean with minimum variance is the sample mean
2. To obtain a formula for the minimum norm generalized inverse that was presented in Sections 13 and 14

In Subsection 22.3 we consider the problem of minimizing a second-degree form given a linear constraint. Obtaining a least square estimator subject to one or more linear constraints is considered in Subsection 22.4.

Subsection 22.4 considers the canonical correlation coefficient obtained as the result of minimizing a bilinear form subject to two constraints.

22.2 AN OVERVIEW OF LAGRANGE MULTIPLIERS

We will give an informal review of the basic ideas about Lagrange multipliers. A few examples of geometrical and/or statistical interest will be presented. This topic is also treated in most calculus textbooks. We first consider general constraints. The statistical applications generally involve linear and quadratic constraints.

Matrix Algebra for Linear Models, First Edition. Marvin H. J. Gruber.
© 2014 John Wiley & Sons, Inc. Published 2014 by John Wiley & Sons, Inc.

Consider a continuous function of three variables for now. Assume that all of its partial derivatives exist. We would like to maximize or minimize $w = f(x,y,z)$ given a constraint of the form $g(x,y,z) = c$. Assume that all of the partial derivatives of g exist and consider neighborhoods of points where $g_z \neq 0$. For such points, as a result of the implicit function theorem, we may solve the constraint equation to obtain $z = h(x,y)$. Then we have

$$w = f(x, y, h(x, y))$$

so that where f has a maximum or minimum

$$w_x = f_x + f_z z_x = 0$$
$$w_y = f_y + f_z z_y = 0.$$

Also

$$g_x + g_z z_x = 0$$
$$g_y + g_z z_y = 0$$

and

$$z_x = -\frac{g_x}{g_z}, \quad z_y = -\frac{g_y}{g_z}.$$

Then

$$w_x = f_x - \frac{f_z}{g_z} g_x = 0$$

$$w_y = f_y - \frac{f_z}{g_z} g_y = 0$$

$$w_z = f_z - \frac{f_z}{g_z} g_z = 0.$$

The quantity

$$\lambda = \frac{f_z}{g_z}$$

is called a Lagrange multiplier. Thus, there exists a quantity λ where all of the maxima and minima of $w = f(x,y,z)$ satisfy the equations

$$f_x - \lambda g_x = 0$$
$$f_y - \lambda g_y = 0$$
$$f_z - \lambda g_z = 0.$$

This result may be generalized to a function of n variables where there are, say, r constraints. Consider the function of n variables

$$w = f(x_1, x_2, \ldots, x_n)$$

together with the r constraints of the form

$$g_i(x_1, x_2, \ldots, x_n) = c_i, \quad i = 1, 2, \ldots, r.$$

Then there exist numbers $\lambda_1, \lambda_2, \ldots, \lambda_r$ such that the values where f is maximized or minimized satisfy the equations

$$f_{x_i} - \sum_{j=1}^{n} \lambda_j g_{j_{x_i}} = 0, \quad i = 1, 2, \ldots, n.$$

Although the results stated above hold for all "nice" functions, the examples we will consider will be linear and quadratic polynomials. These are the cases of interest in regression analysis.

We will now give some examples of general and of statistical interest. The first example is a special case of minimizing a quadratic form subject to a quadratic constraint. The optimization problem for the ridge regression estimator to be obtained later is a more sophisticated example of such a problem.

Example 22.1 Closest and Farthest Point to the Origin on a Sphere

Find the point on the sphere

$$(x-1)^2 + (y-1)^2 + (z-1)^2 = 27$$

closest to the origin.

The square of the distance from the origin is

$$w = x^2 + y^2 + z^2. \qquad (*)$$

Thus for

$$H = x^2 + y^2 + z^2 - \lambda((x-1)^2 + (y-1)^2 + (z-1)^2 - 27),$$
$$H_x = 2x - 2\lambda(x-1) = 0$$
$$H_y = 2y - 2\lambda(y-1) = 0$$
$$H_z = 2z - 2\lambda(z-1) = 0,$$

so that

$$x = \frac{\lambda}{1-\lambda}, \ y = \frac{\lambda}{1-\lambda}, \ z = \frac{\lambda}{1-\lambda}.$$

Substituting into the equation for the sphere, we have

$$3\left(\frac{2\lambda-1}{1-\lambda}\right)^2 = 27,$$

so we have the equations

$$\frac{2\lambda-1}{1-\lambda} = 3 \quad \text{and} \quad \frac{2\lambda-1}{1-\lambda} = -3$$

so

$$\lambda = \frac{4}{5} \quad \text{and} \quad \lambda = 2.$$

The resulting points are $(4,4,4)$ and $(-2,-2,-2)$. These points on the sphere are either at a maximum or minimum distance from the origin. Substitution into (*) tells us that the point on the sphere that is closest to the origin is $(-1,-1,-1)$ and that the farthest point from the origin is $(4,4,4)$.   □

The next example turns out to be a special case of the Gauss–Markov theorem to be discussed in Section 23.

Example 22.2 The Minimum Variance Unbiased Estimator

Recall that an estimator is unbiased if its expectation is equal to the parameter that it is estimating. Let $x_1, x_2, \ldots, x_n$ be a random sample from a population with mean μ and variance σ^2. Let $x' = (x_1, x_2, \ldots, x_n)$ and $a' = (a_1, a_2, \ldots, a_n)$. We want to find the minimum variance unbiased estimator (MVUE) of the mean in the form

$$\hat{\mu} = a'x.$$

We need to minimize

$$\mathrm{Var}\,(\hat{\mu}) = a'a\sigma^2 = \sum_{i=1}^{n} a_i^2 \sigma^2$$

subject to the condition that the estimator is unbiased, namely,

$$\mathrm{E}\,(\hat{\theta}) = a'\mu 1_n = \left(\sum_{i=1}^{n} a_i\right)\mu = \mu.$$

Thus, we need to minimize

$$v = \sum_{i=1}^{n} a_i^2$$

subject to the constraint

$$\sum_{i=1}^{n} a_i = 1.$$

To do this, we employ the method of Lagrange multipliers

$$H = \sum_{i=1}^{n} a_i^2 - \lambda \left(\sum_{i=1}^{n} a_i - 1 \right).$$

Now

$$H_{a_i} = 2a_i - \lambda = 0.$$

Then

$$a_1 = a_2 = \ldots = a_n = \frac{\lambda}{2}.$$

Then

$$\frac{n\lambda}{2} = 1 \quad \text{so} \quad \lambda = \frac{2}{n} \quad \text{and} \quad a_i = \frac{1}{n}.$$

The MVUE is then

$$\hat{\mu} = \overline{x},$$

the sample mean. The Gauss–Markov theorem to be taken up in Section 23 states that for a general linear model, the least square estimator is MVUE. This is a special case for the model

$$y_i = \mu + \varepsilon_i, \quad 1 \leq i \leq n. \qquad \square$$

The next example illustrates a case where more than one constraint is used. Examples with more than one constraint will come up in the derivation of constrained least square estimators subject to multiple hypotheses.

Example 22.3 Point on a Line that Is Closest to the Origin

Find the point that is closest to the origin that lies on the line of intersection of the planes $x + y + z = 3$ and $2x + y + 2z = 5$.

We have

$$
\begin{aligned}
w &= x^2 + y^2 + z^2 - \lambda_1(x+y+z-3) - \lambda_2(2x+y+2z-5), \\
w_x &= 2x - \lambda_1 - 2\lambda_2 = 0 \\
w_y &= 2y - \lambda_1 - \lambda_2 = 0 \\
w_z &= 2z - \lambda_1 - 2\lambda_2 = 0.
\end{aligned}
$$

Then

$$
x = \frac{\lambda_1}{2} + \lambda_2, \ y = \frac{\lambda_1}{2} + \frac{\lambda_2}{2}, \ z = \frac{\lambda_1}{2} + \lambda_2. \tag{**}
$$

Substitution of (**) into the equations of the two planes and some algebra yields

$$
\begin{aligned}
3\lambda_1 + 5\lambda_2 &= 6 \\
5\lambda_1 + 9\lambda_2 &= 10,
\end{aligned}
$$

which has the solution $\lambda_1 = 2$, $\lambda_2 = 0$, so after substitution in (**), we see that the closest point is $(1,1,1)$. □

Example 22.4 The Minimum Norm Generalized Inverse

We will show how to obtain a minimum norm generalized inverse by solving the constrained optimization problem of minimizing $\|a\|^2 = a'a$ subject to the constraint $Xa = b$.

Let

$$
L = \frac{1}{2}a'a - \lambda'(Xa - b), \tag{22.1}
$$

where a is an m-dimensional vector and λ is an n-dimensional vector of Lagrange multipliers. Find the derivative of L with respect to the components of a and set it equal to zero. Thus,

$$
\frac{\partial L}{\partial a} = a' - \lambda'X = 0 \tag{22.2}
$$

and

$$
a = X'\lambda. \tag{22.3}
$$

Substitute a in (22.3) into the constraint $Xa = b$ to obtain

$$
Xa = XX'\lambda = b. \tag{22.4}
$$

Then from (22.4) and a result established in Section 12,

$$\lambda = (XX')^- b + (I - (XX')(XX')^-)z \tag{22.5}$$

for any vector z. Now

$$a = X'\lambda = X'(XX')^- b + (X' - X'(XX')^- (XX'))Z = X'(XX')^- b.$$

In Section 14, we defined the minimum norm generalized inverse as

$$X_m^- = X'(XX')^-. \qquad \square$$

22.3   MINIMIZING A SECOND-DEGREE FORM WITH RESPECT TO A LINEAR CONSTRAINT

First, we minimize a form (21.1) subject to a constraint $R'c = f$, where R is an $n \times s$ matrix and f is an n-dimensional vector. We then illustrate how to obtain the constrained least square estimator given the truth of a linear hypothesis as a special case. In Section 23 prove the Gauss–Markov theorem and obtain the linear Bayes estimator.

Let's see how to minimize the form with respect to the constraint. To this end let λ represent an s-dimensional vector of Lagrange multipliers. Now consider

$$h(c, \lambda) = c'Wc + c'd + e - \lambda'(R'c - f). \tag{22.6}$$

Then

$$\frac{\partial h}{\partial c} = 2Wc + d - R\lambda = 0$$
$$\frac{\partial h}{\partial \lambda} = R'c - f = 0. \tag{22.7}$$

Assume for now that W is positive definite. Then from the first equation in (22.7),

$$c = \frac{1}{2}W^{-1}(R\lambda - d). \tag{22.8}$$

From the second equation of (22.7),

$$R'c = f, \tag{22.9}$$

and thus, from (22.8)

$$R'c = \frac{1}{2}R'W^{-1}(R\lambda - d) = f. \tag{22.10}$$

Solving (22.10) for λ, we get

$$\begin{aligned}\lambda &= 2(R'W^{-1}R)^{-1}f + (R'W^{-1}R)^{-1}R'W^{-1}d \\ &= (R'W^{-1}R)^{-1}(2f + R'W^{-1}d).\end{aligned} \tag{22.11}$$

Substituting (22.11) into (22.8) optimal

$$c_{opt} = \frac{1}{2}W^{-1}(R(R'W^{-1}R)^{-1}(2f + R'W^{-1}d) - d). \tag{22.12}$$

Then

$$\begin{aligned}f(c_{opt}) &= \frac{1}{4}(2f' + d'W^{-1}R)(R'W^{-1}R)^{-1}(2f + R'W^{-1}d) \\ &\quad - \frac{1}{4}d'W^{-1}d + e.\end{aligned} \tag{22.13}$$

The first term corresponds to the difference between the minimum value for the constrained estimator and the unconstrained estimator, while the two terms following it are the minimum of the unconstrained estimator that was obtained in (21.5).

When W is of non-full rank, the inverse of W may be replaced by a generalized inverse.

Example 22.5 Minimizing a Second-Degree Form with Respect to a Linear Constraint

Let $v = [x\ y\ z]$

$$g(x,y,z) = [x \quad y \quad z]\begin{bmatrix} 2 & 1 & 1 \\ 1 & 2 & 1 \\ 1 & 1 & 2 \end{bmatrix}\begin{bmatrix} x \\ y \\ z \end{bmatrix} + [3 \quad 4 \quad 5]\begin{bmatrix} x \\ y \\ z \end{bmatrix} + 6.$$

Let $x + 2y + 3z = 6$ be a constraint equation. Then

$$L(v,\lambda) = g(v) - \lambda(x + 2y + 3z - 6)$$

$$\frac{\partial L}{\partial v} = 2[x \quad y \quad z]\begin{bmatrix} 2 & 1 & 1 \\ 1 & 2 & 1 \\ 1 & 1 & 2 \end{bmatrix} + [3 \quad 4 \quad 5] - \lambda[1 \quad 2 \quad 3] = 0$$

$$\frac{\partial L}{\partial \lambda} = x + 2y + 3z - 6 = 0.$$

The solution to the matrix equation for $\partial L/\partial v=0$ is [x y z]=$[-(\lambda/4)$ $(1/2)+(\lambda/4)$ $-1+$ $(3\lambda/4)]$, and substitution into $\partial L/\partial\lambda=0$ gives $\lambda=-4$. Then the function is minimized subject to the constraint when $v = [-1 \quad \frac{1}{2} \quad 2]$. The minimum value of $g(v)$ is 45/2. The reader may verify that the unconstrained minimum value is 5/2, which is less than the constrained minimum, which is what we would expect.   □

22.4   THE CONSTRAINED LEAST SQUARE ESTIMATOR

Suppose we wish to obtain the least square estimator subject to the constraint $H'\beta=m$. As was done for the unconstrained least square estimator in (21.8), let $W=X'X$, $c=b$, $d=-2X'Y$, $e=0$. In addition, let $f=m$ and $R=H$. The minimization problem is solved by considering

$$h(\beta,\lambda) = \beta'X'X\beta - 2\beta'X'Y - \lambda'(H'\beta - m)$$
$$= (Y - X\beta)'(Y - X\beta) - Y'Y - \lambda'(H'\beta - m). \tag{22.14}$$

Then we have

$$\frac{\partial h}{\partial \beta} = 2X'X\beta - 2X'Y - H\lambda = 0$$
$$\frac{\partial h}{\partial \lambda} = H'\beta - m = 0, \tag{22.15}$$

and solving the first equation in (22.15) for β,

$$\hat{\beta} = (X'X)^- X'Y + (X'X)^- H\lambda. \tag{22.16}$$

Then, from (22.16)

$$H'\hat{\beta} = H'(X'X)^- X'Y + H'(X'X)^- H\lambda = m. \tag{22.17}$$

Solving (22.17) for λ, we have

$$\lambda = [H'(X'X)^- H]^- (m - H'(X'X)^- X'Y). \tag{22.18}$$

The unconstrained least square estimator is $b=(X'X)^-X'Y$. After substitution of (22.18) into (22.16), we obtain the constrained least square estimator

$$\hat{\beta}_c = b - (X'X)^- H[H'(X'X)^- H]^- (H'b - m). \tag{22.19}$$

Substitution of the quantities defined at the beginning of this subsection yields the same result.

In general, for an estimator of the form $A\hat{\theta}$, where A is a matrix and $\hat{\theta}$ is a column vector,

$$E(A\hat{\theta}) = AE(\hat{\theta}) \quad \text{and} \quad D(A\hat{\theta}) = AD(\hat{\theta})A'. \tag{22.20}$$

Thus, for the unconstrained least square estimator,

$$E(b) = (X'X)^- X'X\beta = UU'\beta \quad \text{and} \quad D(b) = \sigma^2(X'X)^- X'X(X'X)^-. \tag{22.21}$$

As a result, for the estimable parametric functions,

$$E(p'b) = p'UU'\beta = p'\beta \tag{22.22a}$$

and

$$
\begin{aligned}
D(p'b) &= p'(X'X)^- (X'X)(X'X)^- p\sigma^2 \\
&= d'U'(X'X)^- U\Lambda U'(X'X)^- Ud\sigma^2 \\
&= d'\Lambda^{-1}\Lambda\Lambda^{-1}d\sigma^2 = d'\Lambda^{-1}d\sigma^2 \\
&= d'U'U\Lambda^{-1}U'Ud\sigma^2 = p'(X'X)^+ p\sigma^2.
\end{aligned}
\tag{22.22b}
$$

For the estimable parametric functions of the constrained least square estimator, assume that the constraints are also estimable so that $H = UM$ for some matrix M. Then since

$$
\begin{aligned}
E(H'b) &= E(M'U'b) = M'E(U'(X'X)^- UU'X'X\beta) \\
&= M'\Lambda^{-1}\Lambda U'\beta = M'U'\beta = H'\beta
\end{aligned}
$$

and the fact that the estimator is derived assuming that $H'\beta = m$, we have that

$$E(p'\hat{\beta}_c) = p'\beta.$$

Now from (22.18)

$$
\begin{aligned}
\hat{\beta}_c &= [I - (X'X)^- H[H'(X'X)^- H]^- H']b \\
&\quad + (X'X)^- H[H'(X'X)^- H]^- m.
\end{aligned}
\tag{22.23}
$$

Only the first term of (22.23) is a random variable. Thus,

$$
\begin{aligned}
D(\hat{\beta}_c) &= \sigma^2(X'X)^- - 2\sigma^2(X'X)^- H[H'(X'X)^- H]^- H'(X'X)^- \\
&\quad + \sigma^2(X'X)^- H[H'(X'X)^- H]^- H'(X'X)^- H[H'(X'X)^- H]^- H'(X'X)^-.
\end{aligned}
\tag{22.24}
$$

If the generalized inverse of $H'(X'X)^-H$ is reflexive, then

$$D(\hat{\beta}_c) = \sigma^2(X'X)^- - \sigma^2(X'X)^-H[H'(X'X)^-H]^-H'(X'X)^-. \qquad (22.25)$$

Comparing (22.25) with (22.22b), it is easily seen that in the sense of the Loewner ordering, the dispersion of the constrained least square estimator is less than that of the unconstrained least square estimator.

Example 22.6 One-Way ANOVA with Four Treatments

We wish to find the constrained least square estimator for one-way ANOVA with five observations of four treatments. The linear model setup would be

$$Y = \begin{bmatrix} 1_5 & 1_5 & 0 & 0 & 0 \\ 1_5 & 0 & 1_5 & 0 & 0 \\ 1_5 & 0 & 0 & 1_5 & 0 \\ 1_5 & 0 & 0 & 0 & 1_5 \end{bmatrix} \begin{bmatrix} \mu \\ \tau_1 \\ \tau_2 \\ \tau_3 \\ \tau_4 \end{bmatrix} + \varepsilon.$$

Then

$$\hat{\beta} = (X'X)^-X'Y,$$

$$\begin{bmatrix} \hat{\mu} \\ \hat{\tau}_1 \\ \hat{\tau}_2 \\ \hat{\tau}_3 \\ \hat{\tau}_4 \end{bmatrix} = \begin{bmatrix} 0 & 0 & 0 & 0 & 0 \\ 0 & \frac{1}{5} & 0 & 0 & 0 \\ 0 & 0 & \frac{1}{5} & 0 & 0 \\ 0 & 0 & 0 & \frac{1}{5} & 0 \\ 0 & 0 & 0 & 0 & \frac{1}{5} \end{bmatrix} \begin{bmatrix} Y_{..} \\ Y_{1.} \\ Y_{2.} \\ Y_{3.} \\ Y_{4.} \end{bmatrix} = \begin{bmatrix} 0 \\ \bar{Y}_{1.} \\ \bar{Y}_{2.} \\ \bar{Y}_{3.} \\ \bar{Y}_{4.} \end{bmatrix}.$$

We first consider the constraint $H'\beta = 0$, where $H' = \begin{bmatrix} 0 & 1 & 1 & -2 & 0 \end{bmatrix}$.
Then

$$(X'X)^-H = \begin{bmatrix} 0 \\ \frac{1}{5} \\ \frac{1}{5} \\ -\frac{2}{5} \\ 0 \end{bmatrix}$$

and

$$H'(X'X)^-H = \begin{bmatrix} \frac{6}{5} \end{bmatrix}$$

so that

$$[H'(X'X)^- H]^{-1} = \left[\tfrac{5}{6}\right].$$

Thus, the reduced least square estimators are

$$\hat{\mu}^r = 0,$$

$$\hat{\tau}_1^r = \bar{Y}_{1.} - \frac{1}{6}(\bar{Y}_{1.} + \bar{Y}_{2.} - 2\bar{Y}_{3.})$$

$$= \frac{5}{6}\bar{Y}_{1.} - \frac{1}{6}\bar{Y}_{2.} + \frac{1}{3}\bar{Y}_{3.},$$

$$\hat{\tau}_2^r = \bar{Y}_{2.} - \frac{1}{6}(\bar{Y}_{1.} + \bar{Y}_{2.} - 2\bar{Y}_{3.})$$

$$= -\frac{1}{6}\bar{Y}_{1.} + \frac{5}{6}\bar{Y}_{2.} + \frac{1}{3}\bar{Y}_{3.},$$

$$\hat{\tau}_3^r = \bar{Y}_{3.} + \frac{2}{6}(\bar{Y}_{1.} + \bar{Y}_{2.} - 2\bar{Y}_{3.})$$

$$= \frac{1}{3}\bar{Y}_{1.} + \frac{1}{3}\bar{Y}_{2.} + \frac{1}{3}\bar{Y}_{3.},$$

$$\hat{\tau}_4^r = \bar{Y}_{4.}.$$

Observe that

$$\hat{\tau}_1^r + \hat{\tau}_2^r - 2\hat{\tau}_3^r = 0.$$

The estimators satisfy the constraint. That is a check on the correctness of the calculation.

Suppose now we consider two constraints

$$H_0\ \tau_2 - \tau_3 = 0,\ 3\tau_1 - \tau_2 - \tau_3 - \tau_4 = 0.$$

Then

$$H' = \begin{bmatrix} 0 & 0 & 1 & -1 & 0 \\ 0 & 3 & -1 & -1 & -1 \end{bmatrix}.$$

The reader may verify the details of the following matrix calculations:

$$(X'X)^- H = \begin{bmatrix} 0 & 0 \\ 0 & \frac{3}{5} \\ \frac{1}{5} & -\frac{1}{5} \\ -\frac{1}{5} & -\frac{1}{5} \\ 0 & -\frac{1}{5} \end{bmatrix},\ [H'(X'X)^- H]^{-1} = \begin{bmatrix} \frac{5}{2} & 0 \\ 0 & \frac{5}{12} \end{bmatrix},$$

and

$$
\begin{bmatrix} \hat{\mu}^r \\ \hat{\tau}_1^r \\ \hat{\tau}_2^r \\ \hat{\tau}_3^r \\ \hat{\tau}_4^r \end{bmatrix} = \begin{bmatrix} 0 \\ \overline{Y}_{1.} \\ \overline{Y}_{2.} \\ \overline{Y}_{3.} \\ \overline{Y}_{4.} \end{bmatrix} - \begin{bmatrix} 0 & 0 \\ 0 & \frac{1}{4} \\ \frac{1}{2} & -\frac{1}{12} \\ -\frac{1}{2} & -\frac{1}{12} \\ 0 & -\frac{1}{12} \end{bmatrix} \begin{bmatrix} \overline{Y}_{2.} - \overline{Y}_{3.} \\ 3\overline{Y}_{1.} - \overline{Y}_{2.} - \overline{Y}_{3.} - \overline{Y}_{4.} \end{bmatrix}
$$

$$
= \begin{bmatrix} 0 \\ \overline{Y}_{..} \\ \frac{1}{12}(3\overline{Y}_{1.} + 5\overline{Y}_{2.} + 5\overline{Y}_{3.} - \overline{Y}_{4.}) \\ \frac{1}{12}(3\overline{Y}_{1.} + 5\overline{Y}_{2.} + 5\overline{Y}_{3.} - \overline{Y}_{4.}) \\ \frac{1}{12}(3\overline{Y}_{1.} - \overline{Y}_{2.} - \overline{Y}_{3.} + 11\overline{Y}_{4.}) \end{bmatrix}.
$$

$\square$

22.5   CANONICAL CORRELATION

We follow, for the most part, the derivation of Yanai et al. (2011). Consider two matrices of observations

$$
X = [x_1 \quad x_2 \quad \dots \quad x_p] \quad \text{and} \quad Y = [y_1 \quad y_2 \quad \dots \quad y_q].
$$

The objective is to find two linear combinations of the column vectors of X and Y, respectively, so that the correlation coefficient is maximized. Let a be a p-dimensional vector and b be a q-dimensional vector. Then the linear combinations can be written Xa and Yb, and we want to maximize

$$
r = \frac{a'X'Yb}{\sqrt{a'X'Xa}\sqrt{b'Y'Yb}}. \tag{22.26}
$$

Without loss of generality, it may be assumed that

$$
a'X'Xa = 1 \quad \text{and} \quad b'Y'Yb = 1. \tag{22.27}
$$

We now maximize $a'X'Yb$ subject to the constraints in (22.27).
   We now have a maximization problem with two constraints.
   Thus, we write down

$$
L(a, b, \lambda_1, \lambda_2) = a'X'Yb - \frac{\lambda_1}{2}(a'X'Xa - 1) - \frac{\lambda_2}{2}(b'Y'Yb - 1). \tag{22.28}
$$

Differentiating L with respect to a and then with respect to b and setting the results equal to zero yield

$$\frac{\partial L}{\partial a} = X'Yb - \lambda_1 X'Xa = 0 \qquad (22.29a)$$

and

$$\frac{\partial L}{\partial b} = Y'Xa - \lambda_2 Y'Yb = 0. \qquad (22.29b)$$

Premultiply (22.29a) by a' and (22.29b) by b'. Then

$$a'X'Yb = \lambda_1 a'X'Xa = \lambda_1$$

and

$$b'Y'Xa = \lambda_2 b'Y'Yb = \lambda_2.$$

Thus, $\lambda_1 = \lambda_2 = \sqrt{\lambda}$ maximizes the correlation between the linear combinations of the column matrix for the two sets of data. Pre-multiply Equation (22.29a) by $X(X'X)^-$ and Equation (22.29) by $Y(Y'Y)^-$ to obtain

$$X(X'X)^- X'Yb = \sqrt{\lambda}Xa \qquad (22.30a)$$

and

$$Y(Y'Y)^- Y'Xa = \sqrt{\lambda}Yb. \qquad (22.30b)$$

Substitute the right-hand side of (22.30b) divided by $\sqrt{\lambda}$ into (22.30a) and make a similar substitution of the right-hand side of (22.30b) and obtain

$$X(X'X)^- X'Y(Y'Y)^- Y'Xa = \lambda Xa \qquad (22.31a)$$

and

$$Y(Y'Y)^- Y'X(X'X)^- X'Yb = \lambda Yb. \qquad (22.31b)$$

Thus, Xa is the eigenvector of $X(X'X)^- X'Y(Y'Y)^- Y'$ and Yb is the eigenvector of $Y(Y'Y)^- Y'X(X'X)^- X$, and the canonical correlation coefficient λ is the largest eigenvalue of both of these matrices.

Example 22.7 Canonical Correlation of Grades on Periodic Examinations and the Final Examination

These data reproduced from Gruber (2010) are the grades of six students on four-hour exams and the final exam from a course taught by the author on differential equations and multivariable calculus. We will find the canonical correlation between the first two-hour exams and the last two-hour exams and the final examination. The data follows.

Exam 1	Exam 2	Exam 3	Exam 4	Final
95	85	97	83	91
69	70	70	72	68
89	84	96	85	88
66	58	100	70	73
77	94	100	86	74
49	71	94	80	71

After standardizing the data, that means for each column subtracting the mean and dividing by the standard deviation, we get the X and Y matrices.

Thus,

$$
X = \begin{bmatrix}
1.25003 & 0.6143 \\
-0.31001 & -0.53751 \\
0.89002 & 0.53751 \\
-0.49001 & -1.45895 \\
0.17000 & 1.30538 \\
-1.51003 & -0.46072
\end{bmatrix}, \quad
Y = \begin{bmatrix}
0.36455 & 0.53906 & 1.41131 \\
-1.99827 & -1.07812 & -0.99315 \\
0.27713 & 0.83309 & 1.09769 \\
0.62719 & -1.37215 & -0.47044 \\
0.62719 & 0.98011 & -0.36590 \\
0.10210 & 0.09801 & -0.67952
\end{bmatrix},
$$

$$
p_X = X(X'X)^- X'
$$

$$
= \begin{bmatrix}
0.3194 & -0.0612 & 0.2227 & -0.6823 & -0.0137 & -0.3990 \\
-0.0612 & 0.0579 & -0.0547 & 0.1590 & -0.1438 & 0.0427 \\
0.2227 & -0.0547 & 0.1584 & -0.0855 & 0.0285 & -0.2695 \\
-0.06824 & 0.1590 & -0.0855 & 0.4762 & -0.4601 & 0.0214 \\
-0.1370 & -0.1438 & 0.0285 & -0.4601 & 0.4650 & 0.1241 \\
-0.3990 & 0.0427 & -0.2695 & -0.0214 & 0.1241 & 0.5230
\end{bmatrix},
$$

and

$$
p_Y = Y(Y'Y)^- Y'
$$

$$
= \begin{bmatrix}
0.4321 & -0.1775 & 0.3041 & -0.0818 & -0.2312 & -0.2475 \\
-0.1775 & 0.7952 & -0.1456 & -0.1918 & -0.2693 & -0.0275 \\
0.3041 & -0.1456 & 0.2701 & -0.2501 & -0.0437 & -0.1375 \\
-0.0818 & -0.1918 & -0.2501 & 0.7996 & -0.2309 & 0.0305 \\
-0.2312 & -0.2693 & -0.0437 & -0.2309 & 0.5187 & 0.2616 \\
-0.2475 & -0.0275 & -0.1375 & -0.0305 & 0.2616 & 0.1843
\end{bmatrix}.
$$

The eigenvalues of $p_x p_y$ and $p_y p_x$ are 0.9819, 0.7587,0,0,0. The largest is 0.9819, the first canonical correlation coefficient.   □

22.6 SUMMARY

Several constrained optimization problems of statistical interest were considered. These included constrained least square estimation, the formula for a minimum norm generalized inverse and canonical correlation. The development of these ideas required a review of Lagrange multipliers.

EXERCISES

22.1 Verify that the estimates in the second part of Example 22.6 satisfy the constraints.

22.2 Redo Example 22.6 for the constraints H_0 $\tau_1 - \tau_2 = 0$, $\tau_1 + \tau_2 - 3\tau_3 + \tau_4 = 0$.

22.3 Show that for the variance given in 22.24 that in the Loewner ordering the variance of $H'\hat{\beta}_c$ is less than that of $H'b$.

22.4 In Example 22.6

$$(X'X)^+ = \frac{1}{125} \begin{bmatrix} 4 & 1 & 1 & 1 & 1 \\ 1 & 19 & -6 & -6 & -6 \\ 1 & -6 & 19 & -6 & -6 \\ 1 & -6 & -6 & 19 & -6 \\ 1 & -6 & -6 & -6 & 19 \end{bmatrix}.$$

Using this Moore–Penrose inverse in place of the generalized inverse in Example 22.6, obtain the unconstrained and constrained least square estimators for the basic estimable parametric functions $\hat{\mu} + \hat{\tau}_i$, $i = 1,2,3,4$. Observe that the unconstrained least square estimators are the same for both generalized inverses but the constrained estimators are not.

22.5 Show that the minimum value of $w = (x - x_0)^2 + (y - y_0)^2 + (z - z_0)^2$ subject to the constraint $ax + by + cz = d$ is

$$w_{min} = \frac{(d - ax_0 - by_0 - cz_0)^2}{a^2 + b^2 + c^2},$$

and then obtain the well-known formula for the distance from a point to a plane.

22.6 Find the point on the circle $(x - 2)^2 + (y - 2)^2 = 1$ nearest the origin. What is the distance from the point to the origin?

22.7 Consider the linear model

$$
\begin{bmatrix} Y_{11} \\ Y_{12} \\ Y_{21} \\ Y_{22} \end{bmatrix} = \begin{bmatrix} 1_2 & 0 \\ 0 & 1_2 \end{bmatrix} \begin{bmatrix} \beta_1 \\ \beta_2 \end{bmatrix} + \varepsilon.
$$

A. Show that the least square estimator is $\hat{\beta}_i = \overline{Y}_{i.}$, $i = 1, 2$.

B. Minimize $\beta_1^2 + \beta_2^2$ subject to the constraint $(\beta_1 - \overline{Y}_{1.})^2 + (\beta_2 - \overline{Y}_{2.})^2 = C^2$.

22.8 Here are grades in an engineering statistics class taught by the author.

Exam 1	Exam 2	Exam 3	Final
91	100	74	77
64	58	76	77
71	89	55	65
81	78	73	76
86	94	50	64
77	80	73	69
64	71	68	72
51	59	52	56
61	46	52	59
76	89	66	76
92	91	64	84
67	66	69	81

Find the first canonical correlation coefficient between the first two exams and the third exam and the final.

SECTION 23

THE GAUSS–MARKOV THEOREM

23.1 INTRODUCTION

Statisticians often want to find the most efficient estimator of a certain type relative to some measure of efficiency. An estimator with smaller variance is considered more efficient. An unbiased estimator is one whose expected value is that of the parameter that is being estimated. In this section we show how to find the unbiased estimator that is a linear combination of the observed values and has minimum variance, the minimum variance unbiased estimator (MVUE). We consider two cases, first when the parameters of the linear model are constants (Subsection 23.2) and second when they are random variables (Subsection 23.3).

23.2 THE GAUSS–MARKOV THEOREM AND THE LEAST SQUARE ESTIMATOR

The Gauss–Markov theorem shows that the MVUE is the least square estimator. Example 22.2 was a special case of this result for the linear model

$$y = 1_n \mu + \varepsilon$$

because the least square estimator in this case was

$$\hat{\mu} = \overline{y}.$$

Matrix Algebra for Linear Models, First Edition. Marvin H. J. Gruber.
© 2014 John Wiley & Sons, Inc. Published 2014 by John Wiley & Sons, Inc.

As was the case for Example 22.2, the proof of the Gauss–Markov theorem consists of finding the linear estimator $L'Y$ of linear combinations of the regression parameters $p'\beta$ by minimizing the variance of $L'Y$ subject to the constraint imposed by the condition of unbiasedness. The formal statement and proof of the Gauss–Markov theorem is given below. Assume that the β parameters in the linear regression model are constants.

Theorem 23.1 (Gauss–Markov). Consider the classical linear regression model

$$Y = X\beta + \varepsilon, \tag{23.1}$$

where X may be of full or non-full rank. Assume that the usual assumptions about the error term of a linear regression model hold true; that is,

$$E(\varepsilon) = 0 \quad \text{and} \quad D(\varepsilon) = \sigma^2 I. \tag{23.2}$$

Then:

1. The linear estimator $L'Y$ is unbiased for $p'\beta$ if and only if $p' = L'X$. This is the condition for $p'\beta$ to be estimable.
2. The least square estimator is the MVUE for the estimable parametric functions.

Proof. First notice that

$$E(L'Y) = L'E(Y) = L'X\beta. \tag{23.3}$$

In order for $L'Y$ to be unbiased for $p'\beta$,

$$E(L'Y) = p'\beta. \tag{23.4}$$

Comparing (23.3) and (23.4), we obtain the necessary and sufficient condition that $p' = L'X$. This establishes the first part of the theorem.

Using (23.2) the problem becomes that of minimizing

$$\text{Var}(L'Y) = \sigma^2 L'L \tag{23.5}$$

subject to the constraint $X'L - p = 0$. Then we have

$$H = L'L\sigma^2 + \lambda'(X'L - p). \tag{23.6}$$

Differentiating (23.6) with respect to the vectors L and λ and setting the result equal to zero,

$$\frac{\partial H}{\partial L} = 2L\sigma^2 + X\lambda = 0$$

$$\frac{\partial H}{\partial \lambda} = X'L - p = 0. \tag{23.7}$$

Multiply the first equation in (23.7) by X' so that

$$2X'L\sigma^2 + X'X\lambda = 0. \tag{23.8}$$

Let G be any generalized inverse of $X'X$. Then from (23.8) and the condition for unbiasedness,

$$\lambda = -2GX'L\sigma^2 = -2Gp\sigma^2. \tag{23.9}$$

Substitution of λ in the first equation of (23.7) yields

$$2L\sigma^2 - 2XGp\sigma^2 = 0 \tag{23.10}$$

so that optimum

$$L_{opt} = XGp \tag{23.11}$$

and

$$p'b = L'_{opt}Y = p'GX'Y. \tag{23.12}$$

∎

Since the estimator is unbiased if and only if the parametric function is estimable, we have that for some t

$$\begin{aligned} Var(p'b) &= p'GX'XGp\sigma^2 = t'XGX'XGXt\sigma^2 \\ &= t'X(X'X)^+ X'X(X'X)^+ Xt\sigma^2 \\ &= p'(X'X)^+ p\sigma^2 \end{aligned}$$

because XGX' is independent of the choice of generalized inverse.

23.3   THE MODIFIED GAUSS–MARKOV THEOREM AND THE LINEAR BAYES ESTIMATOR

In Subsection 23.2, we assumed that the regression parameters were constants. We will now assume that the regression parameters are random variables that satisfy the prior assumptions

$$E(\beta) = \theta \quad \text{and} \quad D(\beta) = F, \tag{23.13}$$

where θ is an m-dimensional vector of parameters and F is a positive-semi-definite matrix. For this development no other distributional assumptions about β will be needed. Because β is now a random variable, the assumptions in (23.2) are modified to be

$$E(\varepsilon \mid \beta) = 0 \quad \text{and} \quad D(\varepsilon \mid \beta) = \sigma^2 I. \tag{23.14}$$

Consider an estimator $\hat{\beta}$ of β. We say that $\hat{\beta}$ is unbiased for β in the extended sense if $E(\hat{\beta}) = \theta$. Our objective is to find the minimum variance unbiased in the extended sense estimator for the parametric functions $p'\beta$ of the form

$$p'\hat{\beta} = a + L'Y, \tag{23.15}$$

where a is a scalar and L is an n-dimensional column vector.

To this end, we minimize the variance

$$v = \text{Var}(p'\beta - a - L'Y) \tag{23.16}$$

subject to the extended unbiasedness condition

$$E(p'\beta - a - L'Y) = 0. \tag{23.17}$$

Calculation of the variance in (23.16) and the expectation in (23.17) reduces the problem to that of minimizing

$$v = p'Fp + L'(XFX' + \sigma^2 I)L - 2p'FX'L \tag{23.18}$$

subject to the constraint

$$p'\theta - a - L'X\theta = 0. \tag{23.19}$$

It turns out that the use of Lagrange multipliers is not necessary here. Solving (23.19) for a, we obtain

$$a = p'\theta - L'X\theta. \tag{23.20}$$

Differentiating (23.18) and setting the result equal to zero, we have

$$\frac{\partial v}{\partial L} = 2L'(XFX' + \sigma^2 I) - 2p'FX' = 0. \tag{23.21}$$

Solving (23.21) for L yields

$$L' = FX'(XFX' + \sigma^2 I)^{-1}.$$ (23.22)

The resulting linear Bayes estimator is

$$p'\hat{\beta} = p'\theta + p'FX'(XFX' + \sigma^2 I)^{-1}(Y - X\theta).$$ (23.23)

Thus, we have

Theorem 23.2 The MVUE in the extended sense is the linear Bayes estimator. ■

For the case where the prior distribution is normal with mean and dispersion given in (23.23) and also assuming the population is normal, the linear Bayes estimator is the mean of the posterior distribution that would be obtained using Bayes theorem.

An estimator that is algebraically equivalent to (23.23) expressed in terms of the least square estimator may be derived by obtaining the MVUE in the extended case in the form

$$p'\hat{\beta} = a + c'b.$$ (23.24)

This is equivalent to solving the optimization problem of minimizing the variance

$$v = \text{Var}(p'\beta - a - c'b),$$ (23.25)

where b is the least square estimator, subject to the unbiasedness condition

$$E(p'\beta - a - c'b) = 0.$$ (23.26)

After obtaining the variance in (23.25) and the expectation in (23.26) the problem reduces to minimizing

$$v = p'Fp + c'[UU'FUU' + \sigma^2 (X'X)^+]c - 2p'FUU'c$$ (23.27a)

subject to

$$p'\theta - a - cUU'\theta = 0.$$ (23.27b)

Solving (23.27) for a gives

$$a = (p' - c'UU')\theta.$$ (23.28)

Differentiating (23.27a) with respect to the vector c, setting the result equal to zero, and transposing terms to isolate c, we have

$$[UU'FUU' + \sigma^2(X'X)^+]c = FUU'p.$$ (23.29)

Then

$$c' = p'FUU[UU'FUU' + \sigma^2(X'X)^+]^+. \tag{23.30}$$

Substitution of (23.28) and (23.30) into (23.24) yields

$$p'\hat{\beta} = p'\theta + p'FUU'[UU'FUU' + \sigma^2(X'X)^+]^+(b - \theta). \tag{23.31}$$

We now establish the equivalence of (23.23) and (23.31) for the full-rank case. We then show how to establish the equivalence for the non-full-rank case for the estimable parametric functions using the reparameterization of the non-full-rank model to a full-rank model. Note that in the full-rank case, $UU' = I$, so (23.31) becomes

$$p'\hat{\beta} = p'\theta + p'F[F + \sigma^2(X'X)^{-1}]^{-1}(b - \theta). \tag{23.32}$$

Recall the identity (3.24) that was established in Subsection 3.4:

$$(A + BCB')^{-1} = A^{-1} - A^{-1}B(C^{-1} + B'A^{-1}B)^{-1}B'A^{-1}. \tag{23.33}$$

Let $A = \sigma^2 I$, $B = X$, $C = F$. Then employing (23.33),

$$FX'(XFX' + \sigma^2 I)^{-1} = \frac{1}{\sigma^2}FX' - \frac{1}{\sigma^2}FX'X\left(F^{-1} + X'X\frac{1}{\sigma^2}I\right)^{-1}X'\frac{1}{\sigma^2}I \tag{23.34}$$
$$= (\sigma^2 F^{-1} + X'X)^{-1}X'.$$

Then (23.23) becomes

$$p'\hat{\beta} = p'\theta + p'(X'X + \sigma^2 F^{-1})^{-1}X'(Y - X\theta) \tag{23.35}$$
$$= p'\theta + p'F((X'X)^{-1}\sigma^2 + F)^{-1}(b - \theta)$$

using the result of Exercise 3.5.

The equivalence of (23.23) and (23.35) for the non-full-rank model may be established using the reparameterization of the non-full-rank model to a full-rank model described in Subsection 10.1. Recall that the reparameterized model was

$$Y = XU\gamma + \varepsilon, \tag{23.36}$$

where $\gamma = U'\beta$. The assumptions on the γ parameters are from

$$E(\gamma) = U'\theta \quad \text{and} \quad D(\gamma) = U'FU. \tag{23.37}$$

We consider estimable parametric functions and assume that $U'FU$ is positive definite. For estimable parametric functions, there exists a d so that $p = Ud$. Also recall that using the singular value decomposition, it is easily shown that $X = XUU'$. Then we have

$$
\begin{aligned}
p'\hat{\beta} &= d'U'\theta + d'(U'FU)U'X'(XU(U'FU)U'X' + \sigma^2 I)^{-1}(Y - (XU)U'\theta) \\
&= d'U'\theta + d'U'FU((U'X'XU)^{-1}\sigma^2 + U'FU)^{-1}(U'b - U'\theta) \\
&= p'\theta + p'F((X'X)^+ \sigma^2 + UU'FUU')^+(b - \theta).
\end{aligned}
\tag{23.38}
$$

Example 23.1 The Linear Bayes Estimator for the Variance Component Model

This example is also discussed in Gruber (2010). Consider the linear model

$$
Y = [1_a \otimes 1_n \quad I_a \otimes 1_n]\begin{bmatrix} \mu \\ \tau \end{bmatrix} + \varepsilon.
$$

Suppose that in the prior assumptions (23.23) $\theta = 0$ and $F = \sigma_\tau^2 I$. Rewrite the linear model in the form

$$
Y - (1_a \otimes 1_n)\mu = (I_a \otimes 1_n)\tau + \varepsilon.
$$

For each μ a form of the linear Bayes estimator is given by

$$
p'\hat{\tau} = p'\left(nI_a + \frac{\sigma^2}{\sigma_a^2}I \right)^{-1} (I_a \otimes 1_n')(Y - (1_a \otimes 1_n)\mu).
$$

A modified Bayes estimator may be obtained by replacing μ by $\bar{Y}...$ Then a modified Bayes estimator would be

$$
p'\hat{\tau} = p'\left(nI_a + \frac{\sigma^2}{\sigma_a^2}I \right)^{-1} (I_a \otimes 1_n')(Y - \bar{Y}..).
$$

The individual components would be

$$
\begin{aligned}
\hat{\tau}_i &= \frac{n\sigma_\tau^2}{\sigma^2 + n\sigma_\tau^2}(\bar{Y}_{i.} - \bar{Y}..) \\
&= \bar{Y}.. + \left(1 - \frac{n\sigma^2}{\sigma^2 + n\sigma_\tau^2} \right)(\bar{Y}_{i.} - \bar{Y}..), \quad 1 \le i \le a.
\end{aligned}
$$

$\square$

Example 23.2 The Best Linear Unbiased Predictor (BLUP)

This example is also discussed in Gruber (2010). The derivation is similar to that of Bulmer (1980). Consider a linear model of the form

$$Y = X\beta + Z\delta + \varepsilon, \tag{23.39}$$

where the β parameters are random variables that satisfy the assumptions in (23.13). The error term ε satisfies (23.14). The δ parameter is constant.

Thus, we have a mixed model where some of the parameters are random variables and some of them are constants.

Let $\tilde{Y} = X\beta + \varepsilon$, where $\tilde{Y} = Y - Z\delta$. The linear Bayes estimator is

$$p'\hat{\beta} = p'FX'(XFX' + \sigma^2 I)^{-1}\tilde{Y}. \tag{23.40}$$

Rewrite the model (23.39) in the form

$$Y = Z\delta + \tilde{\varepsilon}, \tag{23.41}$$

where $\tilde{\varepsilon} = X\beta + \varepsilon$. When $\theta = 0$,

$$E(\tilde{\varepsilon}) = 0 \quad \text{and} \quad D(\tilde{\varepsilon}) = XFX' + \sigma^2 I.$$

The weighted least square estimator for the model (23.41) is given by

$$\hat{\delta} = [Z'(XFX' + \sigma^2 I)^{-1}Z]^{-1}Z'(XFX' + \sigma^2 I)^{-1}Y. \tag{23.42}$$

Substitute (23.42) into (23.40). Then the best linear unbiased predictor (BLUP) consists of the estimators in (23.40) and (23.42). □

23.4 SUMMARY

Two main results were obtained in this section. The first was the Gauss–Markov theorem that states that the MVUE for a linear regression model is the least square estimator. The second is that the linear Bayes estimator is an MVUE in the extended sense. The results were illustrated for the variance component model and in the derivation of the BLUP.

EXERCISES

23.1 Establish the formula for the variance in (23.18).

23.2 Show that if θ is unknown and it is assumed that $p' = L'X$, the problem reduces to the optimization problem for the Gauss–Markov theorem about the least square estimator.

23.3 Assume θ is unknown. Show that the estimator of the form $c'b$ that minimizes

$$v = c'(X'X)^{+}c$$

subject to the constraint

$$p' - cUU' = 0$$

is in fact $p'b$.

23.4 Show that for estimable parametric functions, a minimum variance unbiased in the extended sense estimator of the form

$$p'\widehat{\beta} = a + (p - c)'b$$

is

$$p'\widehat{\beta} = p'b - p'(X'X)^{+}(UU'FUU' + \sigma^{2}(X'X)^{+})'(b - \theta)\sigma^{2},$$

and establish that it is algebraically equivalent to (23.38).

23.5 Fill in the details of the derivation of (23.34) and (23.35).

23.6 Fill in the details for (23.38).

23.7 **A.** For the optimization problem in (23.16) and (23.17), establish that

$$v_{\min} = p'Fp - p'FX'(XFX + \sigma^{2}I)^{-1}XFp.$$

B. For the optimization problem in (23.25) and (23.26), establish that for estimable parametric functions,

$$v_{\min} = p'Fp - p'UU'FUU'[UU'FUU' + \sigma^{2}(X'X)^{+}]^{+}UU'FUU'p.$$

Establish that the results obtained in A and B are algebraically equivalent.

C. Establish that for estimable parametric functions, a form equivalent to the result in B is

$$v_{\min} = p'(X'X)^{+}p\sigma^{2} - p'(X'X)^{+}(UU'FUU' + (X'X)^{+}\sigma^{2})^{+}(X'X)^{+}p\sigma^{4}.$$

The results in this exercise show that the minimum variance is:

1. Less than the variance of the prior parameter
2. Less than the variance of the least square estimator

23.8 Show that for $\theta=0$ and $G=(UU'FUU')^+\sigma^2$, the linear Bayes estimator takes the form

$$p'\hat{\beta} = p'(X'X+G)^+ X'Y.$$

23.9 For the least square estimator $p'\hat{\beta} = p'(X'X)^-X'Y$, where the parametric function is not estimable:

A. What is the bias of $p'\hat{\beta}$?

B. Show that if G is a reflexive generalized inverse,

$$Var(p'\hat{\beta}) = p'(X'X)^-p\sigma^2.$$

The solution to this problem is contained in Gruber (2010, pp. 273–275).

23.10 For the model in (18.7), assume that

1. $E(\varepsilon|\alpha,\beta)=0$, $D(\varepsilon|\alpha,\beta)=\sigma^2I$;
2. $E(\alpha)=0$, $D(\alpha)=\sigma_\alpha^2I$; and
3. $E(\beta)=0$, $D(\beta)=\sigma_\beta^2I$

and that α and β are independent. Show that the modified Bayes estimator is

$$\hat{\alpha}_i = \left(1-\frac{\sigma^2}{\sigma^2+\sigma_\alpha^2}\right)(\bar{Y}_{i.} - \bar{Y}..), \quad 1\leq i\leq a.$$

$$\hat{\beta}_j = \left(1-\frac{\sigma^2}{\sigma^2+\sigma_\beta^2}\right)(\bar{Y}_{.j} - \bar{Y}..), \quad 1\leq j\leq b$$

23.11 Assume that the β parameters satisfy the assumptions $E(\beta)=0$, $D(\beta)=\sigma^2F^{-1}$. Assume that X is of full rank. Find the vector h such that

$$m = E(h'X'Y - p'\beta)(h'X'Y - p'\beta)'$$

is minimized. Show that the optimum estimator is

$$p'\hat{\beta} = p'(X'X+\sigma^2F^{-1})^{-1}X'Y$$

and that the minimum m is

$$m_{min} = p'(X'X+\sigma^2F)^{-1}p\sigma^2.$$

SECTION 24

RIDGE REGRESSION-TYPE ESTIMATORS

24.1 INTRODUCTION

Ridge-type estimators have been proposed as one possible way of dealing with the multicollinearity problem. In this final section, we will consider how they arise as the solution to a constrained optimization problem and as a special case of the linear Bayes estimator. We also study their efficiencies from both the Bayesian and the frequentist point of view. These derivations will furnish a nice application of the matrix methods that have been developed in this book.

First, in Subsections 24.2 and 24.3 we derive the generalized ridge regression estimator as a special case of the more general problem of minimizing a second-degree form with respect to a quadratic constraint. We show how it is also a special case of the mixed estimator of Theil and Goldberger (1961). We then show how for special choices of the biasing parameter matrix different ridge estimators in the literature may be obtained. In Subsections 24.4 and 24.5, we compare the mean square error (MSE) of the ridge-type estimators and the least square estimator with and without averaging over a prior distribution.

24.2 MINIMIZING A SECOND-DEGREE FORM WITH RESPECT TO A QUADRATIC CONSTRAINT

We will consider the minimization of a second-degree form with respect to a single quadratic constraint. This will enable us to show how the ridge-type estimators arise from specializations of the minimization problem. We also show a connection with

Matrix Algebra for Linear Models, First Edition. Marvin H. J. Gruber.
© 2014 John Wiley & Sons, Inc. Published 2014 by John Wiley & Sons, Inc.

the linear Bayes estimator. Thus, we will see how ridge-type estimators can be formulated from both the Bayesian and the frequentist point of view.

Consider minimizing the general second-degree form

$$f(c) = c'Wc + c'd + e \tag{24.1}$$

subject to a constraint of the form

$$g(c) = c'Ac + c'g + h = 0. \tag{24.2}$$

Assume that A and W are positive-semi-definite matrices.

Again employing the method of Lagrange multipliers, we differentiate the function

$$
\begin{aligned}
L(c) &= f(c) + \lambda m(c) \\
&= c'Wc + c'd + e + \lambda(c'Ac + c'g + h) \\
&= c'(W + \lambda A)c + c'(d + \lambda g) + (e + \lambda h)
\end{aligned} \tag{24.3}
$$

and set the result equal to zero. Thus,

$$\frac{\partial L}{\partial c} = 2(W + \lambda A)c + d + \lambda g = 0. \tag{24.4}$$

Solving (24.4) for the vector c, we obtain

$$c = -\frac{1}{2}(W + \lambda A)^{+}(d + \lambda g). \tag{24.5}$$

We will show how to specialize this general form to get the biased ridge-type estimators in the next subsection.

24.3   THE GENERALIZED RIDGE REGRESSION ESTIMATORS

Suppose that in the above development $c = \hat{\beta}$, $A = X'X$, $d = H\theta$, $W = H$, and $g = X'Xb$. Then the generalized ridge regression estimator is

$$\hat{\beta} = [H + \lambda X'X]^{+}[H\theta + \lambda X'Xb]. \tag{24.6}$$

We write the estimator in a slightly different form. Let $G = (1/\lambda)H$. Then

$$\hat{\beta} = [G + X'X]^{+}[G\theta + X'Xb]. \tag{24.7}$$

There are a number of interesting things about (24.7). First, there is a matrix R such that $G = R'R$. Suppose that θ is the least square estimator with respect to the linear model

$$r = R\beta + \eta. \tag{24.8}$$

Then (24.7) is a mixed estimator with respect to an augmented linear model

$$\begin{bmatrix} Y \\ r \end{bmatrix} = \begin{bmatrix} X \\ R \end{bmatrix} \beta + \begin{bmatrix} \varepsilon \\ \eta \end{bmatrix},$$

where the R matrix in (24.8) represents additional observations.

Second, (24.7) is a Bayes estimator with respect to a normal prior and a linear Bayes estimator with respect to any prior with mean θ and dispersion $F = \sigma^2 (UU'GUU')^+$.

Third, for $\theta = 0$ and different choices of G, (24.7) reduces to some of the different ridge-type estimators considered in the literature. For general G and zero prior mean, we have

$$\hat{\beta} = (X'X + G)^+ X'Y, \tag{24.9}$$

the ridge regression estimator of C. R. Rao (1975). When $G = kI$ with k a positive constant, we have

$$\hat{\beta} = (X'X + kI)^{-1} X'Y, \tag{24.10}$$

the ordinary ridge regression estimator of Hoerl and Kennard (1970). For $G = kUU'$ where U is the column orthogonal matrix in the SVD of $X'X$ with k a positive constant, the estimator

$$\hat{\beta} = (X'X + kUU')^+ X'Y \tag{24.11}$$

is the k-s estimator of Baye and Parker (1984), where s is the rank of X. When X is of full rank and $G = UKU'$, the estimator

$$\hat{\beta} = (X'X + UKU')^{-1} X'Y \tag{24.12}$$

is the generalized ridge estimator of Hoerl and Kennard (1970). For $G = X'X$,

$$\hat{\beta} = \frac{1}{1+k} b, \tag{24.13}$$

the contraction estimator of Mayer and Willke (1973).

When

$$G = (1-d)k(I + dk(X'X)^{-1})^{-1} \tag{24.14}$$

with $0 < d < 1$, after some algebra (see Exercise 24.1), we obtain the two-parameter Liu estimator of Ozkale and Kaciranlar (2007),

$$\hat{\beta} = (X'X + kI)^{-1} (X'Y + kdb). \tag{24.15}$$

When $k = 1$, the original Liu (1993) estimator results. When $d = 0$, (24.15) reduces to the ordinary ridge estimator of Hoerl and Kennard. Each of these estimators may be viewed in two ways:

1. As the solution to a constrained optimization problem where both the quantity to be minimized and the constraint is a quadratic form
2. As a linear Bayes estimator and for the case of normal priors and normal populations a Bayes estimator with prior mean zero and dispersion $F = (UU'GUU')^+\sigma^2$.

24.4   THE MEAN SQUARE ERROR OF THE GENERALIZED RIDGE ESTIMATOR WITHOUT AVERAGING OVER THE PRIOR DISTRIBUTION

The MSE averaging over the prior distribution is the minimum variance obtained in the derivation of the Bayes estimators. For the ridge-type estimators, this MSE is less than that of the least square estimator. The MSE without averaging over the prior distribution of the Bayes estimators including the ridge-type estimators is less than that of the least square estimator only for a range of parameters. The main objective in this section is to determine this range of parameters.

We first show that the MSE may be decomposed into the sum of the variance and the squared bias. To this end, consider a parametric function of an estimator $p'\hat{\beta}$. The MSE of the parametric functions is given by

$$\text{MSE}(p'\hat{\beta}) = p'E(\hat{\beta} - \beta)(\hat{\beta} - \beta)'p. \tag{24.16}$$

This MSE may be shown to be the sum of two quantities, the variance and the square of the bias. Observe that after addition and subtraction of $E(\hat{\beta})$,

$$
\begin{aligned}
\text{MSE}(p'\hat{\beta}) &= p'E(\hat{\beta} - E(\hat{\beta}) + E(\hat{\beta}) - \beta)(\hat{\beta} - E(\hat{\beta}) + E(\hat{\beta}) - \beta)'p \\
&= p'E(\hat{\beta} - E(\hat{\beta}))(\hat{\beta} - E(\hat{\beta}))'p + p'E(E(\hat{\beta}) - \beta)(E(\hat{\beta}) - \beta)'p \\
&\quad + p'E(\hat{\beta} - E(\hat{\beta}))(E(\hat{\beta}) - \beta)'p + p'(E(\hat{\beta}) - \beta)E(\hat{\beta} - E(\hat{\beta}))'p.
\end{aligned}
\tag{24.17}
$$

The last two terms of (24.17) are zero because $E(\hat{\beta} - E(\hat{\beta})) = E(\hat{\beta}) - E(\hat{\beta}) = 0$. Thus,

$$
\begin{aligned}
\text{MSE}(p'\hat{\beta}) &= p'E(\hat{\beta} - E(\hat{\beta}))(\hat{\beta} - E(\hat{\beta}))'p + p'E(E(\hat{\beta}) - \beta)(E(\hat{\beta}) - \beta)'p \\
&= \text{Var}(p'\hat{\beta}) + (\text{Bias}(p'\hat{\beta}))^2.
\end{aligned}
\tag{24.18}
$$

Consider the generalized ridge regression estimator (24.9). Then

$$
\begin{aligned}
\text{Var}(p'\hat{\beta}) &= p'(X'X + G)^{-1}\text{Var}(X'Y)(X'X + G)^{-1}p \\
&= p'(X'X + G)^{-1}(X'X)(X'X + G)^{-1}p\sigma^2.
\end{aligned}
\tag{24.19}
$$

Now

$$E(\hat{\beta}) = (X'X+G)^{-1}X'X\beta$$

so that the bias

$$\begin{aligned}\beta - E(\hat{\beta}) &= \beta - (X'X+G)^{-1}X'X\beta \\ &= (X'X+G)^{-1}G\beta\end{aligned} \tag{24.20}$$

and

$$(\text{Bias}(p'\hat{\beta}))^2 = p'(X'X+G)^{-1}G\beta\beta'G(X'X+G)^{-1}p. \tag{24.21}$$

The MSE of $p'\hat{\beta}$ is then the sum of (24.20) and (24.21). Thus,

$$\text{MSE}(p'\hat{\beta}) = p'(X'X+G)^{-1}(\sigma^2(X'X)+G\beta\beta'G)(X'X+G)^{-1}p. \tag{24.22a}$$

For the non-full-rank case, we have, when $G=UAU'$ with A a nonsingular matrix (see Exercise 24.3),

$$\text{MSE}(p'\hat{\beta}) = p'(X'X+G)^+(\sigma^2(X'X)+G\beta\beta'G)(X'X+G)^+p. \tag{24.22b}$$

Theorem 24.1 gives the region for the β parameters where the ridge estimator has a smaller MSE than the least square estimator assuming that X is of full rank and G is nonsingular.

Theorem 24.1 The MSE of the parametric functions of the generalized ridge estimator (24.9) is smaller than that of the least square estimator if and only if β lies in the ellipsoid

$$\beta'(2G^{-1}+(X'X)^{-1})^{-1}\beta \le \sigma^2. \tag{24.23}$$

Proof. The MSE of the parametric functions of the ridge estimator is less than that of the least square estimator if and only if in the sense of the Loewner ordering

$$(X'X+G)^{-1}(\sigma^2(X'X)+G\beta\beta'G)(X'X+G)^{-1} \le \sigma^2(X'X)^{-1}.$$

The ordering is preserved when both sides of the expression are multiplied by the same matrix so that

$$\begin{aligned}(\sigma^2(X'X)+G\beta\beta'G) &\le (X'X+G)\sigma^2(X'X)^{-1}(X'X+G) \\ &= \sigma^2 X'X + 2\sigma^2 G + \sigma^2 G(X'X)^{-1}G\end{aligned}$$

or

$$G\beta\beta'G \le 2\sigma^2 G + \sigma^2 G(X'X)^{-1}G. \tag{24.24}$$

Multiplication of both sides of (24.24) by G^{-1} gives

$$\beta\beta' \le \sigma^2(2G^{-1} + (X'X)^{-1}). \tag{24.25}$$

The result of the theorem follows if we apply the result of Theorem 7.3.   ■

A similar ellipsoid may be derived in the non-full-rank case (see Exercise 24.5).

If we wanted to compare two ridge-type estimators for the MSE in (24.14), it would be difficult to find an ellipsoid where one estimator is smaller than another. Instead of the measure of efficiency in (24.16), let us consider the individual MSE (IMSE) of $\hat{\gamma}_i = U_i'\hat{\beta}, 1 \le i \le s$ where U_i is the ith column of U as discussed in Subsection 10.1. We will compare the IMSE of the $\hat{\gamma}_i$. We have

$$\text{IMSE}(\hat{\gamma}_i) = U_i' E(\hat{\beta} - \beta)(\hat{\beta} - \beta)'U_i = E(\hat{\gamma}_i - \gamma_i)^2. \tag{24.26}$$

If for the generalized ridge estimator (24.9) $G = UKU'$ where K is a diagonal matrix, then

$$\hat{\gamma} = U'\hat{\beta} = U'(X'X + UKU')^+ X'Y = U'U(\Lambda + K)^{-1}U'U\Lambda^{1/2}Z$$
$$= (\Lambda + K)^{-1}\Lambda^{1/2}Z = (\Lambda + K)^{-1}\Lambda\Lambda^{-1/2}Z.$$

The individual coordinates are

$$\hat{\gamma}_i = \frac{\lambda_i}{\lambda_i + k_i}g_i, \quad 1 \le i \le s. \tag{24.27}$$

Using the singular value decomposition on the matrices in (24.22b), we get that the IMSE of $\hat{\gamma}_i$ is

$$\text{IMSE}(\hat{\gamma}_i) = E(\hat{\gamma}_i - \gamma_i)^2 = \frac{\sigma^2\lambda_i + k_i^2\gamma_i^2}{(\lambda_i + k_i)^2}. \tag{24.28}$$

We now state a theorem that compares the IMSE of two ridge-type estimators.

Theorem 24.2 Suppose that $k_{i1} \ge k_{i2}$. Then

$$E(\hat{\gamma}_{i1} - \gamma_i)^2 \le E(\hat{\gamma}_{i2} - \gamma_i)^2 \tag{24.29}$$

if and only if

$$|\gamma_i| \le \sigma\sqrt{\frac{2\lambda_i + k_{i1} + k_{i2}}{\lambda_i(k_{i1} + k_{i2}) + 2k_{i1}k_{i2}}}. \tag{24.30}$$

Proof. Inequality (24.29) holds true if and only if

$$\frac{\sigma^2\lambda_i + k_{i1}^2\gamma_i^2}{(\lambda_i + k_{i1})^2} \le \frac{\sigma^2\lambda_i + k_{i2}^2\gamma_i^2}{(\lambda_i + k_{i2})^2}. \tag{24.31}$$

Then (24.31) holds true if and only if

$$[k_{i1}^2(\lambda_i + k_{i2})^2 - k_{i2}^2(\lambda_i + k_{i1})^2]\gamma_i^2 \le \sigma^2\lambda_i[(\lambda_i + k_{i1})^2 - (\lambda_i + k_{i2})^2]$$

or if

$$[k_{i1}(\lambda_i + k_{i2}) + k_{i2}(\lambda_i + k_{i1})(k_{i1} - k_{i2})]\lambda_i\gamma_i^2$$
$$\le \sigma^2\lambda_i(2\lambda_i + k_{i1} + k_{i2})(k_{i1} - k_{i2}),$$

giving

$$\gamma_i^2 \le \frac{2\lambda_i + k_{i1} + k_{i2}}{\lambda_i(k_{i1} + k_{i2}) + 2k_{i1}k_{i2}}. \tag{24.32}$$

Taking the square root of both sides of (24.32) yields the result.   ∎

The following corollaries of Theorem 24.2 are not hard to establish.

Corollary 24.1 A ridge-type estimator has a smaller IMSE than the least square estimator if and only if

$$|\gamma_i| \le \sigma\sqrt{\frac{2}{k_i} + \frac{1}{\lambda_i}}. \tag{24.33}$$

Proof. In Theorem 24.2 let $k_{i1} = k_i$ and $k_{i2} = 0$.   ∎

Corollary 24.2 Let M be a matrix of the form $U\Delta U'$ where Δ is a positive-semi-definite diagonal matrix. If $k_{i1} \ge k_{i2}$, $1 \le i \le s$ and (24.30) holds true for $1 \le i \le s$, then the weighted MSE

$$E(\hat{\beta}_1 - \beta)'M(\hat{\beta}_1 - \beta) \le E(\hat{\beta}_2 - \beta)'M(\hat{\beta}_2 - \beta) \tag{24.34}$$

where

$$p'\hat{\beta}_j = (X'X + UK_jU')^+X'Y, \quad j = 1, 2$$

with K_j, $j = 1, 2$ is a positive-definite diagonal matrix with elements k_{ij}, $j = 1, 2$.   ∎

Some choices of the matrix M include UU', $X'X$, and I for the full-rank case.

24.5   THE MEAN SQUARE ERROR AVERAGING OVER THE PRIOR DISTRIBUTION

Since the estimators we have found are unbiased in the extended sense, the variances in Exercise 23.7 are the MSE averaging over the prior distribution. The MSE in Exercise 23.7c indicates that the average MSE (AMSE) of the Bayes estimator is less than that of the least square estimator. That means that the MSE without averaging over the prior of the Bayes or ridge estimator is less than that of the least square estimator for at least one value of the β parameter. Since the MSE is a continuous function of the β parameters, this would also be true for a range of values.

We noted that ridge estimator (24.9) is a Bayes estimator where the prior dispersion $F = (UU'GUU')^+\sigma^2$. For the full-rank case, $F = \sigma^2 G^{-1}$. To obtain the average MSE, we find the expectation of (24.22b) and obtain the AMSE as

$$\begin{aligned} \text{AMSE} &= p'(X'X+G)^{-1}(\sigma^2(X'X)+\sigma^2 G)(X'X+G)^{-1}p \\ &= p'(X'X+G)^{-1}p. \end{aligned} \tag{24.35}$$

For the above choice of prior dispersion, (24.35) is equivalent to the expressions obtained in Exercise 23.7. The AMSE of two ridge-type estimators with respect to their own priors may be compared by means of the following theorem.

Theorem 24.3 Assume that $p'\hat{\beta}_1$ is the ridge-type estimator derived with respect to the prior distribution with dispersion $F_1 = \sigma^2 G_1^{-1}$ and that $p'\hat{\beta}_2$ is the ridge-type estimator derived with respect to the prior distribution with dispersion $F_2 = \sigma^2 G_2^{-1}$. If $G_1 \geq G_2$, then

$$\text{AMSE}(p'\hat{\beta}_1) \leq \text{AMSE}(p'\hat{\beta}_2).$$

Proof. The result follows from Theorem 8.8.   ■

24.6   SUMMARY

We showed two of the main methods of deriving ridge-type estimators. First, we derived them as the solution to a constrained optimization problem. Then we derived them as a special case of the linear Bayes estimator. We compared their efficiencies with the least square estimator with and without averaging over a prior distribution.

EXERCISES

24.1 Do the necessary matrix manipulation to show that for the choice of G in (24.14), the generalized ridge estimator reduces to (24.15).

24.2 Show that (24.15) is linear Bayes (and Bayes if the prior and population are normal) for the prior parameters

$$\theta = 0 \quad \text{and} \quad F = \frac{\sigma^2}{(1-d)k}(I + dk(X'X)^{-1}).$$

24.3 Suppose that X is of non-full rank and $G = UAU'$ where A is positive definite. Show that for estimable parametric functions the formula for the MSE of the generalized ridge estimator is given by (24.22b).

24.4 **A.** Consider two generalized ridge estimators $p'\hat{\beta}_1$ and $p'\hat{\beta}_1$ with ridge matrices G_1 and G_2. Assume that $G_i = UD_iU'$, $i = 1,2$ where the D_i are positive-definite diagonal matrices.
Show that if $G_2 \le G_1$ in the sense of the Loewner ordering, then

$$\text{Var}(p'\hat{\beta}_1) \le \text{Var}(p'\hat{\beta}_2).$$

B. Interpret the condition in A for estimators (24.9)–(24.15). Show that for two estimators of the same kind, the one with the larger k has the smaller variance. Then for the same k, derive conditions for comparison of the variance of the different estimators. There are 21 possible comparisons.

24.5 Show that the general ellipsoid for comparing a Bayes estimator to a least square estimator for the non-full-rank case when $U'FU$ is positive definite is

$$(\beta - \theta)'(2UU'FUU' + \sigma^2(X'X)^+)^+(\beta - \theta) \le 1.$$

How would you specialize this result to that of (24.23)?

24.6 Give the form of the ellipsoids in (24.23) for each of the estimators in (24.9)–(24.15).

24.7 **A.** Consider the estimators (24.13) with parameters k_1 and k_2 where $k_1 \ge k_2$. Show that the MSE of the parametric functions of the estimator associated with k_1 is smaller than the MSE of the parametric functions associated with k_2 if and only if

$$\beta'(X'X)\beta \le \frac{(k_1 + k_2 + 2)\sigma^2}{k_1 + k_2 + 2k_1k_2}.$$

B. Assume that X is an orthogonal matrix, so $X'X = I$. Consider the ordinary ridge regression estimator (24.10) with parameters k_1 and k_2 where $k_1 \ge k_2$. Show that the MSE of the parametric functions of the estimator associated with k_1 is smaller than the MSE of the parametric functions associated with k_2 if and only if

$$\beta'\beta \le \frac{\sigma^2(k_1 + k_2 + 2)}{k_1 + k_2 + 2k_1 k_2}.$$

24.8 Show that the optimal value of k_i in (24.28) is σ^2/γ_i^2. What is the minimum value of the IMSE? Since the quantities in the optimum k_i are unknown, how would you suggest estimating them?

24.9 Consider the contraction estimator of Mayer and Willke (24.13). What value of k in terms of the β parameters minimizes the predictive MSE

$$PMSE = E(\hat{\beta} - \beta)'X'X(\hat{\beta} - \beta)$$

24.10 Consider a setup with r linear models

$$Y_i = X\beta_i + \varepsilon_i, \quad 1 \le i \le r,$$

where ε_i are uncorrelated. For each model, consider the estimator

$$p'\hat{\beta}_i = (X'X + G)^{-1} X'Y_i$$

and

$$b_i = (X'X)^{-1} X'Y_i.$$

To keep things simple, assume X is of full rank. Assume $r > r(X)$. Find the vector c that minimizes

$$M = \sum_{i=1}^{r} E(c'b_i - p'\beta_i)(c'b_i - p'\beta_i)',$$

and do appropriate algebra to show that

$$p'\hat{\beta}_i = p'\left(X'X + r\sigma^2 \left(\sum_{i=1}^{r} \beta_i \beta_i'\right)^{-1}\right)^{-1} X'Y_i.$$

Thus, we have a theoretically optimum G. Again since the quantities are known, suggest an estimator based on the sample.

ANSWERS TO SELECTED EXERCISES

PART I

Section 1

1.1 $AB = \begin{bmatrix} 10 & 18 \\ -14 & 6 \end{bmatrix}$, $BA = \begin{bmatrix} 15 & 17 & 2 \\ -13 & -3 & -6 \\ 22 & 22 & 4 \end{bmatrix}$

1.6 $P'D_1 PP'D_2 P = P'D_1 D_2 P = P'D_2 D_1 P = P'D_2 PP'D_1 P$

1.7 $A = \dfrac{1}{2}(A + A') + \dfrac{1}{2}(A - A')$

1.9 $A'B' = (BA)' = (AB)' = B'A'$

1.14 Symmetry of XX'

$$(XX')' = X''X' = XX'$$

1.15 Observe that

$$AB = \begin{bmatrix} a_{11} & a_{12} \\ a_{21} & a_{22} \end{bmatrix} \begin{bmatrix} b_{11} & b_{12} \\ b_{21} & b_{22} \end{bmatrix}$$
$$= \begin{bmatrix} a_{11}b_{11} + a_{12}b_{21} & a_{11}b_{12} + a_{12}b_{22} \\ a_{21}b_{11} + a_{22}b_{21} & a_{21}b_{12} + a_{22}b_{22} \end{bmatrix}$$

Matrix Algebra for Linear Models, First Edition. Marvin H. J. Gruber.
© 2014 John Wiley & Sons, Inc. Published 2014 by John Wiley & Sons, Inc.

Now for example

$$a_{11}b_{11} + a_{12}b_{21} + a_{11}b_{12} + a_{12}b_{22} = a_{11}(b_{11} + b_{12}) + a_{12}(b_{21} + b_{22}) = a_{11} + a_{12} = 1.$$

In a similar way the summation to 1 may be established for the remaining row and the columns.

1.16 B. $X'X = \begin{bmatrix} 4 & 2 & 2 & 2 & 2 \\ 2 & 2 & 0 & 1 & 1 \\ 2 & 0 & 2 & 1 & 1 \\ 2 & 1 & 1 & 2 & 0 \\ 2 & 1 & 1 & 0 & 2 \end{bmatrix}$

1.17 B. $\begin{bmatrix} \frac{13}{54} & \frac{41}{162} & \frac{83}{324} & \frac{1}{4} \end{bmatrix}$

1.18 C. $\pi^{(3)} = \begin{bmatrix} \frac{5}{16} & \frac{51}{256} & \frac{295}{1024} & \frac{205}{1024} \end{bmatrix}$

Section 2

2.1 A. 0 **B.** 3072 **C.** 4 **D.** 4 **E.** $-1/27$

2.6 A. $x = 23/5, y = 11/5$ **B.** $x = -168/85, y = 181/85, z = 58/85$
C. System cannot be solved by Cramer's rule because of zero determinant. If x is taken as arbitrary we get $y = -1 + x$, $z = 2$, $w = 5 - x$. System has infinitely many solutions. If, for example, we choose $x = 4$ then $y = 3$, $z = 2, w = 1$.

2.7 A. $\frac{1}{2}(7 + \sqrt{17}), \frac{1}{2}(7 - \sqrt{17})$

2.9 Observe that

$$\det(A - \lambda I) = \begin{vmatrix} a_{11} - \lambda & a_{12} \\ a_{21} & a_{22} - \lambda \end{vmatrix} = 0$$

Expanding it we get

$$(a_{11} - \lambda)(a_{22} - \lambda) - a_{12}a_{21} = 0$$

or

$$\lambda^2 - (a_{11} + a_{12})\lambda + (a_{11}a_{22} - a_{21}a_{12}) = 0$$

which is the same as

$$\lambda^2 - (\text{tr}A)\lambda + \det A = 0$$

A. The roots of this equation using the quadratic formula are

$$\lambda = \frac{\text{tr}A \pm \sqrt{(\text{tr}A)^2 - 4\det A}}{2}$$

For the roots to be real the expression under the radical sign must be positive or zero. Hence the necessary and sufficient condition is

$$\frac{(\text{tr}A)^2}{\det A} \geq 4.$$

Observe that

$$h = (\text{tr}A)^2 - 4\det A = (a_{11} + a_{22})^2 - 4(a_{11}a_{22} - a_{12}a_{21})$$
$$= (a_{11} - a_{22})^2 + 4a_{12}a_{21}.$$

If A is symmetric then h is a sum of squares. If a_{12} and a_{21} have the same sign both parts of h are positive so the quadratic equation has real roots.

Section 3

3.1 A. $\begin{bmatrix} \frac{2}{3} & -\frac{1}{3} \\ -\frac{1}{3} & \frac{2}{3} \end{bmatrix}$ **B.** $\begin{bmatrix} \frac{\sqrt{3}}{10} & -\frac{1}{10} \\ \frac{1}{10} & \frac{\sqrt{3}}{10} \end{bmatrix}$ **C.** $\begin{bmatrix} \cos\theta & -\sin\theta \\ \sin\theta & \cos\theta \end{bmatrix}$ **D.** $\begin{bmatrix} \frac{3}{4} & -\frac{1}{4} & -\frac{1}{4} \\ -\frac{1}{4} & \frac{3}{4} & -\frac{1}{4} \\ -\frac{1}{4} & -\frac{1}{4} & \frac{3}{4} \end{bmatrix}$

G. $\dfrac{1}{(1-\rho)(1+3\rho)} \begin{bmatrix} 1+2\rho & -\rho & -\rho & -\rho \\ -\rho & 1+2\rho & -\rho & -\rho \\ -\rho & -\rho & 1+2\rho & -\rho \\ -\rho & -\rho & -\rho & 1+2\rho \end{bmatrix}$

3.4 A. $R_\alpha R_\beta = \begin{bmatrix} \cos\alpha\cos\beta - \sin\alpha\sin\beta & \cos\alpha\sin\beta + \sin\alpha\cos\beta \\ -\sin\alpha\cos\beta - \cos\alpha\sin\beta & -\sin\alpha\sin\beta + \cos\alpha\cos\beta \end{bmatrix}$

$$= \begin{bmatrix} \cos(\alpha+\beta) & \sin(\alpha+\beta) \\ -\sin(\alpha+\beta) & \cos(\alpha+\beta) \end{bmatrix} = R_{\alpha+\beta}$$

3.7 B. The matrix $X = 1_n$. Then $X'X = 1'_n 1_n = n, (X'X)^{-1} = [1/n]$, and $X'Y = \sum_{i=1}^{n} y_i$. Then $b_0 = \frac{1}{n}\sum_{i=1}^{n} y_i = \bar{y}$.

Section 4

4.1 $a \ne 0$ and $a + bn \ne 0$. $x = \dfrac{c}{a}$, $y = \dfrac{ad - bc}{a(a + bn)}$

4.3 $I_n - \dfrac{g}{(g-1)n} J_n$

4.4 $H_6 = \begin{bmatrix} \frac{1}{\sqrt{6}} & \frac{1}{\sqrt{6}} & \frac{1}{\sqrt{6}} & \frac{1}{\sqrt{6}} & \frac{1}{\sqrt{6}} & \frac{1}{\sqrt{6}} \\ \frac{1}{\sqrt{2}} & -\frac{1}{\sqrt{2}} & 0 & 0 & 0 & 0 \\ \frac{1}{\sqrt{6}} & \frac{1}{\sqrt{6}} & -\frac{2}{\sqrt{6}} & 0 & 0 & 0 \\ \frac{1}{\sqrt{12}} & \frac{1}{\sqrt{12}} & \frac{1}{\sqrt{12}} & -\frac{3}{\sqrt{12}} & 0 & 0 \\ \frac{1}{\sqrt{20}} & \frac{1}{\sqrt{20}} & \frac{1}{\sqrt{20}} & \frac{1}{\sqrt{20}} & -\frac{4}{\sqrt{20}} & 0 \\ \frac{1}{\sqrt{30}} & \frac{1}{\sqrt{30}} & \frac{1}{\sqrt{30}} & \frac{1}{\sqrt{30}} & \frac{1}{\sqrt{30}} & -\frac{5}{\sqrt{30}} \end{bmatrix}$

4.6 Observe that if d_i are the diagonal elements of D

$$AD = \begin{bmatrix} a_{11}d_1 & a_{12}d_2 & \cdots & a_{1n}d_n \\ a_{21}d_1 & a_{22}d_2 & \cdots & a_{2n}d_n \\ \vdots & \vdots & & \vdots \\ a_{n1}d_1 & a_{n2}d_2 & \cdots & a_{nn}d_n \end{bmatrix}.$$

Then $\det AD = d_1 d_2 \ldots d_n \det A = \det D \det A = \det A \det D$.

4.8 A. $A \otimes B = \begin{bmatrix} 1 & 0 & 1 & 0 \\ 0 & 1 & 0 & 1 \\ 1 & 0 & 1 & 0 \\ 0 & 1 & 0 & 1 \end{bmatrix}$, $B \otimes A = \begin{bmatrix} 1 & 1 & 0 & 0 \\ 1 & 1 & 0 & 0 \\ 0 & 0 & 1 & 1 \\ 0 & 0 & 1 & 1 \end{bmatrix}$

B. $A \otimes B = \begin{bmatrix} -1 & 2 & 4 & -6 & 12 & 24 \\ 4 & -1 & 6 & 24 & -6 & 36 \\ -2 & 4 & 8 & -4 & 8 & 16 \\ 8 & -2 & 12 & 16 & -4 & 24 \end{bmatrix}$

$B \otimes A = \begin{bmatrix} -1 & -6 & 2 & 12 & 4 & 24 \\ -2 & -4 & 4 & 8 & 8 & 16 \\ 4 & 24 & -1 & -6 & 6 & 36 \\ 8 & 16 & -2 & -4 & 12 & 24 \end{bmatrix}$

4.9   We have

$$A \otimes (B \otimes C) = (A \otimes B) \otimes C$$

$$= \begin{bmatrix} 1 & 0 & 1 & 0 \\ 1 & 0 & 1 & 0 \\ 0 & 1 & 0 & 1 \\ 0 & 1 & 0 & 1 \\ 1 & 0 & 1 & 0 \\ 1 & 0 & 1 & 0 \\ 0 & 1 & 0 & 1 \\ 0 & 1 & 0 & 1 \end{bmatrix}$$

4.10   Observe that

$$(P \otimes Q)(P \otimes Q)' = (P \otimes Q)(P' \otimes Q') = (PP' \otimes QQ') = I.$$

Also

$$(P \otimes Q)'(P \otimes Q) = (P' \otimes Q')(P \otimes Q) = (P'P \otimes Q'Q) = I.$$

4.18   **A.** There are a number of correct answers. One possibility is to pre multiply A by

$$\begin{bmatrix} 1 & 0 \\ 0 & -2 \end{bmatrix}\begin{bmatrix} 1 & 1 \\ 0 & 1 \end{bmatrix}\begin{bmatrix} 1 & 0 \\ -3 & 1 \end{bmatrix}\begin{bmatrix} \frac{1}{4} & 0 \\ 0 & 1 \end{bmatrix} = \begin{bmatrix} -\frac{1}{2} & 1 \\ \frac{3}{2} & -2 \end{bmatrix} = A^{-1}$$

Then

$$A = \begin{bmatrix} 4 & 0 \\ 0 & 1 \end{bmatrix}\begin{bmatrix} 1 & 0 \\ 3 & 1 \end{bmatrix}\begin{bmatrix} 1 & -1 \\ 0 & 1 \end{bmatrix}\begin{bmatrix} 1 & 0 \\ 0 & -\frac{1}{2} \end{bmatrix}$$

B. Pre-multiply by

$$P = \begin{bmatrix} 1 & 0 \\ -\frac{b}{a} & 1 \end{bmatrix}$$

and post-multiply by

$$Q = \begin{bmatrix} 1 & -\frac{b}{a} \\ 0 & 1 \end{bmatrix}\begin{bmatrix} \frac{1}{a} & 0 \\ 0 & \frac{a}{a^2-b^2} \end{bmatrix} = \begin{bmatrix} \frac{1}{a} & -\frac{b}{a^2-b^2} \\ 0 & \frac{a}{a^2-b^2} \end{bmatrix}$$

to get the identity matrix. Then

$$A^{-1} = QP = \frac{1}{a^2 - b^2}\begin{bmatrix} a & -b \\ -b & a \end{bmatrix}$$

and

$$A = \begin{bmatrix} 1 & 0 \\ \frac{b}{a} & 1 \end{bmatrix} \begin{bmatrix} a & 0 \\ 0 & \frac{b^2-a^2}{a} \end{bmatrix} \begin{bmatrix} 1 & \frac{b}{a} \\ 0 & 1 \end{bmatrix}$$

C. $C^{-1} = \begin{bmatrix} 1 & 0 & 0 \\ 0 & 1 & 0 \\ 0 & 1 & 1 \end{bmatrix} \begin{bmatrix} 1 & 0 & 0 \\ 0 & -1 & 0 \\ 0 & 1 & 1 \end{bmatrix} \begin{bmatrix} 1 & 1 & 0 \\ 0 & 1 & 0 \\ 0 & 0 & 1 \end{bmatrix} \begin{bmatrix} 1 & 0 & 0 \\ 0 & 1 & -1 \\ 0 & 0 & 1 \end{bmatrix} \cdot$

$$\begin{bmatrix} 1 & 0 & 0 \\ 0 & 1 & 0 \\ 0 & 0 & \frac{1}{2} \end{bmatrix} \begin{bmatrix} 1 & 0 & 0 \\ -1 & 1 & 0 \\ 0 & 1 & 1 \end{bmatrix} \begin{bmatrix} 1 & 0 & 0 \\ -1 & 1 & 0 \\ 0 & 0 & 1 \end{bmatrix} \begin{bmatrix} 0 & 0 & 1 \\ 0 & 1 & 0 \\ 1 & 0 & 0 \end{bmatrix}$$

$$= \begin{bmatrix} -\frac{1}{2} & \frac{1}{2} & \frac{1}{2} \\ \frac{1}{2} & -\frac{1}{2} & \frac{1}{2} \\ \frac{1}{2} & \frac{1}{2} & -\frac{1}{2} \end{bmatrix}$$

$$C = \begin{bmatrix} 0 & 0 & 1 \\ 0 & 1 & 0 \\ 1 & 0 & 0 \end{bmatrix} \begin{bmatrix} 1 & 0 & 0 \\ 1 & 1 & 0 \\ 0 & 0 & 1 \end{bmatrix} \begin{bmatrix} 1 & 0 & 0 \\ 0 & 1 & 0 \\ 0 & -1 & 1 \end{bmatrix} \cdot$$

$$\begin{bmatrix} 1 & 0 & 0 \\ 0 & 1 & 0 \\ 0 & 0 & 2 \end{bmatrix} \begin{bmatrix} 1 & 0 & 0 \\ 0 & 1 & 1 \\ 0 & 0 & 1 \end{bmatrix} \begin{bmatrix} 1 & -1 & 0 \\ 0 & 1 & 0 \\ 0 & 0 & 1 \end{bmatrix} \begin{bmatrix} 1 & 0 & 0 \\ 0 & -1 & 0 \\ 0 & 1 & 1 \end{bmatrix} \cdot$$

$$\begin{bmatrix} 1 & 0 & 0 \\ 0 & 1 & 0 \\ 0 & -1 & 1 \end{bmatrix}$$

D. Premultiply the matrix in D by

$$P = \begin{bmatrix} 1 & 0 & 0 \\ 0 & 1 & 0 \\ 0 & -\frac{1}{4} & 1 \end{bmatrix} \begin{bmatrix} 1 & 0 & 0 \\ -3 & 1 & 0 \\ 0 & 0 & 0 \end{bmatrix} \begin{bmatrix} 1 & 0 & 0 \\ 0 & 1 & 0 \\ -2 & 0 & 1 \end{bmatrix} = \begin{bmatrix} 1 & 0 & 0 \\ -3 & 1 & 0 \\ -\frac{5}{4} & -\frac{1}{4} & 1 \end{bmatrix}$$

$$Q = \begin{bmatrix} 1 & -2 & 0 \\ 0 & 1 & 0 \\ 0 & 0 & 1 \end{bmatrix} \begin{bmatrix} 1 & 0 & -3 \\ 0 & 1 & 0 \\ 0 & 0 & 1 \end{bmatrix} \begin{bmatrix} 1 & 0 & 0 \\ 0 & 1 & -2 \\ 0 & 0 & 1 \end{bmatrix} \begin{bmatrix} 1 & 0 & 0 \\ 0 & -\frac{1}{4} & 0 \\ 0 & 0 & -\frac{1}{3} \end{bmatrix}$$

$$= \begin{bmatrix} 1 & \frac{1}{2} & -\frac{1}{3} \\ 0 & -\frac{1}{4} & \frac{2}{3} \\ 0 & 0 & -\frac{1}{3} \end{bmatrix}$$

$$D^{-1} = QP = \begin{bmatrix} -\frac{1}{12} & \frac{7}{12} & -\frac{1}{3} \\ -\frac{1}{12} & -\frac{5}{12} & \frac{2}{3} \\ \frac{5}{12} & \frac{1}{12} & -\frac{1}{3} \end{bmatrix}$$

$$D = \begin{bmatrix} 1 & 0 & 0 \\ 0 & 1 & 0 \\ 2 & 0 & 1 \end{bmatrix} \begin{bmatrix} 1 & 0 & 0 \\ 3 & 1 & 0 \\ 0 & 0 & 1 \end{bmatrix} \begin{bmatrix} 1 & 0 & 0 \\ 0 & 1 & 0 \\ 0 & \frac{1}{4} & 1 \end{bmatrix} \begin{bmatrix} 1 & 0 & 0 \\ 0 & -4 & 0 \\ 0 & 0 & -3 \end{bmatrix}.$$

$$\begin{bmatrix} 1 & 0 & 0 \\ 0 & 1 & 2 \\ 0 & 0 & 1 \end{bmatrix} \begin{bmatrix} 1 & 0 & 3 \\ 0 & 1 & 0 \\ 0 & 0 & 1 \end{bmatrix} \begin{bmatrix} 1 & 2 & 0 \\ 0 & 1 & 0 \\ 0 & 0 & 1 \end{bmatrix}$$

4.21 **B.** One way is to simply differentiate the elements of A^2. Then

$$\frac{\partial A^2}{\partial \theta} = \begin{bmatrix} -2\sin 2\theta & 2\cos 2\theta \\ -2\cos 2\theta & -2\sin 2\theta \end{bmatrix}.$$

The other way is to use the fact that

$$\frac{\partial A^2}{\partial \theta} = 2A\frac{\partial A}{\partial \theta}$$

Then

$$\frac{\partial A^2}{\partial \theta} = 2\begin{bmatrix} \cos\theta & \sin\theta \\ -\sin\theta & \cos\theta \end{bmatrix}\begin{bmatrix} -\sin\theta & \cos\theta \\ -\cos\theta & -\sin\theta \end{bmatrix}$$

$$= \begin{bmatrix} -4\cos\theta\sin\theta & 2(\cos^2\theta - \sin^2\theta) \\ -2(\cos^2\theta - \sin^2\theta) & -4\cos\theta\sin\theta \end{bmatrix}$$

$$= \begin{bmatrix} -2\sin 2\theta & 2\cos 2\theta \\ -2\cos 2\theta & -2\sin 2\theta \end{bmatrix}.$$

4.23 Recall that

$$A \otimes B = a_{ij}B, \quad i = 1, 2, \ldots, m, \quad j = 1, 2, \ldots, n$$

Then

$$\frac{\partial(A \otimes B)}{\partial \alpha} = \frac{\partial(a_{ij}B)}{\partial \alpha} = a_{ij}\frac{\partial B}{\partial \alpha} + \frac{\partial a_{ij}}{\partial \alpha}B, \quad i = 1, 2, \ldots, m, \quad j = 1, 2, \ldots, m$$

$$= A \otimes \frac{\partial B}{\partial \alpha} + \frac{\partial A}{\partial \alpha} \otimes B$$

4.25 **A.** One possible choice is

$$P = \begin{bmatrix} 1 & 0 \\ -\frac{3}{7} & 1 \end{bmatrix}, Q = \begin{bmatrix} \frac{1}{7} & -\frac{3}{40} \\ 0 & \frac{7}{40} \end{bmatrix}$$

B. A possible choice is

$$P = \begin{bmatrix} 1 & 0 & 0 \\ -2 & 1 & 0 \\ -\frac{1}{3} & -\frac{1}{3} & 1 \end{bmatrix}, Q = \begin{bmatrix} 1 & \frac{2}{3} & -\frac{1}{4} \\ 0 & -\frac{1}{3} & -\frac{1}{4} \\ 0 & 0 & \frac{3}{4} \end{bmatrix}$$

Section 5

5.1 The subspaces would be those with the one dimensional basis consisting of the single vectors $(1,0,0)$, $(0,1,0)$ and $(0,0,1)$.

5.4 Showing that it is a vector space consists of a straightforward verification that the axioms are satisfied. By definition e^x and xe^x span the vector space. To show linear independence observe that if

$$c_1 e^x + c_2 x e^x \equiv 0$$

we have

$$(c_1 + c_2 x)e^x \equiv 0.$$

Since $e^x \neq 0$,

$$c_1 + c_2 x \equiv 0 \text{ for all } x.$$

This is impossible unless $c_1 = 0, c_2 = 0$.

5.5 If we assume that for any vectors v, w, and x

$$(av + bw, x) = a(v,x) + b(w,x)$$

we have that

$$\begin{aligned} (av + bw, cx + dy) &= a(v, cx + dy) + b(w, cx + dy) \\ &= a(cx + dy, v) + b(cx + dy, w) \\ &= ac(x,v) + ad(y,v) + bc(x,w) + bd(y,w) \\ &= ac(v,x) + ad(v,y) + bc(w,x) + bd(w,y) \\ &= ac(v,x) + bc(w,x) + ad(v,y) + bd(w,y). \end{aligned}$$

5.8 Observe that by the Cauchy–Schwarz inequality

$$\rho^2 = \frac{(\text{cov}(X,Y))^2}{\text{VarX VarY}} = \frac{(X,Y)^2}{\|X\|^2 \|Y\|^2} \leq 1.$$

The result then follows.

5.9   The equality $\dim(U \oplus V) = \dim U + \dim V$ holds true when $U \cap V = \{0\}$. We know that $\dim(U \oplus V) \leq m + n$. Let $x \in U \oplus V$. Suppose that a basis of U is $u_1, u_2, \ldots u_m$ and a basis of V is $v_1, v_2, \ldots, v_n$. Then

$$x = \sum_{i=1}^{m} c_i u_i + \sum_{j=1}^{n} d_j v_j$$

The elements $u_1, u_2, \ldots, u_m,\ v_1, v_2, \ldots, v_n$ must be linearly independent. Otherwise either the u's or the v's are linearly dependent or a non-zero vector containing only u_i s could be expressed in one containing only v_I s contradicting the fact that $U \cap V = \{0\}$.

5.11   **C.** The vectors g_1, g_2 and g_3 are linearly independent because the only solution to the system of equations

$$2c_1 + c_2 + c_3 = 0$$
$$c_1 + 2c_2 + c_3 = 0$$
$$c_1 + c_2 + 2c_3 = 0$$

is $c_1 = c_2 = c_3 = 0$.
We have

$$e_1 = \frac{3}{4}g_1 - \frac{1}{4}g_2 - \frac{1}{4}g_3$$

$$e_2 = -\frac{1}{4}g_1 + \frac{3}{4}g_2 - \frac{1}{4}g_3$$

$$e_3 = -\frac{1}{4}g_1 - \frac{1}{4}g_2 + \frac{3}{4}g_3$$

and

$$f_1 = -\frac{1}{2}g_1 + \frac{1}{2}g_2 + \frac{1}{2}g_3$$

$$f_2 = \frac{1}{2}g_1 - \frac{1}{2}g_2 + \frac{1}{2}g_3$$

$$f_3 = \frac{1}{2}g_1 + \frac{1}{2}g_2 - \frac{1}{2}g_3$$

The transformation from e to g has matrix

$$\begin{bmatrix} 2 & 1 & 1 \\ 1 & 2 & 1 \\ 1 & 1 & 2 \end{bmatrix} \text{ with inverse } \frac{1}{4}\begin{bmatrix} 3 & -1 & -1 \\ -1 & 3 & -1 \\ -1 & -1 & 3 \end{bmatrix}.$$

The transformation from f to g has matrix

$$\begin{bmatrix} 0 & 1 & 1 \\ 1 & 0 & 1 \\ 1 & 1 & 0 \end{bmatrix} \text{ with inverse } \frac{1}{2}\begin{bmatrix} -1 & 1 & 1 \\ 1 & -1 & 1 \\ 1 & 1 & -1 \end{bmatrix}.$$

5.15 The matrix for D is

$$\begin{bmatrix} 0 & -1 \\ 1 & 0 \end{bmatrix}.$$

The matrix for G is

$$\begin{bmatrix} 0 & 1 \\ -1 & 0 \end{bmatrix}.$$

Section 6

6.1 **A.** 4 **B.** 3 **C.** 1 **E.** 2

6.2 **A.** $x = 7/4, y = -1/4, z = -1/2, w = 0$ **B.** inconsistent

  C. Let z and w be arbitrary. Then $x = 2 - 3/4$, $z - 1/4$, w, $y = 2 - 1/4$, and $z - 3/4$ w.

6.3 This is a rank 2 matrix. The normal equations would be

$$4\hat{\mu} + 2\hat{\alpha}_1 + 2\hat{\alpha}_2 + 2\hat{\beta}_1 + 2\hat{\beta}_2 = Y_{..}$$
$$2\hat{\mu} + 2\hat{\alpha}_1 + \hat{\beta}_1 + \hat{\beta}_2 = Y_{1.}$$
$$2\hat{\mu} + 2\hat{\alpha}_2 + \hat{\beta}_1 + \hat{\beta}_2 = Y_{2.}$$
$$2\hat{\mu} + \hat{\alpha}_1 + \hat{\alpha}_2 + 2\hat{\beta}_1 = Y_{.1}$$
$$2\hat{\mu} + \hat{\alpha}_1 + \hat{\alpha}_2 + 2\hat{\beta}_2 = Y_{.2}$$

The coefficient matrix and the augmented matrix are of rank 2 so there are infinitely many solutions for the least square estimator. One possibility is

$$\hat{\mu} = \overline{Y}.., \hat{\alpha}_1 = \overline{Y}_{1.} - \overline{Y}.., \hat{\alpha}_2 = \overline{Y}_{2.} - \overline{Y}.., \hat{\beta}_1 = \overline{Y}_{.1} - \overline{Y}.., \hat{\beta}_2 = \overline{Y}_{.2} - \overline{Y}...$$

6.6 **B.** Assume matrix A has rank r and matrix B has rank s. Then let the vectors $a_1, a_2, \ldots, a_r$ be a basis of C(A) and $b_1, b_2, \ldots, b_s$ be a basis for C(B). Then a basis for

$$C\begin{pmatrix} A & 0 \\ 0 & B \end{pmatrix} \text{ consists of the vectors}$$

$$\begin{bmatrix} a_1 \\ 0 \end{bmatrix}, \begin{bmatrix} a_2 \\ 0 \end{bmatrix}, \dots, \begin{bmatrix} a_r \\ 0 \end{bmatrix}, \begin{bmatrix} 0 \\ b_1 \end{bmatrix}, \begin{bmatrix} 0 \\ b_2 \end{bmatrix}, \dots, \begin{bmatrix} 0 \\ b_s \end{bmatrix}.$$

because they span $C \begin{pmatrix} A & 0 \\ 0 & B \end{pmatrix}$. However, if they were not linearly independent either the a vectors or the b vectors would be dependent and either A would have rank less than r or B would have rank less than s.

6.8   **A.** 1   **B.** 1   **C.** 2   **D.** 3

6.10 $A_1 = \begin{bmatrix} 1 & 2 \\ 3 & 4 \end{bmatrix}, C = \begin{bmatrix} 0 \\ 3 \end{bmatrix}$

PART II

Section 7

7.1   **A.** Eigenvalues 100, 50 Eigenvectors $\begin{bmatrix} -\frac{3}{5} \\ \frac{4}{5} \end{bmatrix}, \begin{bmatrix} \frac{4}{5} \\ \frac{3}{5} \end{bmatrix}$

Spectral decomposition

$$\begin{bmatrix} 68 & -24 \\ -24 & 82 \end{bmatrix} = 100 \begin{bmatrix} -\frac{3}{5} \\ \frac{4}{5} \end{bmatrix} \begin{bmatrix} -\frac{3}{5} & \frac{4}{5} \end{bmatrix} + 50 \begin{bmatrix} \frac{4}{5} \\ \frac{3}{5} \end{bmatrix} \begin{bmatrix} \frac{4}{5} & \frac{3}{5} \end{bmatrix}$$

$$= \begin{bmatrix} -\frac{3}{5} & \frac{4}{5} \\ \frac{4}{5} & \frac{3}{5} \end{bmatrix} \begin{bmatrix} 100 & 0 \\ 0 & 50 \end{bmatrix} \begin{bmatrix} -\frac{3}{5} & \frac{4}{5} \\ \frac{4}{5} & \frac{3}{5} \end{bmatrix}$$

B. Eigenvalues 3,1. Eigenvectors $\begin{bmatrix} \frac{1}{\sqrt{2}} \\ \frac{1}{\sqrt{2}} \end{bmatrix}, \begin{bmatrix} -\frac{1}{\sqrt{2}} \\ \frac{1}{\sqrt{2}} \end{bmatrix}$.

$$\begin{bmatrix} 2 & 1 \\ 1 & 2 \end{bmatrix} = 3 \begin{bmatrix} \frac{1}{\sqrt{2}} \\ \frac{1}{\sqrt{2}} \end{bmatrix} \begin{bmatrix} \frac{1}{\sqrt{2}} & \frac{1}{\sqrt{2}} \end{bmatrix} + \begin{bmatrix} -\frac{1}{\sqrt{2}} \\ \frac{1}{\sqrt{2}} \end{bmatrix} \begin{bmatrix} -\frac{1}{\sqrt{2}} & \frac{1}{\sqrt{2}} \end{bmatrix}$$

$$= \begin{bmatrix} \frac{1}{\sqrt{2}} & -\frac{1}{\sqrt{2}} \\ \frac{1}{\sqrt{2}} & \frac{1}{\sqrt{2}} \end{bmatrix} \begin{bmatrix} 3 & 0 \\ 0 & 1 \end{bmatrix} \begin{bmatrix} \frac{1}{\sqrt{2}} & \frac{1}{\sqrt{2}} \\ -\frac{1}{\sqrt{2}} & \frac{1}{\sqrt{2}} \end{bmatrix}$$

C. Eigenvalues 7, 4, 4 Chosing orthogonal eigenvectors $\begin{bmatrix} \frac{1}{\sqrt{3}} \\ \frac{1}{\sqrt{3}} \\ \frac{1}{\sqrt{3}} \end{bmatrix}, \begin{bmatrix} \frac{1}{\sqrt{6}} \\ \frac{1}{\sqrt{6}} \\ -\frac{1}{\sqrt{2}} \end{bmatrix}, \begin{bmatrix} \frac{1}{\sqrt{2}} \\ -\frac{1}{\sqrt{2}} \\ 0 \end{bmatrix}$.

We can write

$$
\begin{bmatrix} 5 & 1 & 1 \\ 1 & 5 & 1 \\ 1 & 1 & 5 \end{bmatrix} = 7 \begin{bmatrix} \frac{1}{\sqrt{3}} \\ \frac{1}{\sqrt{3}} \\ \frac{1}{\sqrt{3}} \end{bmatrix} \begin{bmatrix} \frac{1}{\sqrt{3}} & \frac{1}{\sqrt{3}} & \frac{1}{\sqrt{3}} \end{bmatrix} + 4 \begin{bmatrix} \frac{1}{\sqrt{6}} \\ \frac{1}{\sqrt{6}} \\ -\frac{2}{\sqrt{6}} \end{bmatrix} \begin{bmatrix} \frac{1}{\sqrt{6}} & \frac{1}{\sqrt{6}} & -\frac{2}{\sqrt{6}} \end{bmatrix}
$$

$$
+ 4 \begin{bmatrix} -\frac{1}{\sqrt{2}} \\ \frac{1}{\sqrt{2}} \\ 0 \end{bmatrix} \begin{bmatrix} -\frac{1}{\sqrt{2}} & \frac{1}{\sqrt{2}} & 0 \end{bmatrix} = \begin{bmatrix} \frac{1}{\sqrt{3}} & \frac{1}{\sqrt{6}} & -\frac{1}{\sqrt{2}} \\ \frac{1}{\sqrt{3}} & \frac{1}{\sqrt{6}} & \frac{1}{\sqrt{2}} \\ \frac{1}{\sqrt{3}} & -\frac{2}{\sqrt{6}} & 0 \end{bmatrix} \begin{bmatrix} 7 & 0 & 0 \\ 0 & 4 & 0 \\ 0 & 0 & 4 \end{bmatrix} \begin{bmatrix} \frac{1}{\sqrt{3}} & \frac{1}{\sqrt{3}} & \frac{1}{\sqrt{3}} \\ \frac{1}{\sqrt{6}} & \frac{1}{\sqrt{6}} & -\frac{2}{\sqrt{6}} \\ -\frac{1}{\sqrt{2}} & \frac{1}{\sqrt{2}} & 0 \end{bmatrix}
$$

D. Eigenvalues 7/5,4/5,4/5

$$
\begin{bmatrix} 1 & \frac{1}{5} & \frac{1}{5} \\ \frac{1}{5} & 1 & \frac{1}{5} \\ \frac{1}{5} & \frac{1}{5} & 1 \end{bmatrix} = 7/5 \begin{bmatrix} \frac{1}{\sqrt{3}} \\ \frac{1}{\sqrt{3}} \\ \frac{1}{\sqrt{3}} \end{bmatrix} \begin{bmatrix} \frac{1}{\sqrt{3}} & \frac{1}{\sqrt{3}} & \frac{1}{\sqrt{3}} \end{bmatrix} + 4/5 \begin{bmatrix} \frac{1}{\sqrt{6}} \\ \frac{1}{\sqrt{6}} \\ -\frac{2}{\sqrt{6}} \end{bmatrix} \begin{bmatrix} \frac{1}{\sqrt{6}} & \frac{1}{\sqrt{6}} & -\frac{2}{\sqrt{6}} \end{bmatrix}
$$

$$
+ 4/5 \begin{bmatrix} -\frac{1}{\sqrt{2}} \\ \frac{1}{\sqrt{2}} \\ 0 \end{bmatrix} \begin{bmatrix} -\frac{1}{\sqrt{2}} & \frac{1}{\sqrt{2}} & 0 \end{bmatrix} = \begin{bmatrix} \frac{1}{\sqrt{3}} & \frac{1}{\sqrt{6}} & -\frac{1}{\sqrt{2}} \\ \frac{1}{\sqrt{3}} & \frac{1}{\sqrt{6}} & \frac{1}{\sqrt{2}} \\ \frac{1}{\sqrt{3}} & -\frac{2}{\sqrt{6}} & 0 \end{bmatrix} \begin{bmatrix} \frac{7}{5} & 0 & 0 \\ 0 & \frac{4}{5} & 0 \\ 0 & 0 & \frac{4}{5} \end{bmatrix} \begin{bmatrix} \frac{1}{\sqrt{3}} & \frac{1}{\sqrt{3}} & \frac{1}{\sqrt{3}} \\ \frac{1}{\sqrt{6}} & \frac{1}{\sqrt{6}} & -\frac{2}{\sqrt{6}} \\ -\frac{1}{\sqrt{2}} & \frac{1}{\sqrt{2}} & 0 \end{bmatrix}
$$

E. Eigenvalues 12, 3,3,0. Orthogonal eigenvectors $\begin{bmatrix} \frac{3}{\sqrt{12}} \\ \frac{1}{\sqrt{12}} \\ \frac{1}{\sqrt{12}} \\ \frac{1}{\sqrt{12}} \end{bmatrix}, \begin{bmatrix} 0 \\ \frac{1}{\sqrt{6}} \\ \frac{1}{\sqrt{6}} \\ -\frac{2}{\sqrt{6}} \end{bmatrix}, \begin{bmatrix} 0 \\ \frac{1}{\sqrt{12}} \\ -\frac{1}{\sqrt{12}} \\ 0 \end{bmatrix}, \begin{bmatrix} \frac{1}{2} \\ -\frac{1}{2} \\ -\frac{1}{2} \\ -\frac{1}{2} \end{bmatrix}$

$$
\begin{bmatrix} 9 & 3 & 3 & 3 \\ 3 & 3 & 0 & 0 \\ 3 & 0 & 3 & 0 \\ 3 & 0 & 0 & 3 \end{bmatrix} = 12 \begin{bmatrix} \frac{3}{\sqrt{12}} \\ \frac{1}{\sqrt{12}} \\ \frac{1}{\sqrt{12}} \\ \frac{1}{\sqrt{12}} \end{bmatrix} \begin{bmatrix} \frac{3}{\sqrt{12}} & \frac{1}{\sqrt{12}} & \frac{1}{\sqrt{12}} & \frac{1}{\sqrt{12}} \end{bmatrix}
$$

$$
+ 3 \begin{bmatrix} 0 & 0 \\ \frac{1}{\sqrt{6}} & \frac{1}{\sqrt{12}} \\ \frac{1}{\sqrt{6}} & -\frac{1}{\sqrt{12}} \\ -\frac{2}{\sqrt{6}} & 0 \end{bmatrix} \begin{bmatrix} 0 & \frac{1}{\sqrt{6}} & \frac{1}{\sqrt{6}} & -\frac{2}{\sqrt{6}} \\ 0 & \frac{1}{\sqrt{2}} & -\frac{1}{\sqrt{2}} & 0 \end{bmatrix}
$$

$$
= \begin{bmatrix} \frac{3}{\sqrt{12}} & 0 & 0 \\ \frac{1}{\sqrt{12}} & \frac{1}{\sqrt{6}} & \frac{1}{\sqrt{2}} \\ \frac{1}{\sqrt{12}} & \frac{1}{\sqrt{6}} & -\frac{1}{\sqrt{2}} \\ \frac{1}{\sqrt{12}} & -\frac{2}{\sqrt{6}} & 0 \end{bmatrix} \begin{bmatrix} 12 & 0 & 0 \\ 0 & 3 & 0 \\ 0 & 0 & 3 \end{bmatrix} \begin{bmatrix} \frac{3}{\sqrt{12}} & \frac{1}{\sqrt{12}} & \frac{1}{\sqrt{12}} & \frac{1}{\sqrt{12}} \\ 0 & \frac{1}{\sqrt{6}} & \frac{1}{\sqrt{6}} & -\frac{2}{\sqrt{6}} \\ 0 & \frac{1}{\sqrt{2}} & -\frac{1}{\sqrt{2}} & 0 \end{bmatrix}
$$

7.2 **A.** Positive semi-definite but not positive definite

Eigenvalues 10,0

Eigenvectors

$$\begin{bmatrix} \frac{1}{\sqrt{10}} \\ \frac{1}{\sqrt{10}} \end{bmatrix}, \begin{bmatrix} -\frac{3}{\sqrt{10}} \\ \frac{1}{\sqrt{10}} \end{bmatrix} \begin{bmatrix} 1 & 3 \\ 3 & 9 \end{bmatrix} = 10 \begin{bmatrix} \frac{1}{\sqrt{10}} \\ \frac{3}{\sqrt{10}} \end{bmatrix} \begin{bmatrix} \frac{1}{\sqrt{10}} & \frac{3}{\sqrt{10}} \end{bmatrix}$$

B. Not positive semi- definite Eigenvalues 15, −5

Eigenvectors

$$\begin{bmatrix} \frac{1}{\sqrt{2}} \\ \frac{1}{\sqrt{2}} \end{bmatrix} \begin{bmatrix} \frac{1}{\sqrt{2}} \\ \frac{1}{\sqrt{2}} \end{bmatrix} \begin{bmatrix} 5 & 10 \\ 10 & 5 \end{bmatrix} = 15 \begin{bmatrix} \frac{1}{\sqrt{2}} \\ \frac{1}{\sqrt{2}} \end{bmatrix} \begin{bmatrix} \frac{1}{\sqrt{2}} & \frac{1}{\sqrt{2}} \end{bmatrix} - 5 \begin{bmatrix} -\frac{1}{\sqrt{2}} \\ \frac{1}{\sqrt{2}} \end{bmatrix} \begin{bmatrix} -\frac{1}{\sqrt{2}} & \frac{1}{\sqrt{2}} \end{bmatrix}$$

$$= \begin{bmatrix} \frac{1}{\sqrt{2}} & -\frac{1}{\sqrt{2}} \\ \frac{1}{\sqrt{2}} & \frac{1}{\sqrt{2}} \end{bmatrix} \begin{bmatrix} 15 & 0 \\ 0 & -5 \end{bmatrix} \begin{bmatrix} \frac{1}{\sqrt{2}} & \frac{1}{\sqrt{2}} \\ -\frac{1}{\sqrt{2}} & \frac{1}{\sqrt{2}} \end{bmatrix}$$

C. Positive definite Eigenvalues 1.5, 0.5 Eigenvectors $\begin{bmatrix} \frac{1}{\sqrt{2}} \\ -\frac{1}{\sqrt{2}} \end{bmatrix}, \begin{bmatrix} \frac{1}{\sqrt{2}} \\ \frac{1}{\sqrt{2}} \end{bmatrix}$

$$\begin{bmatrix} 1 & -0.5 \\ -0.5 & 1 \end{bmatrix} = 1.5 \begin{bmatrix} \frac{1}{\sqrt{2}} \\ -\frac{1}{\sqrt{2}} \end{bmatrix} \begin{bmatrix} \frac{1}{\sqrt{2}} & -\frac{1}{\sqrt{2}} \end{bmatrix} + 0.5 \begin{bmatrix} \frac{1}{\sqrt{2}} \\ \frac{1}{\sqrt{2}} \end{bmatrix} \begin{bmatrix} \frac{1}{\sqrt{2}} & \frac{1}{\sqrt{2}} \end{bmatrix}$$

$$= \begin{bmatrix} \frac{1}{\sqrt{2}} & \frac{1}{\sqrt{2}} \\ -\frac{1}{\sqrt{2}} & \frac{1}{\sqrt{2}} \end{bmatrix} \begin{bmatrix} 1.5 & 0 \\ 0 & 0.5 \end{bmatrix} \begin{bmatrix} \frac{1}{\sqrt{2}} & -\frac{1}{\sqrt{2}} \\ \frac{1}{\sqrt{2}} & \frac{1}{\sqrt{2}} \end{bmatrix}$$

7.3 Positive definite if $-(1/2) < r < 1$. Positive semidefinite if $-(1/2) \le r < 1$.

7.5 **A.** Suppose that there are constants c_1 and c_2 where

$$c_1 \begin{bmatrix} 2 \\ 3 \end{bmatrix} + c_2 \begin{bmatrix} 1 \\ 2 \end{bmatrix} = \begin{bmatrix} 0 \\ 0 \end{bmatrix}$$

Then we must have

$$2c_1 + c_2 = 0$$
$$3c_1 + 2c_2 = 0$$

The only solutions to this system of equations is $c_1 = 0$, $c_2 = 0$, hence the two vectors are linearly independent.

B. Eigenvalues $4 + \sqrt{17}, 4 - \sqrt{17}$ Observe that there is one positive and one negative eigenvalue.

7.6   Eigenvalues 75,0

7.7   Suppose that there exist constants $c_1, c_2, \ldots, c_n$ where

$$c_1 p_1 + c_2 p_2 + \cdots c_n p_n = 0$$

For each $i = 1, 2, \ldots, n$ the inner product

$$(p_i, c_1 p_1 + c_2 p_2 + \cdots c_n p_n) = c_1 (p_i, p_1) + c_2 (p_i, p_2) + \cdots c_i (p_i, p_i)$$
$$+ \cdots c_n (p_i, p_n) = c_i = 0.$$

This establishes independence.

7.9   Not positive semidefinite. This is a skew symmetric matrix where $A' = -A$. The eigenvalues and eigenvectors are complex numbers.

Eigenvalues $i\sqrt{3}, -i\sqrt{3}, 0$ Eigenvectors $\begin{bmatrix} -\frac{1}{2} - i\frac{\sqrt{3}}{2} \\ \frac{1}{2} - i\frac{\sqrt{3}}{2} \\ 1 \end{bmatrix}, \begin{bmatrix} -\frac{1}{2} + i\frac{\sqrt{3}}{2} \\ \frac{1}{2} + i\frac{\sqrt{3}}{2} \\ 1 \end{bmatrix}, \begin{bmatrix} 1 \\ -1 \\ 1 \end{bmatrix}$

7.10   **A.** Eigenvalues 9,4 $P = \begin{bmatrix} \frac{1}{\sqrt{10}} & \frac{1}{\sqrt{5}} \\ \frac{3}{\sqrt{10}} & \frac{2}{\sqrt{5}} \end{bmatrix}, Q = \begin{bmatrix} -\frac{2}{\sqrt{5}} & -\frac{3}{\sqrt{10}} \\ \frac{1}{\sqrt{5}} & \frac{1}{\sqrt{10}} \end{bmatrix}$

   B. Eigenvalues 3,2,1 $P = \begin{bmatrix} \frac{1}{\sqrt{3}} & \frac{1}{\sqrt{2}} & 0 \\ \frac{1}{\sqrt{3}} & -\frac{1}{\sqrt{2}} & \frac{1}{\sqrt{2}} \\ \frac{1}{\sqrt{3}} & 0 & -\frac{1}{\sqrt{2}} \end{bmatrix}, Q = \begin{bmatrix} \frac{1}{\sqrt{3}} & -\frac{2}{\sqrt{6}} & -\frac{1}{\sqrt{6}} \\ \frac{1}{\sqrt{3}} & \frac{1}{\sqrt{6}} & -\frac{1}{\sqrt{6}} \\ \frac{1}{\sqrt{3}} & \frac{1}{\sqrt{6}} & \frac{2}{\sqrt{6}} \end{bmatrix}.$

7.11   Not positive semi-definite

Eigenvalues $2, \frac{1}{2}(1 + \sqrt{3}), \frac{1}{2}(1 - \sqrt{3}),$

$$P = \begin{bmatrix} 0 & \frac{\sqrt{3}}{\sqrt{8}} & -\frac{\sqrt{3}}{\sqrt{8}} \\ \frac{1}{\sqrt{2}} & -\frac{2}{\sqrt{8}} & -\frac{2}{\sqrt{8}} \\ \frac{1}{\sqrt{2}} & \frac{1}{\sqrt{8}} & \frac{1}{\sqrt{8}} \end{bmatrix}, Q = \begin{bmatrix} 0 & \frac{\sqrt{3}}{\sqrt{5}} & -\frac{\sqrt{3}}{\sqrt{5}} \\ \frac{1}{\sqrt{5}} & -\frac{1}{\sqrt{5}} & -\frac{1}{\sqrt{5}} \\ \frac{2}{\sqrt{5}} & \frac{1}{\sqrt{5}} & \frac{1}{\sqrt{5}} \end{bmatrix}$$

Section 8

8.1   **A.** Characteristic equation for both A and B $10 - 7\lambda + \lambda^2 = 0$, Eigenvalues $\lambda = 5, 2$. Normalized eigenvectors of A

$$\begin{bmatrix} \frac{2}{\sqrt{5}} \\ \frac{1}{\sqrt{5}} \end{bmatrix}, \begin{bmatrix} -\frac{1}{\sqrt{2}} \\ \frac{1}{\sqrt{2}} \end{bmatrix}$$

Orthonormal Eigenvalues of B

$$\begin{bmatrix} \frac{\sqrt{2}}{\sqrt{3}} \\ \frac{1}{\sqrt{3}} \end{bmatrix}, \begin{bmatrix} -\frac{1}{\sqrt{3}} \\ \frac{\sqrt{2}}{\sqrt{3}} \end{bmatrix}$$

B. Normalized eigenvectors of A'

$$\begin{bmatrix} \frac{1}{\sqrt{2}} \\ \frac{1}{\sqrt{2}} \end{bmatrix}, \begin{bmatrix} -\frac{1}{\sqrt{5}} \\ \frac{2}{\sqrt{5}} \end{bmatrix}$$

C. $H'Hx = HHx = H\lambda x = \lambda Hx = \lambda^2 x$.

D. The eigenvalues of A'A are $15 \pm 5\sqrt{5}$. The eigenvalues of A^2 are 25,4 the squares of 5 and 2 which are the eigenvalues of A.

8.2 **A.** Eigenvalues of A 11, 5 with multiplicity 2. Eigenvalues of B 8,6

Normalized eigenvectors of A   Normalized Eigenvectors of B

$$\begin{bmatrix} \frac{1}{\sqrt{3}} \\ \frac{1}{\sqrt{3}} \\ \frac{1}{\sqrt{3}} \end{bmatrix}, \begin{bmatrix} -\frac{1}{\sqrt{2}} \\ 0 \\ \frac{1}{\sqrt{2}} \end{bmatrix}, \begin{bmatrix} -\frac{1}{\sqrt{2}} \\ \frac{1}{\sqrt{2}} \\ 0 \end{bmatrix} \quad \begin{bmatrix} \frac{1}{\sqrt{2}} \\ \frac{1}{\sqrt{2}} \end{bmatrix}, \begin{bmatrix} -\frac{1}{\sqrt{2}} \\ \frac{1}{\sqrt{2}} \end{bmatrix}$$

B. Eigenvalues of $A \otimes B$ 88, 66, 40 with multiplicity 2, 30 with multiplicity 2.

Normalized Eigenvectors

$$\begin{bmatrix} \frac{1}{\sqrt{6}} \\ \frac{1}{\sqrt{6}} \\ \frac{1}{\sqrt{6}} \\ \frac{1}{\sqrt{6}} \\ \frac{1}{\sqrt{6}} \\ \frac{1}{\sqrt{6}} \end{bmatrix}, \begin{bmatrix} -\frac{1}{\sqrt{6}} \\ \frac{1}{\sqrt{6}} \\ -\frac{1}{\sqrt{6}} \\ \frac{1}{\sqrt{6}} \\ -\frac{1}{\sqrt{6}} \\ \frac{1}{\sqrt{6}} \end{bmatrix}, \begin{bmatrix} -\frac{1}{2} \\ -\frac{1}{2} \\ 0 \\ 0 \\ \frac{1}{2} \\ \frac{1}{2} \end{bmatrix}, \begin{bmatrix} -\frac{1}{2} \\ -\frac{1}{2} \\ \frac{1}{2} \\ \frac{1}{2} \\ 0 \\ 0 \end{bmatrix}, \begin{bmatrix} \frac{1}{2} \\ -\frac{1}{2} \\ 0 \\ 0 \\ -\frac{1}{2} \\ \frac{1}{2} \end{bmatrix}, \begin{bmatrix} \frac{1}{2} \\ -\frac{1}{2} \\ -\frac{1}{2} \\ \frac{1}{2} \\ 0 \\ 0 \end{bmatrix}$$

C. $\det(A \otimes B) = 8,363,520,000$

$$\text{Trace}(A \otimes B) = 294$$

8.4 **A.** Let p be any vector and let $q = M'p$ Then $p'MAM'p = q'Aq \geq 0$ because A is positive semi-definite.

B. follows from A.

8.6 **A.** Eigenvalues 2.8, 0.1 with multiplicity 2.

One set of eigenvectors is

$$\begin{bmatrix} \frac{1}{\sqrt{3}} \\ \frac{1}{\sqrt{3}} \\ \frac{1}{\sqrt{3}} \end{bmatrix}, \begin{bmatrix} \frac{1}{\sqrt{2}} \\ -\frac{1}{\sqrt{2}} \\ 0 \end{bmatrix}, \begin{bmatrix} \frac{1}{\sqrt{6}} \\ \frac{1}{\sqrt{6}} \\ -\frac{2}{\sqrt{6}} \end{bmatrix}$$

B. By Cayley–Hamilton theorem

$$\Sigma^{-1} = \frac{1}{0.028}(\Sigma^2 - 3\Sigma + 0.57I)$$

$$\Sigma^{-1} = \begin{bmatrix} 6.78571 & -3.21429 & -3.21429 \\ -3.21429 & 6.78571 & -3.21429 \\ -3.21429 & -3.21429 & 6.78571 \end{bmatrix}$$

8.7 If E is idempotent then $(I-E)^2 = I - 2E + E^2 = I - 2E + E = I - E$. Also $E(I-E) = E - E^2 = E - E = 0$.

8.11 **A.** $B^2 - A^2 = \begin{bmatrix} 0 & \varepsilon \\ \varepsilon & 2\varepsilon + \varepsilon^2 \end{bmatrix}$, $\det(B^2 - A^2) = -\varepsilon^2 < 0$. So $B^2 - A^2$ is not positive semi-definite.

Section 9

9.3 **A.** $\begin{bmatrix} \frac{1}{2} & \frac{1}{2} \\ \frac{1}{2} & \frac{1}{2} \\ \frac{1}{2} & -\frac{1}{2} \\ \frac{1}{2} & -\frac{1}{2} \end{bmatrix} \begin{bmatrix} \sqrt{6} & 0 \\ 0 & \sqrt{2} \end{bmatrix} \begin{bmatrix} \frac{2}{\sqrt{6}} & \frac{1}{\sqrt{6}} & \frac{1}{\sqrt{6}} \\ 0 & \frac{1}{\sqrt{2}} & -\frac{1}{\sqrt{2}} \end{bmatrix}$

B. $\begin{bmatrix} -\frac{1}{2} & -\frac{1}{2} \\ -\frac{1}{2} & -\frac{1}{2} \\ \frac{1}{2} & -\frac{1}{2} \\ \frac{1}{2} & -\frac{1}{2} \end{bmatrix} \begin{bmatrix} 12 & 0 \\ 0 & 8 \end{bmatrix} \begin{bmatrix} -\frac{1}{2} & -\frac{1}{2} & \frac{1}{2} & \frac{1}{2} \\ \frac{1}{2} & \frac{1}{2} & \frac{1}{2} & \frac{1}{2} \end{bmatrix}$

C. $\begin{bmatrix} \frac{1}{2} & -\frac{1}{2} & -\frac{1}{2} \\ \frac{1}{2} & \frac{1}{2} & -\frac{1}{2} \\ \frac{1}{2} & -\frac{1}{2} & \frac{1}{2} \\ \frac{1}{2} & \frac{1}{2} & \frac{1}{2} \end{bmatrix} \begin{bmatrix} 2 & 0 & 0 \\ 0 & \sqrt{2} & 0 \\ 0 & 0 & \sqrt{2} \end{bmatrix} \begin{bmatrix} \frac{1}{2} & \frac{1}{2} & \frac{1}{2} & \frac{1}{2} \\ 0 & 0 & -\frac{1}{\sqrt{2}} & \frac{1}{\sqrt{2}} \\ -\frac{1}{\sqrt{2}} & \frac{1}{\sqrt{2}} & 0 & 0 \end{bmatrix}$

D. $\begin{bmatrix} \frac{1}{\sqrt{3}} & -\frac{1}{\sqrt{2}} \\ \frac{1}{\sqrt{3}} & \frac{1}{\sqrt{2}} \\ \frac{1}{\sqrt{3}} & 0 \end{bmatrix} \begin{bmatrix} 2 & 0 \\ 0 & 1 \end{bmatrix} \begin{bmatrix} \frac{1}{\sqrt{3}} & \frac{1}{\sqrt{3}} & \frac{1}{\sqrt{3}} \\ -\frac{1}{\sqrt{2}} & \frac{1}{\sqrt{2}} & 0 \end{bmatrix}$

9.4 By the SVD we have $A'A = U\Lambda U' = A^2$ and $AA' = S'\Lambda S = A^2$ because A is symmetric. Then $U = S'$. The SVD takes the form $A = U\Delta U'$ where $\Delta = \Lambda^{1/2}$. For the factorization $B = U\Delta^{1/2}, B' = \Delta^{1/2}U'$.

9.5 **A.** One factorization is

$$
B = \begin{bmatrix} \frac{\sqrt{6}}{2} & \frac{\sqrt{2}}{2} \\ \frac{\sqrt{6}}{2} & \frac{\sqrt{2}}{2} \\ \frac{\sqrt{6}}{2} & -\frac{\sqrt{2}}{2} \\ \frac{\sqrt{6}}{2} & -\frac{\sqrt{2}}{2} \end{bmatrix}, C = \begin{bmatrix} \frac{2}{\sqrt{6}} & \frac{1}{\sqrt{6}} & \frac{1}{\sqrt{6}} \\ 0 & \frac{1}{\sqrt{2}} & -\frac{1}{\sqrt{2}} \end{bmatrix}
$$

B. One factorization is

$$
B = \begin{bmatrix} -\sqrt{3} & -\sqrt{2} \\ -\sqrt{3} & -\sqrt{2} \\ \sqrt{3} & -\sqrt{2} \\ \sqrt{3} & -\sqrt{2} \end{bmatrix}, C = B'
$$

C. One factorization is

$$
B = \begin{bmatrix} \frac{1}{2} & -\frac{1}{2} & -\frac{1}{2} \\ \frac{1}{2} & \frac{1}{2} & -\frac{1}{2} \\ \frac{1}{2} & -\frac{1}{2} & \frac{1}{2} \\ \frac{1}{2} & \frac{1}{2} & \frac{1}{2} \end{bmatrix}, C = \begin{bmatrix} 1 & 1 & 1 & 1 \\ 0 & 0 & -1 & 1 \\ -1 & 1 & 0 & 0 \end{bmatrix}
$$

D. A factorization is

$$
B = \begin{bmatrix} \frac{2}{\sqrt{6}} & -\frac{1}{\sqrt{2}} \\ \frac{2}{\sqrt{6}} & \frac{1}{\sqrt{2}} \\ \frac{2}{\sqrt{6}} & 0 \end{bmatrix}, C = B'
$$

This is an example of a factorization of the type in Exercise 9.4.

9.9 **A.** $B = \begin{bmatrix} 7 & 2 \\ 1 & 7 \end{bmatrix} = \begin{bmatrix} 7 & \frac{3}{5} \\ \frac{3}{2} & 7 \end{bmatrix} + \begin{bmatrix} 0 & \frac{1}{2} \\ -\frac{1}{2} & 0 \end{bmatrix}$

B. 1. Eigenvalues $7 \pm \sqrt{2}$, Singular values 8.51783, 5.51783.

2. Eigenvalues 17/2, 11/2 Singular values 17/2, 11/2

3. Eigenvalues $\pm \frac{1}{2}i$, Singular values $\frac{1}{2}, \frac{1}{2}$

Section 10

10.1 A. Reparametized Model

$$Y = \begin{bmatrix} \frac{3}{\sqrt{6}} & -\frac{1}{\sqrt{2}} & 0 \\ \frac{3}{\sqrt{6}} & \frac{1}{\sqrt{2}} & 0 \end{bmatrix} \begin{bmatrix} \gamma_1 \\ \gamma_2 \\ \gamma_3 \end{bmatrix} + \varepsilon$$

where

$$\gamma_1 = \frac{1}{\sqrt{6}}(2\mu + \beta_1 + \beta_2)$$
$$\gamma_2 = \frac{1}{\sqrt{2}}(-\beta_1 + \beta_2)$$
$$\gamma_3 = \frac{1}{\sqrt{3}}(-\mu + \beta_1 + \beta_2)$$

LS Estimator

$$g_1 = \frac{1}{\sqrt{6}}(y_1 + y_2), g_2 = \frac{1}{\sqrt{2}}(-y_1 + y_2), g_3 = 0$$

where

$$Y = \begin{bmatrix} y_1 \\ y_2 \end{bmatrix}$$

10.1 B. Reparametized model

$$Y = \begin{bmatrix} \sqrt{2} & -1 & 0 & 0 & 0 \\ \sqrt{2} & 0 & -1 & 0 & 0 \\ \sqrt{2} & 0 & 1 & 0 & 0 \\ \sqrt{2} & 1 & 0 & 0 & 0 \end{bmatrix} \begin{bmatrix} \gamma_1 \\ \gamma_2 \\ \gamma_3 \\ \gamma_4 \\ \gamma_5 \end{bmatrix} + \varepsilon$$

where

$$\gamma_1 = \frac{1}{2\sqrt{2}}(2\mu + \tau_1 + \tau_2 + \theta_1 + \theta_2)$$
$$\gamma_2 = \frac{1}{2}(-\tau_1 + \tau_2 - \theta_1 + \theta_2)$$
$$\gamma_2 = \frac{1}{2}(-\tau_1 + \tau_2 - \theta_1 + \theta_2)$$
$$\gamma_3 = \frac{1}{2}(-\tau_1 + \tau_2 + \theta_1 - \theta_2)$$
$$\gamma_4 = \frac{1}{\sqrt{3}}(-\mu + \theta_1 + \theta_2)$$
$$\gamma_5 = \frac{1}{2\sqrt{6}}(-2\mu + 3\tau_1 + 3\tau_2 - \theta_1 - \theta_2)$$

LS estimator

$$g_1 = \frac{1}{\sqrt{8}}z_1, g_2 = \frac{1}{\sqrt{2}}z_2, g_3 = \frac{1}{\sqrt{2}}z_3, g_4 = 0$$

Where

$$z_1 = \tfrac{1}{2}y_{..}, z_2 = \tfrac{1}{\sqrt{2}}(-y_{11}+y_{22}), z_3 = \tfrac{1}{\sqrt{2}}(-y_{12}+y_{21})$$

Where

$$Y' = \begin{bmatrix} y_{11} & y_{12} & y_{21} & y_{22} \end{bmatrix}$$

10.2   Covariance Matrix

$$\begin{bmatrix} 1.333571 & 0.521071 & 0.707500 & 0.353929 \\ 0.521071 & 0.522679 & 0.692321 & -0.84607 \\ 0.707500 & 0.692321 & 1.069821 & -1.305536 \\ 0.353929 & -0.84607 & -1.305536 & 4.053679 \end{bmatrix}$$

Its eigenvalues are

$$\lambda_1 = 4.79791, \lambda_2 = 2.04758, \lambda_3 = 0.0852433, \lambda_4 = 0.0490107$$

We see from the ratio of each eigenvalue to the trace of the matrix that

PC1 = −0.182 Germany − 0.239 France − 0.363 Italy + 0.900 Spain accounts for 68.7% of the variation,

PC2 = −0.797 Germany − 0.316 France − 0.433 Italy − 0.275 Spain accounts for 29.3% of the variation,

PC3 = 0.583 Germany − 0.203 France − 0.714 Italy − 0.320 Spain accounts for 1.2% of the variation,

and

PC4 = 0.155 Germany − 0.895 France + 0.412 Italy − 0.68 Spain accounts for only 0.7% of the variation.

10.3   **A.** Food = − 29.5 + 0.706 Apparel − 0.131 Housing − 0.129 Transportation + 0.5 68 Medical Care.

The model accounts for 99.8% of the variation but none of the variables are significant given the others.

B. PC1 = −0.476 Apparel + 0.514 Housing + 0.493 Transportation + 0.516 Medical Care accounts for 93.6% of variation

PC2 = −0.790 Apparel + 0.131 Housing + 0.598 Transportation + 0.028 Medical Care accounts for 6% of the variation

PC3 = 0.362 Apparel + 0.606 Housing − 0.627 Transportation + 0.329 Medical Care accounts for 0.4% of variation

PC4 = −0.132 Apparel + 0.593 Housing + 0.081 Transportation − 0.790 Medical Care accounts for less than 0.1% of variation

PC1 and PC2 are the only components of any significance

Section 11

11.1 A. $Q = \begin{bmatrix} \frac{5}{2\sqrt{37}} & -\frac{1}{2} & -\frac{1}{\sqrt{62}} & -\frac{36}{\sqrt{2294}} \\ \frac{5}{2\sqrt{37}} & \frac{1}{2} & -\frac{6}{\sqrt{62}} & -\frac{1}{\sqrt{2294}} \\ \frac{5}{2\sqrt{37}} & -\frac{1}{2} & 0 & \frac{31}{\sqrt{2294}} \\ \frac{5}{2\sqrt{37}} & \frac{1}{2} & \frac{5}{\sqrt{62}} & -\frac{6}{\sqrt{2294}} \end{bmatrix}$, $D = \begin{bmatrix} \sqrt{74} & 0 \\ 0 & \sqrt{18} \end{bmatrix}$

B. $A = \begin{bmatrix} \frac{1}{\sqrt{2}} & -\frac{1}{\sqrt{2}} \\ \frac{1}{\sqrt{2}} & \frac{1}{\sqrt{2}} \end{bmatrix} \begin{bmatrix} \frac{5}{\sqrt{74}} & 0 \\ 0 & \frac{1}{\sqrt{2}} \end{bmatrix} \begin{bmatrix} \frac{\sqrt{148}}{2} & \frac{\sqrt{148}}{2} \\ -3 & 3 \end{bmatrix}$

$B = \begin{bmatrix} \frac{1}{\sqrt{2}} & -\frac{1}{\sqrt{2}} \\ \frac{1}{\sqrt{2}} & \frac{1}{\sqrt{2}} \end{bmatrix} \begin{bmatrix} \frac{7}{\sqrt{74}} & 0 \\ 0 & \frac{1}{\sqrt{2}} \end{bmatrix} \begin{bmatrix} \frac{\sqrt{148}}{2} & \frac{\sqrt{148}}{2} \\ -3 & 3 \end{bmatrix}$

11.2 A. 1, 5/7
B. 1, 7/5

11.3 The B singular values of A are the same as the square roots of the eigenvalues of $(A'A)(B'B)^{-1}$. Likewise the A singular values of B are the square roots of the eigenvalues of $(B'B)(A'A)^{-1}$. The result then follows because $(A'A)(B'B)^{-1}$ and $(B'B)(A'A)^{-1}$ are inverses.

11.8 Since Q is P orthogonal $Q'PQ = Q'U\Lambda U'Q = Q'U\Lambda^{1/2}\Lambda^{1/2}UQ = (\Lambda^{1/2}UQ)'\Lambda^{1/2}UQ = I$.

11.10 A. $\begin{bmatrix} 16 & 4 \\ 4 & 9 \end{bmatrix} = \begin{bmatrix} 4 & 0 \\ 1 & \sqrt{8} \end{bmatrix} \begin{bmatrix} 4 & 1 \\ 0 & \sqrt{8} \end{bmatrix}$

B. $\begin{bmatrix} 4 & 1 & 1 \\ 1 & 4 & 1 \\ 1 & 1 & 4 \end{bmatrix} = \begin{bmatrix} 2 & 0 & 0 \\ \frac{1}{2} & \frac{\sqrt{15}}{2} & 0 \\ \frac{1}{2} & \frac{\sqrt{15}}{10} & \frac{3\sqrt{10}}{5} \end{bmatrix} \begin{bmatrix} 2 & \frac{1}{2} & \frac{1}{2} \\ 0 & \frac{\sqrt{15}}{2} & \frac{\sqrt{15}}{10} \\ 0 & 0 & \frac{3\sqrt{10}}{5} \end{bmatrix}$

11.12 Since $U = (L^{-1})'Q$ and $S = LL'$ it follows that $Q'L^{-1}LL'(L')^{-1}Q = I$ or $U'SU = I$.

PART III

Section 12

12.1 A. Two possible answers are $\begin{bmatrix} \frac{3}{2} & 0 & -\frac{1}{4} \\ 0 & 0 & 0 \\ -\frac{1}{2} & 0 & \frac{1}{4} \end{bmatrix}$ and $\begin{bmatrix} 2 & -1 & 0 \\ -1 & 1 & 0 \\ 0 & 0 & 0 \end{bmatrix}$. Both matrices are reflexive generalized inverses.

12.4   **C.** $\hat{\mu} = \bar{Y}.., \hat{\tau}_i = \bar{Y}_{i.} - \bar{Y}..$

$\hat{\mu} + \hat{\tau}_i = \bar{Y}_{i.}, i = 1,2,3,4$

12.5 $AHA = A(G + \alpha(I - GA)A = (AG + \alpha(A - AGA))A$

$\qquad = AGA = A$

Section 13

13.3   Verification of Equation (13.13)

Simply verify the four Penrose Axioms.

1. $UNU'UN^{-1}U'UNU' = UNN^{-1}NU' = UNU'$

2. $UN^{-1}U'UNU'UN^{-1}U' = UN^{-1}NN^{-1}U' = UN^{-1}U'$

3. $UNU'UN^{-1}U' = UNN^{-1}U' = UU'$, a symmetric matrix.

4. $UN^{-1}U'UNU' = UN^{-1}NU' = UU'$, also a symmetric matrix.

Verification of Equation (13.14)

Again verify the Penrose Axioms using the defining property of orthogonal matrices $P'P = I$ and the Penrose Axioms on A.

1. $PAP'PA^+P'PAP = PAA^+AP' = PAP'$

2. $PA^+P'PAP'PA^+P' = PA^+AA^+P' = PA^+P'$

3. $(PAP'PA^+P')' = (PAA^+P')' = P(AA^+)'P' = PAA^+P'$

4. $(PA^+P'PAP')' = (PA^+AP')' = P(A^+A)'P' = PA^+AP'$

Verification of Equation (13.15)

Use Block multiplication of the matrices and the fact that A and B satisfy the Penrose axioms.

1. $\begin{bmatrix} A & 0 \\ 0 & B \end{bmatrix}\begin{bmatrix} A^+ & 0 \\ 0 & B^+ \end{bmatrix}\begin{bmatrix} A & 0 \\ 0 & B \end{bmatrix} = \begin{bmatrix} AA^+A & 0 \\ 0 & BB^+B \end{bmatrix} = \begin{bmatrix} A & 0 \\ 0 & B \end{bmatrix}$

2. $\begin{bmatrix} A^+ & 0 \\ 0 & B^+ \end{bmatrix}\begin{bmatrix} A & 0 \\ 0 & B \end{bmatrix}\begin{bmatrix} A^+ & 0 \\ 0 & B^+ \end{bmatrix} = \begin{bmatrix} A^+AA^+ & 0 \\ 0 & B^+BB^+ \end{bmatrix} = \begin{bmatrix} A^+ & 0 \\ 0 & B^+ \end{bmatrix}$

3. $\begin{bmatrix} \begin{bmatrix} A & 0 \\ 0 & B \end{bmatrix}\begin{bmatrix} A^+ & 0 \\ 0 & B^+ \end{bmatrix} \end{bmatrix}' = \begin{bmatrix} AA^+ & 0 \\ 0 & BB^+ \end{bmatrix} = \begin{bmatrix} (AA^+)' & 0 \\ 0 & (BB^+)' \end{bmatrix}$

$\qquad = \begin{bmatrix} (AA^+) & 0 \\ 0 & (BB^+) \end{bmatrix}$

4. $\begin{bmatrix} \begin{bmatrix} A^+ & 0 \\ 0 & B^+ \end{bmatrix}\begin{bmatrix} A & 0 \\ 0 & B \end{bmatrix} \end{bmatrix}' = \begin{bmatrix} A^+A & 0 \\ 0 & B^+B \end{bmatrix}' = \begin{bmatrix} (A^+A)' & 0 \\ 0 & (B^+B)' \end{bmatrix}$

$\qquad = \begin{bmatrix} (A^+A) & 0 \\ 0 & (B^+B) \end{bmatrix}$

13.4

A. 1.

$$\begin{bmatrix} \frac{1}{12} & \frac{1}{12} & \frac{1}{12} & \frac{1}{12} & \frac{1}{12} & \frac{1}{12} \\ \frac{1}{6} & \frac{7}{24} & \frac{1}{6} & -\frac{1}{12} & -\frac{5}{24} & -\frac{1}{12} \\ -\frac{1}{12} & -\frac{5}{24} & -\frac{1}{12} & \frac{1}{6} & \frac{7}{24} & \frac{1}{6} \\ \frac{1}{6} & -\frac{5}{24} & \frac{1}{6} & -\frac{1}{12} & \frac{7}{24} & -\frac{1}{12} \\ -\frac{1}{12} & \frac{7}{24} & -\frac{1}{12} & \frac{1}{6} & -\frac{5}{24} & \frac{1}{6} \end{bmatrix}$$

2.

$$\begin{bmatrix} \frac{1}{24} & \frac{1}{48} & \frac{1}{18} & \frac{1}{48} & \frac{1}{48} \\ \frac{1}{48} & \frac{19}{96} & -\frac{17}{96} & -\frac{5}{96} & \frac{7}{96} \\ \frac{1}{48} & -\frac{17}{96} & \frac{19}{96} & \frac{7}{96} & -\frac{5}{96} \\ \frac{1}{48} & -\frac{5}{96} & \frac{7}{96} & \frac{19}{96} & -\frac{17}{96} \\ \frac{1}{48} & \frac{7}{96} & -\frac{5}{96} & -\frac{17}{96} & \frac{19}{96} \end{bmatrix}$$

3.

$$\begin{bmatrix} \frac{11}{144} & \frac{1}{72} & \frac{11}{144} & -\frac{7}{144} & \frac{1}{72} & -\frac{7}{144} \\ \frac{1}{72} & \frac{19}{72} & \frac{1}{72} & \frac{1}{72} & -\frac{17}{72} & \frac{1}{72} \\ \frac{11}{144} & \frac{1}{72} & \frac{11}{144} & -\frac{7}{144} & \frac{1}{72} & -\frac{7}{144} \\ -\frac{7}{144} & \frac{1}{72} & -\frac{7}{144} & \frac{11}{144} & \frac{1}{72} & \frac{11}{144} \\ \frac{1}{72} & -\frac{17}{72} & \frac{1}{72} & \frac{1}{72} & \frac{19}{72} & \frac{1}{72} \\ -\frac{7}{144} & \frac{1}{72} & -\frac{7}{144} & \frac{11}{144} & \frac{1}{72} & \frac{11}{144} \end{bmatrix}$$

B. There are many other Generalized Inverses. One possibility is

$$\begin{bmatrix} 0 & 0 & 0 & 0 & 0 \\ 0 & 0 & 0 & 0 & 0 \\ 0 & 0 & \frac{3}{4} & -\frac{1}{4} & -\frac{1}{2} \\ 0 & 0 & -\frac{1}{4} & \frac{5}{12} & \frac{1}{6} \\ 0 & 0 & -\frac{1}{2} & \frac{1}{6} & \frac{2}{3} \end{bmatrix}$$

Note that $X'X$ is a 5×5 matrix of rank 3. This generalized inverse was obtained by inverting the 3×3 sub-matrix in the lower right hand corner.

13.7 A. Observe that

$$(X'X + VV')^{-1} = \left[[U \quad V] \begin{bmatrix} \Lambda & 0 \\ 0 & I \end{bmatrix} \begin{bmatrix} U' \\ V' \end{bmatrix} \right]$$

$$= \left[[U \quad V] \begin{bmatrix} \Lambda^{-1} & 0 \\ 0 & I \end{bmatrix} \begin{bmatrix} U' \\ V' \end{bmatrix} \right]$$

$$= U\Lambda^{-1}U' + VV'$$

and $X'X = U\Lambda U'$. Then

$$X'X(X'X + VV')^{-1}X'X = U\Lambda U'(U\Lambda^{-1}U' + VV')U\Lambda U'$$

$$= U\Lambda U'U\Lambda^{-1}U'U\Lambda U' + U\Lambda U'VV'U\Lambda U'$$

$$= U\Lambda U' = X'X.$$

B. Notice that

$$X'X(X'X+VV')^{-1} = U\Lambda U'(U\Lambda^{-1}U'+VV')$$
$$= U\Lambda U'U\Lambda^{-1}U' + U\Lambda U'VV'$$
$$= UU',$$

a symmetric matrix. Moreover

$$(X'X+VV')^{-1}X'X = (U\Lambda^{-1}U'+VV')U\Lambda U'$$
$$= U\Lambda^{-1}U'U\Lambda U' + U\Lambda UVV'U\Lambda U'$$
$$= UU',$$

also a symmetric matrix. However

$$(X'X+VV')^{-1}X'X(X'X+VV')^{-1} = (U\Lambda^{-1}U'+VV')U\Lambda U'(U\Lambda^{-1}U'+VV')$$
$$= U\Lambda^{-1}U'U\Lambda U'U\Lambda^{-1}U'$$
$$= U\Lambda^{-1}U' = (X'X)^{+} \neq (X'X+VV')^{-1}.$$

13.9　　We know that $X^{+}=U\Lambda^{-1/2}S$. Now since $X=S'\Lambda^{1/2}U'$ and $X'=U\Lambda^{1/2}S$, $(X')^{+}=$ $S'\Lambda^{-1/2}U'$. Then $X^{+}X'^{+}=U\Lambda^{-1/2}SS'\Lambda^{-1/2}U'=U\Lambda^{-1}U'=(X'X)^{+}$.

13.12　　The characteristic equation satisfied by $A'A$ is

$$32(A'A)^{2} - 36(A'A)^{3} + 12(A'A)^{4} - (A'A)^{5} = 0$$

Then

$$(A'A)^{+} = \frac{1}{32}(36UU' - 12A'A + (A'A)^{2})$$

$$= \begin{bmatrix} \frac{1}{16} & \frac{1}{32} & \frac{1}{32} & \frac{1}{32} & \frac{1}{32} \\ \frac{1}{32} & \frac{17}{64} & -\frac{15}{64} & \frac{1}{64} & \frac{1}{64} \\ \frac{1}{32} & -\frac{15}{64} & \frac{17}{64} & \frac{1}{64} & \frac{1}{64} \\ \frac{1}{32} & \frac{1}{64} & \frac{1}{64} & \frac{17}{64} & -\frac{15}{64} \\ \frac{1}{32} & \frac{1}{64} & \frac{1}{64} & -\frac{15}{64} & \frac{17}{64} \end{bmatrix}.$$

Section 14

14.2　　If AG is symmetric then $(AG)'=AG$. Then $G'A'=AG$. But then $G'A'A=AGA=A$ because G is a generalized inverse of A. Then we have $A'AG=A'$.

On the other hand assume that $A'AG=A'$. Then $G'A'AG=G'A'$. But $G'A'A=(A'AG)'=A''=A$ so that $G'A'AG=AG$ and $AG=G'A'=(AG)'$. Thus, AG is symmetric.

14.7 **A.** 1. The answer for the minimum norm and the least square generalized inverse depends on the choice of generalized inverse. There is only one correct answer for the Moore Penrose inverse.

$$(XX')^- = \begin{bmatrix} 0 & 0 & 0 & 0 \\ 0 & \frac{3}{8} & \frac{1}{8} & 0 \\ 0 & \frac{1}{8} & \frac{3}{8} & 0 \\ 0 & 0 & 0 & 0 \end{bmatrix} \quad \text{and} \quad (XX')^- = \begin{bmatrix} \frac{1}{4} & 0 & 0 \\ 0 & \frac{1}{4} & 0 \\ 0 & 0 & 0 \end{bmatrix}$$

B. 1. $X_m^- = \begin{bmatrix} 0 & \frac{1}{2} & \frac{1}{2} & 0 \\ 0 & \frac{1}{4} & -\frac{1}{4} & 0 \\ 0 & -\frac{1}{4} & \frac{1}{4} & 0 \end{bmatrix}, X_{LS}^- = \begin{bmatrix} \frac{1}{4} & \frac{1}{4} & \frac{1}{4} & \frac{1}{4} \\ \frac{1}{4} & \frac{1}{4} & -\frac{1}{4} & -\frac{1}{4} \\ 0 & 0 & 0 & 0 \end{bmatrix}$

$$X^+ = \begin{bmatrix} \frac{1}{4} & \frac{1}{4} & \frac{1}{4} & \frac{1}{4} \\ \frac{1}{8} & \frac{1}{8} & -\frac{1}{8} & -\frac{1}{8} \\ -\frac{1}{8} & -\frac{1}{8} & \frac{1}{8} & \frac{1}{8} \end{bmatrix}$$

The Moore Penrose inverse

2. For the minimum norm and least square generalized inverse the answer depends on the choice of $(AA')^-$ and $(A'A)^-$. For

$$(AA')^- = (A'A)^- = \begin{bmatrix} \frac{1}{2} & 0 \\ 0 & 0 \end{bmatrix} \quad \text{and} \quad A_m^- = \begin{bmatrix} -\frac{1}{2} & 0 \\ \frac{1}{2} & 0 \end{bmatrix}, A_{LS}^- = \begin{bmatrix} -\frac{1}{2} & \frac{1}{2} \\ 0 & 0 \end{bmatrix}, A^+ =$$

$$X_m^- X X_{LS}^- = \begin{bmatrix} -\frac{1}{4} & \frac{1}{4} \\ \frac{1}{4} & -\frac{1}{4} \end{bmatrix}$$

14.9 Show that $A^r = A^- A A^-$ is a reflexive generalized inverse. We have $AA^rA = AA^-AA^-A = AA^-A = A$ and $A^rAA^r = A^-AA^-AA^-AA^- = A^-AA^-AA^- = A^-AA^- = A^r$.

Section 15

15.3 **A.** One possible way to diagonalize the matrix is $P = R_3R_2R_1$

where

$$R_1 = \begin{bmatrix} 1 & 0 & 0 & 0 \\ -1 & 1 & 0 & 0 \\ 0 & 0 & 1 & 0 \\ 0 & 0 & -1 & 1 \end{bmatrix}, R_2 = \begin{bmatrix} 1 & 0 & 0 & 0 \\ 0 & 0 & 1 & 0 \\ 0 & 1 & 0 & 0 \\ 0 & 0 & 0 & 1 \end{bmatrix}, R_3 = \begin{bmatrix} 1 & 1 & 0 & 0 \\ 0 & 1 & 0 & 0 \\ 0 & 0 & 1 & 0 \\ 0 & 0 & 0 & 1 \end{bmatrix}.$$

Then

$$P = \begin{bmatrix} 1 & 0 & 1 & 0 \\ 0 & 0 & 1 & 0 \\ -1 & 1 & 0 & 0 \\ 0 & 0 & -1 & 1 \end{bmatrix}$$

and

$$Q = C_1 C_2 C_3$$

where

$$C_1 = \begin{bmatrix} 1 & 0 & 0 \\ 0 & 1 & 0 \\ -1 & -1 & 1 \end{bmatrix}, C_2 = \begin{bmatrix} 1 & 0 & 0 \\ 0 & 1 & 1 \\ 0 & 1 & 1 \end{bmatrix}, C_3 = \begin{bmatrix} 1 & 0 & -1 \\ 0 & 1 & 0 \\ 0 & 1 & 0 \end{bmatrix}.$$

Thus,

$$Q = \begin{bmatrix} 1 & 0 & -1 \\ 0 & 1 & 1 \\ -1 & -1 & 1 \end{bmatrix}.$$

Now

$$PXQ = \begin{bmatrix} 1 & 0 & 0 \\ 0 & -1 & 0 \\ 0 & 0 & 0 \\ 0 & 0 & 0 \end{bmatrix}$$

and

$$X^- = \begin{bmatrix} 1 & 0 & -1 \\ 0 & 1 & 1 \\ -1 & -1 & 1 \end{bmatrix} \begin{bmatrix} 1 & 0 & 0 & 0 \\ 0 & -1 & 0 & 0 \\ 0 & 0 & 0 & 0 \end{bmatrix} \begin{bmatrix} 1 & 0 & 1 & 0 \\ 0 & 0 & 1 & 0 \\ -1 & 1 & 0 & 0 \\ 0 & 0 & -1 & 1 \end{bmatrix}$$

$$= \begin{bmatrix} 1 & 0 & 1 & 0 \\ 0 & 0 & -1 & 0 \\ -1 & 0 & 0 & 0 \end{bmatrix}$$

B. One Possible Solution

$$P = \begin{bmatrix} 1 & 0 & 0 \\ 0 & 1 & 0 \\ 0 & 1 & 1 \end{bmatrix} \begin{bmatrix} 1 & 0 & 0 \\ 0 & 1 & 0 \\ -\frac{1}{2} & 0 & 1 \end{bmatrix} \begin{bmatrix} 1 & 0 & 0 \\ -\frac{1}{2} & 1 & 0 \\ 0 & 0 & 1 \end{bmatrix} = \begin{bmatrix} 1 & 0 & 0 \\ -\frac{1}{2} & 1 & 0 \\ -1 & 1 & 1 \end{bmatrix}$$

$$Q = \begin{bmatrix} 1 & -\frac{1}{2} & 0 \\ 0 & 1 & 0 \\ 0 & 0 & 1 \end{bmatrix} \begin{bmatrix} 1 & 0 & -\frac{1}{2} \\ 0 & 1 & 0 \\ 0 & 0 & 1 \end{bmatrix} \begin{bmatrix} 1 & 0 & 0 \\ 0 & 1 & 1 \\ 0 & 0 & 1 \end{bmatrix} = \begin{bmatrix} 1 & -\frac{1}{2} & -1 \\ 0 & 1 & 1 \\ 0 & 0 & 1 \end{bmatrix}$$

$$\Delta = PMQ = \begin{bmatrix} 2 & 0 & 0 \\ 0 & \frac{1}{2} & 0 \\ 0 & 0 & 0 \end{bmatrix} \quad \text{and} \quad \Delta^- = \begin{bmatrix} \frac{1}{2} & 0 & 0 \\ 0 & 2 & 0 \\ 0 & 0 & 0 \end{bmatrix}.$$

Thus,

$$M^- = Q\Delta^- P = \begin{bmatrix} 1 & -1 & 0 \\ -1 & 2 & 0 \\ 0 & 0 & 0 \end{bmatrix}$$

C. The problem is easier if it is recognized that

$$H = \begin{bmatrix} 4 & 2 \\ 2 & 4 \end{bmatrix} \otimes J_2$$

Now

$$J_2^+ = \frac{1}{4} J_2.$$

Also by one row and one column operation

$$\begin{bmatrix} 1 & 0 \\ -\frac{1}{2} & 1 \end{bmatrix}\begin{bmatrix} 4 & 2 \\ 2 & 4 \end{bmatrix}\begin{bmatrix} 1 & -\frac{1}{2} \\ 0 & 1 \end{bmatrix} = \begin{bmatrix} 4 & 0 \\ 0 & 3 \end{bmatrix}$$

and

$$\begin{bmatrix} 1 & \frac{1}{2} \\ 0 & 1 \end{bmatrix}\begin{bmatrix} \frac{1}{4} & 0 \\ 0 & \frac{1}{3} \end{bmatrix}\begin{bmatrix} 1 & 0 \\ -\frac{1}{2} & 1 \end{bmatrix} = \begin{bmatrix} \frac{1}{3} & -\frac{1}{6} \\ -\frac{1}{6} & \frac{1}{3} \end{bmatrix}.$$

Then

$$H^- = \begin{bmatrix} \frac{1}{3} & -\frac{1}{6} \\ -\frac{1}{6} & \frac{1}{3} \end{bmatrix} \otimes \frac{1}{4} J_2 = \begin{bmatrix} \frac{1}{12} & \frac{1}{12} & -\frac{1}{24} & -\frac{1}{24} \\ \frac{1}{12} & \frac{1}{12} & -\frac{1}{24} & -\frac{1}{24} \\ -\frac{1}{24} & -\frac{1}{24} & \frac{1}{12} & \frac{1}{12} \\ -\frac{1}{24} & -\frac{1}{24} & \frac{1}{12} & \frac{1}{12} \end{bmatrix}$$

It turns out that this is the Moore Penrose Inverse of H. Why does that seem reasonable?

D. Observe that

$$Z = \begin{bmatrix} 1 & -1 \\ -1 & 1 \end{bmatrix} \otimes J_2. \text{ Again } J_2^+ = \frac{1}{4} J_2.$$

Since

$$\begin{bmatrix} 1 & 0 \\ 1 & 1 \end{bmatrix}\begin{bmatrix} 1 & -1 \\ -1 & 1 \end{bmatrix}\begin{bmatrix} 1 & 1 \\ 0 & 1 \end{bmatrix} = \begin{bmatrix} 1 & 0 \\ 0 & 0 \end{bmatrix}$$

and

$$\begin{bmatrix} 1 & 1 \\ 0 & 1 \end{bmatrix}\begin{bmatrix} 1 & 0 \\ 0 & 0 \end{bmatrix}\begin{bmatrix} 1 & 0 \\ 1 & 1 \end{bmatrix} = \begin{bmatrix} 1 & 0 \\ 0 & 0 \end{bmatrix},$$

$$G = \begin{bmatrix} 1 & 0 \\ 0 & 0 \end{bmatrix} \otimes \begin{bmatrix} \frac{1}{4} & \frac{1}{4} \\ \frac{1}{4} & \frac{1}{4} \end{bmatrix} = \begin{bmatrix} \frac{1}{4} & \frac{1}{4} & 0 & 0 \\ \frac{1}{4} & \frac{1}{4} & 0 & 0 \\ 0 & 0 & 0 & 0 \\ 0 & 0 & 0 & 0 \end{bmatrix}.$$

15.4 Observe that

$$X_1'X_1 = \begin{bmatrix} 4 & 2 & 2 \\ 2 & 2 & 0 \\ 2 & 0 & 2 \end{bmatrix}.$$

One partition is $A = [4]$, $B = [2 \quad 2]$, $D = \begin{bmatrix} 2 & 0 \\ 0 & 2 \end{bmatrix}$. Now $A^- = \left[\frac{1}{4}\right]$. Then by

substitution and matrix multiplication

$$Q = \begin{bmatrix} 1 & -1 \\ -1 & 1 \end{bmatrix} \text{ and } Q^- = \begin{bmatrix} \frac{1}{4} & -\frac{1}{4} \\ -\frac{1}{4} & \frac{1}{4} \end{bmatrix}. \text{ Then } (X_1'X_1) = \begin{bmatrix} \frac{1}{4} & 0 & 0 \\ 0 & \frac{1}{4} & -\frac{1}{4} \\ 0 & -\frac{1}{4} & \frac{1}{4} \end{bmatrix}.$$

Now $X_2'X_2 = \begin{bmatrix} 4 & 2 & 2 & 2 & 2 \\ 2 & 2 & 0 & 1 & 1 \\ 2 & 0 & 2 & 1 & 1 \\ 2 & 1 & 1 & 2 & 0 \\ 2 & 1 & 1 & 0 & 2 \end{bmatrix}$. The partition is $A = X_1'X_1$, $B = \begin{bmatrix} 2 & 2 \\ 1 & 1 \\ 1 & 1 \end{bmatrix}$, and

$D = \begin{bmatrix} 2 & 0 \\ 0 & 2 \end{bmatrix}$. Then $A^- = (X_1'X_1)^-$ from the first part of the problem and

$$Q^- = \begin{bmatrix} \frac{1}{4} & -\frac{1}{4} \\ -\frac{1}{4} & \frac{1}{4} \end{bmatrix}. \text{ Then } (X_2'X_2)^+ = \begin{bmatrix} \frac{1}{4} & 0 & 0 & 0 & 0 \\ 0 & \frac{1}{4} & -\frac{1}{4} & 0 & 0 \\ 0 & -\frac{1}{4} & \frac{1}{4} & 0 & 0 \\ 0 & 0 & 0 & \frac{1}{4} & -\frac{1}{4} \\ 0 & 0 & 0 & -\frac{1}{4} & \frac{1}{4} \end{bmatrix}.$$

$$
\text{For } (X_3'X_3) = \begin{bmatrix}
4 & 2 & 2 & 2 & 2 & 1 & 1 & 1 & 1 \\
2 & 2 & 0 & 1 & 1 & 1 & 1 & 0 & 0 \\
2 & 0 & 2 & 1 & 1 & 0 & 0 & 1 & 1 \\
2 & 1 & 1 & 2 & 0 & 1 & 0 & 1 & 0 \\
2 & 1 & 1 & 0 & 2 & 0 & 1 & 0 & 1 \\
1 & 1 & 0 & 1 & 0 & 1 & 0 & 0 & 0 \\
1 & 1 & 0 & 0 & 1 & 0 & 1 & 0 & 0 \\
1 & 0 & 1 & 1 & 0 & 0 & 0 & 1 & 0 \\
1 & 0 & 1 & 0 & 1 & 0 & 0 & 0 & 1
\end{bmatrix}
$$

$A = X_2'X_2,$

$$
B = \begin{bmatrix}
1 & 1 & 1 & 1 \\
1 & 1 & 0 & 0 \\
0 & 0 & 1 & 1 \\
1 & 0 & 1 & 0 \\
0 & 1 & 0 & 1
\end{bmatrix}, D = I_4, Q = \begin{bmatrix}
\frac{1}{4} & -\frac{1}{4} & -\frac{1}{4} & \frac{1}{4} \\
-\frac{1}{4} & \frac{1}{4} & \frac{1}{4} & -\frac{1}{4} \\
-\frac{1}{4} & \frac{1}{4} & \frac{1}{4} & -\frac{1}{4} \\
\frac{1}{4} & -\frac{1}{4} & -\frac{1}{4} & \frac{1}{4}
\end{bmatrix} = \begin{bmatrix} 1 & -1 \\ -1 & 1 \end{bmatrix} \otimes \frac{1}{4}\begin{bmatrix} 1 & -1 \\ -1 & 1 \end{bmatrix}.
$$

The matrix Q is idempotent so it is its own Moore Penrose inverse. We finally get

$$
(X_3'X_3)^{+} = \begin{bmatrix}
\frac{1}{4} & 0 & 0 & 0 & 0 & 0 & 0 & 0 & 0 \\
0 & \frac{1}{4} & -\frac{1}{4} & 0 & 0 & 0 & 0 & 0 & 0 \\
0 & -\frac{1}{4} & \frac{1}{4} & 0 & 0 & 0 & 0 & 0 & 0 \\
0 & 0 & 0 & \frac{1}{4} & -\frac{1}{4} & 0 & 0 & 0 & 0 \\
0 & 0 & 0 & -\frac{1}{4} & \frac{1}{4} & 0 & 0 & 0 & 0 \\
0 & 0 & 0 & 0 & 0 & \frac{1}{4} & -\frac{1}{4} & -\frac{1}{4} & \frac{1}{4} \\
0 & 0 & 0 & 0 & 0 & -\frac{1}{4} & \frac{1}{4} & \frac{1}{4} & -\frac{1}{4} \\
0 & 0 & 0 & 0 & 0 & -\frac{1}{4} & \frac{1}{4} & \frac{1}{4} & -\frac{1}{4} \\
0 & 0 & 0 & 0 & 0 & \frac{1}{4} & -\frac{1}{4} & -\frac{1}{4} & \frac{1}{4}
\end{bmatrix}.
$$

15.5 If we partition H as

$$
A = \begin{bmatrix} 4 & 4 \\ 4 & 4 \end{bmatrix}, B = D = \begin{bmatrix} 2 & 2 \\ 2 & 2 \end{bmatrix}, \text{ then if we use } A^- = \begin{bmatrix} \frac{1}{4} & 0 \\ 0 & 0 \end{bmatrix}, Q = \begin{bmatrix} 3 & 3 \\ 3 & 3 \end{bmatrix},
$$

and $Q^- = \begin{bmatrix} \frac{1}{3} & 0 \\ 0 & 0 \end{bmatrix}$ we get $H^- = \begin{bmatrix} \frac{1}{3} & 0 & -\frac{1}{6} & 0 \\ 0 & 0 & 0 & 0 \\ -\frac{1}{6} & 0 & \frac{1}{3} & 0 \\ 0 & 0 & 0 & 0 \end{bmatrix}.$

Partition Z in a manner similar to H. Then if

$$A^- = \begin{bmatrix} 1 & 0 \\ 0 & 0 \end{bmatrix}, Q = 0.$$

Then you can use any matrix for a generalized inverse of Q so use I_2. Then we get

$$Z^- = \begin{bmatrix} 3 & 0 & 1 & 1 \\ 0 & 0 & 0 & 0 \\ 1 & 0 & 1 & 0 \\ 1 & 0 & 0 & 1 \end{bmatrix}$$

There are other correct answers and solutions to the problem.

15.6 $(A'A)^+ = \begin{bmatrix} \frac{1}{9} & \frac{1}{18} & \frac{1}{18} \\ \frac{1}{18} & \frac{5}{18} & -\frac{2}{9} \\ \frac{1}{18} & -\frac{2}{9} & \frac{5}{18} \end{bmatrix}$ and $(B'B)^+ = \begin{bmatrix} \frac{1}{18} & \frac{1}{36} & \frac{1}{36} \\ \frac{1}{36} & \frac{13}{288} & -\frac{5}{288} \\ \frac{1}{36} & -\frac{5}{288} & \frac{13}{288} \end{bmatrix}$

Section 16

16.1 Let $p'\beta$ and $q'\beta$ be parametric functions. Then there are vectors t and w such that $p' = t'X$ and $q' = w'X$. Then for scalars a and b

$$ap'\beta + bq'\beta = at'X\beta + bw'X\beta = (at' + bw')X\beta$$

establishing that the linear combination is estimable.

16.2 The parametric functions in b and d are estimable but those in a and c are not.

16.3 A least square generalized inverse of $X'X$ obtained using

$$(X'X)^{-ls} = (X'XX'X)^- X'X$$

with

$$(X'XX'X)^- = \frac{1}{36} \begin{bmatrix} 0 & 0 & 0 & 0 \\ 0 & 3 & -1 & -1 \\ 0 & -1 & 3 & -1 \\ 0 & -1 & -1 & 3 \end{bmatrix}$$

is

$$G_{ls} = \begin{bmatrix} 0 & 0 & 0 & 0 \\ \frac{1}{12} & \frac{1}{4} & -\frac{1}{12} & -\frac{1}{12} \\ \frac{1}{12} & -\frac{1}{12} & \frac{1}{4} & -\frac{1}{12} \\ \frac{1}{12} & -\frac{1}{12} & -\frac{1}{12} & \frac{1}{4} \end{bmatrix}.$$

The least square estimators are

$$\hat{\mu} = 0$$

$$\hat{\tau}_1 = \tfrac{1}{12}y_{..} + \tfrac{1}{4}y_{1.} - \tfrac{1}{12}y_{2.} - \tfrac{1}{12}y_{3.}$$

$$\hat{\tau}_2 = \tfrac{1}{12}y_{..} - \tfrac{1}{12}y_{1.} + \tfrac{1}{4}y_{2.} - \tfrac{1}{12}y_{3.}$$

$$\hat{\tau}_3 = \tfrac{1}{12}y_{..} - \tfrac{1}{12}y_{1.} - \tfrac{1}{12}y_{2.} + \tfrac{1}{4}y_{3.}.$$

For example

$$\hat{\mu} + \hat{\tau}_2 = \tfrac{1}{12}y_{..} - \tfrac{1}{12}y_{1.} + \tfrac{1}{4}y_{2.} - \tfrac{1}{12}y_{3.}$$
$$= \tfrac{1}{3}y_{2.} = \bar{y}_{2.}$$

A minimum norm generalized inverse, computed by finding

$$(X'X)^{-mn} = X'X(X'XX'X)^{-}$$

using the same generalized inverse as above

$$G_{mn} = \begin{bmatrix} 0 & \tfrac{1}{12} & \tfrac{1}{12} & \tfrac{1}{12} \\ 0 & \tfrac{1}{4} & -\tfrac{1}{12} & -\tfrac{1}{12} \\ 0 & -\tfrac{1}{12} & \tfrac{1}{4} & -\tfrac{1}{12} \\ 0 & -\tfrac{1}{12} & -\tfrac{1}{12} & \tfrac{1}{4} \end{bmatrix}.$$

The least square estimators are

$$\hat{\mu} = \tfrac{1}{12}y_{1.} + \tfrac{1}{12}y_{2.} + \tfrac{1}{12}y_{3.} = \tfrac{1}{12}y_{..}$$

$$\hat{\tau}_1 = \tfrac{1}{4}y_{1.} - \tfrac{1}{12}y_{2.} - \tfrac{1}{12}y_{3.}$$

$$\hat{\tau}_2 = -\tfrac{1}{12}y_{1.} + \tfrac{1}{4}y_{2.} - \tfrac{1}{12}y_{3.}$$

$$\hat{\tau}_3 = -\tfrac{1}{12}y_{1.} - \tfrac{1}{12}y_{2.} + \tfrac{1}{4}y_{3.}$$

For example

$$\hat{\mu} + \hat{\tau}_3 = \bar{y}_3.$$

There are other correct solutions and answers depending on the choice of generalized inverse of $X'XX'X$.

16.5 A. Two generalized inverses are

$$G_1 = \begin{bmatrix} 0 & 0 & 0 & 0 & 0 \\ 0 & \tfrac{1}{4} & 0 & 0 & 0 \\ 0 & 0 & \tfrac{1}{4} & 0 & 0 \\ 0 & 0 & 0 & \tfrac{1}{8} & -\tfrac{1}{8} \\ 0 & 0 & 0 & -\tfrac{1}{8} & \tfrac{1}{8} \end{bmatrix}, G_2 = \begin{bmatrix} \tfrac{1}{8} & 0 & 0 & 0 & 0 \\ -\tfrac{1}{8} & \tfrac{1}{4} & 0 & 0 & 0 \\ -\tfrac{1}{8} & 0 & \tfrac{1}{4} & 0 & 0 \\ -\tfrac{1}{8} & 0 & 0 & \tfrac{1}{4} & 0 \\ -\tfrac{1}{8} & 0 & 0 & 0 & \tfrac{1}{4} \end{bmatrix}.$$

For G_1 the least square estimators of the individual parameters are

$$\hat{\mu}=0, \hat{\tau}_1 = \tfrac{1}{4}y_{.1.}, \hat{\tau}_2 = \tfrac{1}{4}y_{.2.}, \hat{\beta}_1 = \tfrac{1}{8}y_{1..} - \tfrac{1}{8}y_{2..}, \hat{\beta}_2 = -\tfrac{1}{8}y_{1..} + \tfrac{1}{8}y_{2..}.$$

For G_2

$$\hat{\mu}=\tfrac{1}{8}y_{...}, \hat{\tau}_1 = \tfrac{1}{4}y_{.1.} - \tfrac{1}{8}y_{...}, \hat{\tau}_2 = \tfrac{1}{4}y_{.2.} - \tfrac{1}{8}y_{...}, \hat{\beta}_1 = \tfrac{1}{4}y_{1..} - \tfrac{1}{8}y_{...}, \hat{\beta}_2 = \tfrac{1}{4}y_{2..} - \tfrac{1}{8}y_{...}$$

B. Now consider for example

$$\theta_{12} = \mu + \tau_1 + \beta_2.$$

For $p' = t'X$ we have

$$1 = \sum_{i=1}^{8} t_i$$
$$1 = \sum_{i=1}^{4} t_i$$
$$0 = \sum_{i=5}^{8} t_i$$
$$0 = t_1 + t_2 + t_5 + t_6$$
$$1 = t_3 + t_4 + t_7 + t_8$$

One possible choice of t is

$$t_1 = t_2 = t_5 = t_6 = t_7 = t_8 = 0, t_3 = t_4 = \frac{1}{2}.$$

For β_2 for example

$$0 = \sum_{i=1}^{8} t_i$$
$$0 = \sum_{i=1}^{4} t_i$$
$$0 = \sum_{i=5}^{8} t_i$$
$$0 = t_1 + t_2 + t_5 + t_6$$
$$1 = t_3 + t_4 + t_7 + t_8,$$

an inconsistent system of equations.

16.6 For the generalized inverse

$$G = \begin{bmatrix} 0 & 0 & 0 & 0 \\ 0 & \tfrac{1}{2} & 0 & 0 \\ 0 & 0 & \tfrac{1}{3} & 0 \\ 0 & 0 & 0 & \tfrac{1}{4} \end{bmatrix},$$

$\hat{\mu}=0, \quad \hat{\tau}_i = \bar{y}_{i.}, \quad i=1,2,3 \quad \text{and} \quad \hat{\mu}+\hat{\tau}_i = \bar{y}_{i.}, \quad i=1,2,3.$

PART IV

Section 17

17.1 $w_1 = \sqrt{2}\left(x_1 - \frac{1}{2}x_2\right)$, $w_2 = \sqrt{\frac{3}{2}}x_2$ or $w_1 = \sqrt{\frac{3}{2}}(-x_1 + x_2)$, $w_2 = \frac{1}{\sqrt{2}}(x_1 - x_2)$

17.2 **A.** $y_1^2 + y_2^2$ where $y_1 = \sqrt{\frac{6}{5}}(-2x_1 + x_2)$, $y_2 = \frac{1}{\sqrt{5}}(x_1 + 2x_2)$

  B. y_1^2 where $y_1 = -2x_1 + x_2, y_2 = 0$. The rank of the matrix of this quadratic form is 1.

  C. $z_1^2 + z_2^2 + z_3^2$ where $z_1 = \frac{1}{\sqrt{2}}(x_1 + x_2 + x_3)$, $z_2 = \frac{\sqrt{3}}{2\sqrt{2}}(-x_1 + x_3)$,

$$z_3 = \frac{1}{2\sqrt{2}}(-x_1 + 2x_2 - x_3)$$

  D. $v_1^2 + v_2^2$ where $v_1 = 2x_1 + x_2 + x_3$, $v_2 = -x_2 + x_3$ and $v_3 = 0$.

  E. $w_1^2 + w_2^2$ where $w_1 = \sqrt{3}(x_1 + x_2 + x_3)$, $w_2 = \sqrt{2}(-x_1 + x_2)$, $w_3 = 0$

17.3 For a $\chi^2(2)$ distribution let $r=s=1$. Substitution into (17.5) yields a $\chi^2(2)$ distribution. Assume that we have a $\chi^2(k)$ distribution. By the Lemma the sum of a $\chi^2(k)$ and a $\chi^2(1)$ distribution is a $\chi^2(k+1)$ distribution.

17.5 The matrix $\left(I - \frac{1}{n}J\right)$ is idempotent of rank $n - 1$.

17.7 $a = 1/n$

17.9 The random variable $y = (x - \mu)V^{-1/2} \sim N(0,1)$. Then

$$(x - \mu)'A(x - \mu) = (x - \mu)'V^{-1/2}V^{1/2}AV^{1/2}V^{-1/2}(x - \mu)$$
$$= y'V^{1/2}AV^{1/2}y.$$

The result follows from Theorem 17.5. In particular if $x \sim N(\mu, \sigma^2 I)$ then

$$(x - \mu)'A(x - \mu)/\sigma^2$$

is idempotent by what was proved above using A/σ^2 in place of **A**.

17.10 **B.** $(a - 1)b$.

Section 18

18.1 A. First

$$Q_R^2 = \tilde{X}(\tilde{X}'\tilde{X})^{-1}\tilde{X}'\tilde{X}(\tilde{X}'\tilde{X})^{-1}\tilde{X}' = \tilde{X}(\tilde{X}'\tilde{X})^{-1}\tilde{X}'.$$

Now

$$Q_E^2 = \left(I - \tilde{X}(\tilde{X}'\tilde{X})^{-1}\tilde{X}' - \frac{1}{n}J \right)^2$$

$$= (I - \tilde{X}(\tilde{X}'\tilde{X})^{-1}\tilde{X}')^2 - \left(I - \tilde{X}(\tilde{X}'\tilde{X})^{-1}\tilde{X}' \right)\frac{1}{n}J$$

$$- \frac{1}{n}J\left(I - \tilde{X}(\tilde{X}'\tilde{X})^{-1}\tilde{X}' \right) + \frac{1}{n^2}J^2 .$$

For any idempotent matrix E we have that $I - E$ is idempotent so the first term of Q_E^2 is idempotent.
   Now

$$J\tilde{X} = J\left(Z - \frac{1}{n}JZ \right) = JZ - JZ = 0.$$

Also

$$\frac{1}{n^2}J^2 = \frac{1}{n}J.$$

Then

$$Q_E^2 = I - \tilde{X}(\tilde{X}'\tilde{X})^{-1}\tilde{X}' - \frac{2}{n}J + \frac{1}{n^2}J^2 = Q_E .$$

C. $Q_R Q_E = \tilde{X}(\tilde{X}'\tilde{X})^{-1}\tilde{X}'\left(I - \tilde{X}(\tilde{X}'\tilde{X})^{-1}\tilde{X}' - \frac{1}{n}J \right)$

$$= \tilde{X}(\tilde{X}'\tilde{X})^{-1}\tilde{X}' - \tilde{X}(\tilde{X}'\tilde{X})^{-1}\tilde{X}' - \tilde{X}(\tilde{X}'\tilde{X})^{-1}\tilde{X}'\frac{1}{n}J$$

$$= 0.$$

18.4 A. We have partitioning the matrix with the first column

$$\tilde{X} = \begin{bmatrix} \frac{1}{2} & -\frac{1}{2} & \frac{1}{2} & -\frac{1}{2} \\ \frac{1}{2} & -\frac{1}{2} & -\frac{1}{2} & \frac{1}{2} \\ -\frac{1}{2} & \frac{1}{2} & \frac{1}{2} & -\frac{1}{2} \\ \frac{1}{2} & \frac{1}{2} & -\frac{1}{2} & \frac{1}{2} \end{bmatrix}, \tilde{X}'\tilde{X} = \begin{bmatrix} 1 & -1 & 0 & 0 \\ -1 & 1 & 0 & 0 \\ 0 & 0 & 1 & -1 \\ 0 & 0 & -1 & 1 \end{bmatrix},$$

and

$$(\tilde{X}'\tilde{X})^+ = \begin{bmatrix} \frac{1}{4} & -\frac{1}{4} & 0 & 0 \\ -\frac{1}{4} & \frac{1}{4} & 0 & 0 \\ 0 & 0 & \frac{1}{4} & -\frac{1}{4} \\ 0 & 0 & -\frac{1}{4} & \frac{1}{4} \end{bmatrix}.$$

The least square estimators are

$$\hat{\mu} = \bar{y}.., \hat{\tau}_1 = \tfrac{1}{4}(y_1. - y_2.), \hat{\tau}_2 = \tfrac{1}{4}(y_2. - y_1.), \hat{\beta}_1 = \tfrac{1}{4}(y_{.1} - y_{.2}), \hat{\beta}_2 = \tfrac{1}{4}(y_{.2} - y_{.1}).$$

The model sum of squares is

$$SSR = \tfrac{1}{4}(y_1. - y_2.)^2 + \tfrac{1}{4}(y_{.1} - y_{.2})^2.$$

B. $SS(\beta) = \tfrac{1}{4}(y_{.1} - y_{.2})^2.$

18.5

Source	df	SS	MS	F
Speed limits	5	69791	14192	2.54 < 3.15
Error	24	131648	5485	
Total	29	201439		

There does not appear to be a significant difference in fatalities for the different speed limits at $\alpha = 0.05$. However it is close because the p value is 0.055.

18.6

Source	df	SS	MS	F
States	4	69.693	17.4231	0.70 < 3.38
Speed limits	5	413.817	82.7634	3.32* > 3.15
Error	20	497.920	24.8960	
Total	29	981.430		

There does not appear to be a significant difference in fatalities between states. There is a significant difference between speed limits.

18.8 We have that

$$b'X'Xb = y'XG'X'XGX'y = y'S'\Lambda^{1/2}U'G'U\Lambda U'GU\Lambda^{1/2}Sy$$
$$= y'S'\Lambda^{1/2}\Lambda^{-1}\Lambda\Lambda^{-1}\Lambda^{1/2}Sy = y'S'Sy$$

for any generalized inverse establishing independence.

18.11

Source	df	SS	MS	F
Brand	3	84663	28221	271.14** > 5.4
Error	26	2706	104	
Total	29	87369		

There is highly significant evidence of a difference amongst the brand.

Section 19

19.1

Source	df	SS	MS	F
Clubs	1	9.389	9.389	3.19
Course	2	44.444	22.222	7.55*
Clubs × Course	2	0.445	0.223	0.008
Error	12	35.333	2.944	
Total	17	89.611		

    A. There is no significant difference in the scores between the older and the new clubs.

    B. At $\alpha = 0.05$ there is a significant difference between the scores at the different golf courses.

    C. There is no significant interaction.

19.2

Source	df	SS	MS	F
Hospital	2	665.68	332.84	1.74
Subject (Hospital)	9	1720.67	191.19	146.88**
Error	12	15.62	1.30	
Total	23	2401.97		

There is no significant difference between hospitals in blood coagulation time. However, there is a highly significant difference between subjects nested in hospitals.

Section 20

20.2 **A.** $F = 0.00115$ not significant.

    B. $F = 0.001757$ not significant

20.3 $Q_1 = 22.5005$ $F_1 = 0.6148$ not significant $Q_2 = 17.6331$ $F_2 = 0.4818$ not significant. $Q_1 + Q_2 = 40.13$ the regression sum of squares.

20.4

Source	df	SS	MS	F
Speed limits	3	102364	34121	7.61*
Error	8	35874	4484	
Total	11	138238		

Contrast 1 F = 14.66 highly significant $Q_1 = 65734.0$

Contrast 2 F = 8.13 highly significant $Q_2 = 36,450$

Contrast 3 F = .041 not significant $Q_3 = 182.25$

Qs should add up to 102364. We are close probably due to roundoff error.

20.5 Using the generalized inverse of X'X

$$G_1 = \begin{bmatrix} 0 & 0 & 0 & 0 \\ 0 & \frac{1}{5} & 0 & 0 \\ 0 & 0 & \frac{1}{5} & 0 \\ 0 & 0 & 0 & \frac{1}{5} \end{bmatrix}$$

the least square estimator is

$$b = \begin{bmatrix} 0 \\ 83.2 \\ 80.2 \\ 79.4 \end{bmatrix}$$

$$Q = 34611.2$$

Using the Moore Penrose inverse

$$G_2 = \begin{bmatrix} \frac{3}{80} & \frac{1}{80} & \frac{1}{80} & \frac{1}{80} \\ \frac{1}{80} & \frac{11}{80} & -\frac{1}{16} & -\frac{1}{16} \\ \frac{1}{80} & -\frac{1}{16} & \frac{11}{80} & -\frac{1}{16} \\ \frac{1}{80} & -\frac{1}{16} & -\frac{1}{16} & \frac{11}{80} \end{bmatrix}$$

and

$$b = \begin{bmatrix} 60.7 \\ 22.5 \\ 19.5 \\ 18.7 \end{bmatrix}$$

$$Q = 3681.82$$

PART V

Section 21

21.3 Two possible ways to do the problem will be presented. First the form to be minimized can be shown to be algebraically equivalent to

$$f(\beta) = \beta'X'V^{-1}X\beta - 2\beta'X'V^{-1}Y.$$

By matrix differentiation

$$\frac{\partial f}{\partial \beta} = 2X'V^{-1}X\beta - 2X'V^{-1}Y = 0$$

and

$$\hat{\beta} = (X'V^{-1}X)^{-1}X'V^{-1}Y.$$

Alternatively let $Z = V^{-1/2}Y$ and $W = V^{-1/2}X$. Then it can be shown that

$$f(\beta) = (Y - X\beta)'V^{-1}(Y - X\beta) - Y'V^{-1}Y$$
$$= (Z - W\beta)'(Z - W\beta) - Z'Z$$

From previous considerations $f(\beta)$ is minimized when

$$\hat{\beta} = (W'W)^{-1}W'Y = (X'V^{-1}X)^{-1}X'V^{-1}Y.$$

21.4 We have

$$\hat{\beta} = (Z'V^{-1}Z)^{-1}Z'V^{-1}Y$$

$$= \left[\begin{bmatrix} X' & R' \end{bmatrix} \begin{bmatrix} \frac{1}{\sigma^2}I & 0 \\ 0 & \frac{1}{\tau^2}I \end{bmatrix} \begin{bmatrix} X \\ R \end{bmatrix} \right]^{-1} \begin{bmatrix} X' & R' \end{bmatrix} \begin{bmatrix} \frac{1}{\sigma^2}I & 0 \\ 0 & \frac{1}{\tau^2}I \end{bmatrix} \begin{bmatrix} Y \\ r \end{bmatrix}$$

$$= \left(\frac{X'X}{\sigma^2} + \frac{R'R}{\tau^2} \right)^{-1} \left(\frac{X'Y}{\sigma^2} + \frac{R'r}{\tau^2} \right)$$

$$= (\tau^2 X'X + \sigma^2 R'R)^{-1}(\tau^2 X'Y + \sigma^2 R'r).$$

21.7 **A.** $\begin{bmatrix} x \\ y \end{bmatrix} = \begin{bmatrix} 6/7 \\ -(5/7) \end{bmatrix}$ Minimum value $= 5/7$

   B. All solutions to $2x + y = 1$, Minimum value $= 3$
   C. No minimum value. See the result in 21.8.
   D. No Minimum E. Minimum value $= 6$ for all point on the hyperplane $8x_1 + 8x_2 + 8x_3 + 8x_4 = -4$.

Section 22

22.2
$$\hat{\mu}^r = 0$$
$$\hat{\tau}_1^r = \frac{5}{12}\bar{y}_{1.} + \frac{5}{12}\bar{y}_{2.} + \frac{1}{4}\bar{y}_{3.} - \frac{1}{12}\bar{y}_{4.}$$
$$\hat{\tau}_2^r = \frac{5}{12}\bar{y}_{1.} + \frac{5}{12}\bar{y}_{2.} + \frac{1}{4}\bar{y}_{3.} - \frac{1}{12}\bar{y}_{4.}$$
$$\hat{\tau}_3^r = \frac{1}{4}\bar{y}_{1.} + \frac{1}{4}\bar{y}_{2.} + \frac{1}{4}\bar{y}_{3.} + \frac{1}{4}\bar{y}_{4.}$$
$$\hat{\tau}_4^r = -\frac{1}{12}\bar{y}_{1.} - \frac{1}{12}\bar{y}_{2.} + \frac{1}{12}\bar{y}_{3.} + \frac{11}{12}\bar{y}_{4.}$$

22.4 For $H_0 \ \tau_1 + \tau_2 - 2\tau_3 = 0$

$$\hat{\mu}^r + \hat{\tau}_1^r = \frac{1}{750}\left(750\bar{y}.. + 475\bar{y}_{1.} - 275\bar{y}_{2.} + 100\bar{y}_{3.} - 150\bar{y}_{4.}\right)$$

$$\hat{\mu}^r + \hat{\tau}_2^r = \frac{1}{750}\left(750\bar{y}.. - 275\bar{y}_{1.} + 475\bar{y}_{2.} + 100\bar{y}_{3.} - 150\bar{y}_{4.}\right)$$

$$\hat{\mu}^r + \hat{\tau}^r = \frac{1}{750}\left(750\bar{y}.. + 100\bar{y}_{1.} + 100\bar{y}_{2.} + 100\bar{y}_{3.} - 150\bar{y}_{4.}\right)$$

$$\hat{\mu}^r + \hat{\tau}_4^r = \frac{1}{750}\left(750\bar{y}.. - 150\bar{y}_{1.} - 150\bar{y}_{2.} - 150\bar{y}_{3.} + 600\bar{y}_{4.}\right)$$

For $H_0 \ \tau_2 - \tau_3 = 0, \ 3\tau_1 - \tau_2 - \tau_3 - \tau_4 = 0$

$$\hat{\mu}^r + \hat{\tau}_1^r = \frac{1}{1500}\left(1500\bar{y}.. + 75\bar{y}_{1.} + 75\bar{y}_{2.} + 75\bar{y}_{3.} + 75\bar{y}_{4.}\right)$$

$$\hat{\mu}^r + \hat{\tau}_2^r = \frac{1}{1500}\left(1500\bar{y}.. + 75\bar{y}_{1.} + 325\bar{y}_{2.} + 325\bar{y}_{3.} - 425\bar{y}_{4.}\right)$$

$$\hat{\mu}^r + \hat{\tau}_3^r = \frac{1}{1500}\left(1500\bar{y}.. + 75\bar{y}_{1.} + 325\bar{y}_{2.} + 325\bar{y}_{3.} - 425\bar{y}_{4.}\right)$$

$$\hat{\mu}^r + \hat{\tau}_4^r = \frac{1}{1500}\left(1500\bar{y}.. + 75\bar{y}_{1.} - 425\bar{y}_{2.} - 425\bar{y}_{3.} + 1075\bar{y}_{4.}\right)$$

22.6 $\left(2 - \frac{1}{\sqrt{2}}, 2 - \frac{1}{\sqrt{2}}\right), d = \sqrt{9 - \frac{8}{\sqrt{2}}}$

22.7 $\hat{\beta}_1 = \bar{Y}_{1.}\left(1 - \frac{c}{\sqrt{\bar{Y}_{1.}^2 + \bar{Y}_{2.}^2}}\right), \hat{\beta}_2 = \bar{Y}_{2.}\left(1 - \frac{c}{\sqrt{\bar{Y}_{1.}^2 + \bar{Y}_{2.}^2}}\right)$

22.8 0.367498

Section 23

23.3 Let λ be the vector of Lagrange multipliers. Let

$$T = c'(X'X)^{+}c - (p' - c'UU')\lambda.$$

Differentiate T with respect to c' and set the result equal to zero. Then

$$(X'X)^{+}c + UU'\lambda = 0. \tag{*}$$

Multiplication by $X'X$ gives

$$UU'c + X'X\lambda = 0.$$

Then

$$\lambda = -(X'X)^{+}UU'c = -(X'X)^{+}p. \tag{**}$$

Substitution of λ in (**) into (*) yields

$$(X'X)^{+}c = UU'(X'X)^{+}p = (X'X)^{+}p. \tag{***}$$

Multiply both sides of (***) by $X'X$ to obtain

$$UU'c = UU'p.$$

Then

$$c'b = c'UU'b = p'UU'b = p'b.$$

23.4 In order that the estimator be unbiased in the extended sense

$$E(a + (p - c)'b - p'\theta) = a + (p - c)'UU'\theta - p'\theta = 0$$

and

$$a = (c - p)'UU'\theta + p'\theta$$
$$= c'UU'\theta + p'(I - UU')\theta$$
$$= c'UU'\theta$$

for the estimable parametric functions.
  We want to minimize

$$v = \text{Var}(p'\beta - a - (p - c)'b)$$
$$= p'Fp + (p - c)'(UU'FUU' + \sigma^2(X'X)^{+})(p - c)$$
$$- 2(p - c)'FUU'p.$$

Differentiating with respect to c and setting the result equal to zero yields

$$2(UU'FUU' + \sigma^2 (X'X)^+)(p-c) - 2FUU'p = 0$$

so that

$$\begin{aligned} c' &= p'UU' - p'UU'F(UU'FUU' + \sigma^2 (X'X)^+)^+ \\ &= p'UU' - p'UU'FU(U'FU + \sigma^2 \Lambda^{-1})^{-1} U' \\ &= p'U\Lambda^{-1}U'(U'FU + \sigma^2 \Lambda^{-1})^{-1} U'\sigma^2 \\ &= p'(X'X)^+ (UU'FUU' + \sigma^2 (X'X)^+)^+ \sigma^2 . \end{aligned}$$

The result follows by substitution. The algebraic equivalence can be obtained by manipulation of the terms in the equality above.

23.7 Just substitute the expressions obtained for L and c into the variances that were to be minimized.

Section 24

24.1 We show that the estimator with respect to the given prior is the Liu estimator. Use the form $\hat{\beta} = F(F + \sigma^2 (X'X)^{-1})b$. Then

$$\begin{aligned} p'\hat{\beta} &= p'\left(\frac{\sigma^2}{(1-d)k}\left(I + dk(X'X)^{-1}\right)\right)\left((X'X)^{-1}\sigma^2 + \left(\frac{\sigma^2}{(1-d)k}(I + dk(X'X)^{-1}\right)\right)^{-1} b \\ &= p'(I + dk(X'X)^{-1})(I + k(X'X)^{-1})^{-1} b \\ &= p'(I + k(X'X)^{-1})^{-1} b + dk(X'X)^{-1}(I + k(X'X)^{-1})^{-1} b \\ &= p'(X'X + kI)^{-1} X'Xb + dk(X'X + kI)^{-1} b \\ &= p'(X'X + kI)^{-1}(X'Y + dkb). \end{aligned}$$

24.3 For the less than full rank model reparametized to the full rank model we get

$$MSE(\hat{\gamma}) = (\Lambda + A)^{-1}(\sigma^2 \Lambda^{-1} + \gamma\gamma')(\Lambda + A)^{-1}.$$

For estimable parametric functions there is a vector d where $p = dU$. Then

$$\begin{aligned} MSE(p'\hat{\beta}) &= p'E(\hat{\beta} - \beta)(\hat{\beta} - \beta)'p = d'U'E(\hat{\gamma} - \gamma)(\hat{\gamma} - \gamma)'Ud \\ &= d'(\Lambda + A)^{-1}(\sigma^2 \Lambda^{-1} + \gamma\gamma')(\Lambda + A)^{-1} d \\ &= d'U'U(\Lambda + A)^{-1} U'U(\sigma^2 \Lambda^{-1} + \gamma\gamma')U'U(\Lambda + A)^{-1} U'Ud \\ &= p'(U\Lambda U' + UAU')^+ (\sigma^2 U\Lambda^{-1}U' + U\gamma\gamma'U')(U\Lambda U' + UAU')^+ p \\ &= p'(X'X + G)^+ (\sigma^2 (X'X)^+ + \beta\beta')(X'X + G)^+ p. \end{aligned}$$

24.5 Outline of the solution.

One form of the linear Bayes estimator is

$$p'\hat{\beta} = p'\theta + p'F(UU'FUU' + \sigma^2(X'X)^+)^+(b-\theta)$$

For the estimable parametric functions where there is a d such that $p = Ud$

$$d'\hat{\gamma} = d'U'\theta + d'U'FU(U'FU + \sigma^2\Lambda^{-1})^{-1}(g - U'\theta)$$

Then

$$MSE(d'\hat{\gamma}) = d'U'FU(U'FU + \sigma^2\Lambda^{-1})^{-1}\sigma^2\Lambda^{-1}(U'FU + \sigma^2\Lambda^{-1})^{-1}U'FUd$$
$$+ d'\sigma^2\Lambda^{-1}(U'FU + \sigma^2\Lambda^{-1})^{-1}(\gamma - U'\theta)(\gamma - U'\theta)'$$
$$(U'FU + \sigma^2\Lambda^{-1})^{-1}\sigma^2\Lambda^{-1}d$$

This MSE is less than that of the least square estimator if and only if in the sense of the Lowner ordering we have

$$\sigma^2\Lambda^{-1}(U'FU + \sigma^2\Lambda^{-1})^{-1}(\gamma - U'\theta)(\gamma - U'\theta)'(U'FU + \sigma^2\Lambda^{-1})^{-1}\sigma^2\Lambda^{-1}$$
$$\le \sigma^2\Lambda^{-1} - (I - \sigma^2\Lambda^{-1}(U'FU + \sigma^2\Lambda^{-1})^{-1})\sigma^2\Lambda^{-1}(I - (U'FU + \sigma^2\Lambda^{-1})^{-1}\sigma^2\Lambda^{-1})$$
$$= 2\sigma^2\Lambda^{-1}(U'FU + \sigma^2\Lambda^{-1})^{-1})\sigma^2\Lambda^{-1} - \sigma^2\Lambda^{-1}(U'FU + \sigma^2\Lambda^{-1})^{-1})\sigma^2\Lambda^{-1}(U'FU$$
$$+ \sigma^2\Lambda^{-1})^{-1}\sigma^2\Lambda^{-1}$$

This holds true

$$(\gamma - U'\theta)(\gamma - U'\theta)' \le 2U'FU + \sigma^2\Lambda^{-1}$$

or

$$(\gamma - U'\theta)'(2U'FU + \sigma^2\Lambda^{-1})^{-1}(\gamma - U'\theta) \le 1$$

or

$$(\beta - \theta)'(2UU'FUU' + \sigma^2(X'X)^+)^+(\beta - \theta) \le 1$$

24.8 We have

$$g(k_i) = \frac{\sigma^2\lambda_i + k_i^2\gamma_i^2}{(\lambda_i + k_i)^2}.$$

Then

$$g'(k_i) = \frac{2\lambda_i k_i\gamma_i^2 - 2\lambda_i\sigma^2}{(\lambda_i + k_i)^3} = 0.$$

and

$$k_i = \frac{\sigma^2}{\gamma_2^2}$$

Minimum value is

$$\frac{\sigma^2}{\lambda_i + \sigma^2 / \gamma_i^2}$$

24.9 Optimum $k = \dfrac{\sigma^2 s}{\beta'(X'X)\beta}$.

REFERENCES

Baksalary, J.K. and R. Kala (1983). Partial orderings of matrices one of which is of rank one. Bulletin of the Polish Academy of Science, Mathematics 31:5–7.

Baye, M.R. and D.F. Parker (1984). Combining ridge and principal components regression: A money demand illustration. Communications in Statistics–Theory and Methods A,13: 197–205.

Bulmer, M.G. (1980). The Mathematical Theory of Quantitative Genetics. Oxford University Press, Oxford.

Economic Report of the President 2009.

Economic Report of the President 2010.

Farebrother, R.W. (1976). Further results on the mean square error of ridge regression. Journal of the Royal Statistical Society B, 38:248–50.

Gruber, M.H.J. (1998). Improving Efficiency by Shrinkage: The James-Stein and Ridge Regression Estimators. Marcel Dekker: New York.

Gruber, M.H.J. (2010). Regression Estimators: A Comparative Study. Second edition. Johns Hopkins University Press: Baltimore.

Harville, D.A. (2008). Matrix Algebra from a Statistician's Perspective. Springer: New York.

Hicks, C.R. and K.V. Turner, Jr. (1999). Fundamental Concepts in the Design of Experiments. Fifth edition. Oxford University Press: New York.

Hoerl, A.E. and R.W. Kennard (1970). Ridge regression: Biased estimation for non-orthogonal problems. Technometrics, 12:55–67.

Hogg, R.V.A., A. Craig, and J.W. Mc Kean (2005). Introduction to Mathematical Statistics. Sixth edition. Prentice-Hall: Englewood Cliffs.

Liu, K. (1993). A new class of biased estimate in linear regression. Communications in Statistics–Theory and Methods, 22(2): 393–402.

Mayer, L.W. and T.A. Willke (1973). On biased estimation in linear models. Technometrics, 15:497–508.

Montgomery, D.C. (2005). Design and Analysis of Experiments. Sixth edition. John Wiley & Sons: New York.

Montgomery, D.C. and G.C. Runger (2007). Applied Statistics and Probability for Engineers. Fourth edition. John Wiley & Sons: New York.

Ozkale, M.R. and S. Kaciranlar (2007). The restricted and unrestricted two-parameter estimators. Communications in Statistics–Theory and Methods, 36:2707–2725.

Rao, C.R. (1973). Linear Statistical Inference and Its Applications. Second edition. Wiley: London.

Rao, C.R. (1975). Simultaneous estimation of parameters in different linear models and applications to biometric problems. Biometrics, 31:545–554.

Rao, C.R. and Mitra, S.K. (1971). Generalized Inverse of Matrices and Its Applications. John Wiley & Sons: New York.

Rhode, C.A. (1965). Generalized inverses of partitioned matrices. Journal of the Society of Industrial and Applied Mathematics 13:1033–1035.

Schott, J.R. (2005). Matrix Algebra for Statistics. Second edition. John Wiley & Sons: Hoboken, NJ.

Searle, S.R. (1971). Linear Models. John Wiley & Sons: New York.

Stewart, F. (1963). Introduction to Linear Algebra. Van Nostrand: New York.

Terasvirta, T. (1980). A comparison of mixed and minimax estimators of linear models. Research Report 13. Department of Statistics, University of Helsinki, Helsinki.

Theil, H. and A.S. Goldberger (1961). On pure and mixed estimation in economics. International Economic Review, 2:65–78.

Theobald, C.M. (1974). Generalizations of the mean square error applied to ridge regression. Journal of the Royal Statistical Society B, 36:103–106.

Van Loan, C.F. (1976). Generalizing the singular value decomposition. SIAM Journal on Numerical Analysis, 13(1):76–83.

Wardlaw, W.P. (2005). Row rank equals column rank. American Mathematical Monthly 78(4):316–18.

Yanai, H., K. Takeuchi, and Y. Takane (2011). Projection Matrices, Generalized Inverse Matrices and Singular Value Decomposition. Springer. New York.

FURTHER READING

Ben-Israel, T. and T.N.E. Greville (2003). Generalized Inverses: Theory and Applications. Second edition. Springer: New York.

Putanen, S., G.P.H. Styan, and J. Isotalo (2011). Matrix Tricks for Linear Statistical Models: Our Personal Top Twenty. Springer: Heidelberg.

INDEX

Matrix Algebra for Linear Models, First Edition. Marvin H. J. Gruber.
© 2014 John Wiley & Sons, Inc. Published 2014 by John Wiley & Sons, Inc.

Printed and bound by CPI Group (UK) Ltd, Croydon, CR0 4YY

16/04/2025